Excursions in MODERN MATHEMATICS

Peter Tannenbaum
Robert Arnold

California State University—Fresno

Prentice Hall, Englewood Cliffs, New Jersey 07632

Library of Congress Cataloging-in-Publication Data

Tannenbaum, Peter, (date)
    Excursions in modern mathematics / Peter Tannenbaum and Robert
Arnold.
      p.   cm.
    Includes bibliographical references and index.
    ISBN 0-13-298233-1
    1. Mathematics.   I. Arnold, Robert, mathematician. II. Title.
QA36.T35   1992
510—dc20
                                            91-23215
                                             CIP

Production Editor: *Valerie Zaborski*
Acquisitions Editor: *Priscilla McGeehon*
Editor-in-Chief: *Tim Bozik*
Marketing Manager: *Paul Banks*
Marketing Assistant: *Jeanine Reiter*
Copy Editor: *Carol Dean*
Interior and Cover Design: *Anne T. Bonanno*
Cover Illustration: *Jeanette Adams*
Prepress Buyer: *Paula Massenaro*
Manufacturing Buyer: *Lori Bulwin*
Supplements Editor: *Susan Black*
Editorial Assistant: *Marisol L. Torres*

 © 1992 by Prentice-Hall, Inc.
A Simon and Schuster Company
Englewood Cliffs, New Jersey 07632

Printed in the United States of America
10 9 8 7 6 5 4 3 2 1

ISBN 0-13-298233-1

Prentice-Hall International (UK) Limited, *London*
Prentice-Hall of Australia Pty. Limited, *Sydney*
Prentice-Hall Canada Inc., *Toronto*
Prentice-Hall Hispanoamericana, S.A., *Mexico*
Prentice-Hall of India Private Limited, *New Delhi*
Prentice-Hall of Japan, Inc., *Tokyo*
Simon & Schuster Asia Pte. Ltd., *Singapore*
Editora Prentice-Hall do Brasil, Ltda., *Rio de Janeiro*

To my parents, Nicholas and Anna,
and my wife Sally

PT

To my wife Rachael
and my son Craig

RA

Contents

11 **Symmetry of Motion.** *Mirror, Mirror, off the Wall* **325**

12 **Symmetry of Scale and Fractals.** *Fractally Speaking* **360**

PART IV

Statistics

13 **Collecting Data.** *Censuses, Surveys, and Clinical Studies* **389**

Preface

To most outsiders, modern mathematics is unknown territory. Its borders are protected by dense thickets of technical terms; its landscapes are a mass of indecipherable equations and incomprehensible concepts. Few realize that the world of modern mathematics is rich with vivid images and provocative ideas.

IVARS PETERSON, *The Mathematical Tourist*

Excursions in Modern Mathematics is, as we hope the name might suggest, a collection of "trips" into that vast and alien frontier that many people perceive mathematics to be. While the purpose of this book is quite conventional — it is intended to serve as a textbook for a college-level liberal arts mathematics course — its contents are not. We have made a concerted effort to introduce the reader to a view of mathematics entirely different from the traditional algebra-geometry-trigonometry-finite math curriculum which so many people have learned to dread, fear, and occasionally abhor. The notion that general education mathematics must be dull, unrelated to the real world, highly technical, and deal mostly with concepts that are historically ancient is totally unfounded.

The "excursions" in this book represent a collection of topics chosen to meet a few simple criteria.

■ **Applicability.** The connection between the mathematics presented here and down-to-earth, concrete real-life problems is direct and immediate. The often heard question, "What is this stuff good for?" is a legitimate one and deserves to be met head on. The often heard answer, "Well, you need to learn the material in Math 101 so that you can understand Math 102 which you will need to know if you plan to take Math 201 which will teach you the real applications," is less than persuasive and in many cases reinforces students' convictions that mathematics is remote, labyrinthine, and ultimately useless to them.

- **Accessibility.**   Sophisticated mathematical topics need not always be highly technical and built on layers of concepts. (Most of us are painfully familiar with the experience of missing or misunderstanding a concept in the middle of a math course and consequently having trouble with the rest of the course material.) In general the choice of topics is such that a heavy mathematical infrastructure is not needed. We have found Intermediate Algebra to be a sufficient prerequisite. In the few instances in which more advanced concepts are unavoidable we have endeavored to provide enough background to make the material self-contained. A word of caution — this does not mean that the material is easy! In mathematics, as in many other walks of life, simple and straightforward is not synonymous with easy and superficial.

- **Age.**   Much of the mathematics in this book has been discovered in this century, some as recently as 20 years ago. Modern mathematical discoveries do not have to be only within the grasp of experts.

- **Aesthetics.**   The notion that there is such a thing as beauty in mathematics is surprising to most casual observers. There is an important aesthetic component in mathematics and, just as in art and music (which mathematics very much resembles), it often surfaces in the simplest ideas. A fundamental objective of this book is to develop an appreciation for the aesthetic elements of mathematics. It is not necessary that the reader love everything in the book — it is sufficient that he or she find one topic about which they can say, ''I really enjoyed learning this stuff!'' We believe that anyone coming in with an open mind almost certainly will.

OUTLINE

The material in the book is divided into four independent parts. Each of these parts in turn contains four chapters dealing with interrelated topics.

- **Part I** (Chapters 1 through 4).   *The Mathematics of Social Choice.* This part deals with mathematical applications in social science. How do groups make decisions? How are elections decided? How can power be measured? When there are competing interests among members of a group, how are conflicts resolved in a fair and equitable way?

- **Part II** (Chapters 5 through 8).   *Management Science.* This part deals with methods for solving problems involving the organization and management of complex activities — that is, activities involving either a large number of steps and/or a large number of variables (building a skyscraper, putting a person on the moon, organizing a banquet, scheduling classrooms at a big university, etc.). Efficiency is the name of the game in all these problems. Some limited or precious resource (time, money, raw materials) must be managed in such a way that waste is minimized. We deal with problems of this type (consciously or unconsciously) every day of our lives.

■ **Part III** (Chapters 9 through 12).   *Growth and Symmetry*.
This part deals with nontraditional geometric ideas. What do sunflowers and seashells have in common? How do populations grow? What are the symmetries of a pattern? What is the geometry of natural (as opposed to artificial) shapes? What kind of symmetry lies hidden in a cloud?

■ **Part IV** (Chapters 13 through 16).   *Statistics*.
In one way or another, statistics affects all of our lives. Government policy, insurance rates, our health, and our diet are all governed by statistical laws. This part deals with some of the basic elements of statistics. How are statistical data collected? How are they summarized so that they say something intelligible? How are they interpreted? What are the patterns of statistical data?

EXERCISES

We have endeavored to write a book that is flexible enough to appeal to a wide range of readers in a variety of settings. The exercises, in particular, have been designed to convey the depth of the subject matter by addressing a broad spectrum of levels of difficulty — from the routine to the challenging. For convenience we have classified them into three levels of difficulty: **Walking** (these are straightforward applications of the concepts discussed in the chapter, and it is intended that all readers be able to do them); **Jogging** (these are exercises that are not difficult per se, but require a little thinking on the part of the reader); and **Running** (these are the really challenging exercises and are intended for those readers who like a little challenge in life).

TEACHING EXTRAS

Because this course can be taught in a myriad of styles, we have included material which will increase the text's flexibility. Bibliography listings at the end of every chapter can be used by students and instructors as a source of additional readings which provide more indepth or tangentially related information.

In addition, Prentice Hall and the *New York Times* are sponsoring "A Contemporary View," a program which will provide both students and instructors with access to information which further illustrates the relevance and currency of mathematics. Through this program, at the instructor's request, each student will receive a supplement containing recent articles from the *New York Times* which pertain in one way or another to the themes of our excursions in this book. These articles may be used to prompt classroom discussion, to suggest topics for term papers and research projects, or merely to show that the contents of this course have relevance to a well-informed person. Instructors are encouraged to call their Prentice Hall representative for more information.

To ease the instructor's transition to a new book, we have prepared an instructor's manual based on our own experience using the manuscript in class. The instructor's manual contains a test bank consisting of four hundred multiple choice questions as well as answers to all exercises (including worked out solutions to many), helpful notes about teaching emphasis, suggested topic ordering, etc.

A FINAL WORD

This book grew out of the conviction that a textbook of this type should provide much more than just a battery of technical skills. Our ultimate purpose is to instill in the reader something that is more subjective but at the same time more long-lasting — an appreciation of mathematics as a discipline and an exposure to a few of its global paradigms: algorithms as solutions to mathematical problems, the concept of recursion, the idea of complexity and the trade-offs that it entails, the difference between conceptually difficult and procedurally complicated, the idea that mathematics is one of those rare disciplines that can define its own boundaries (impossibility theorems), etc. Last, but not least, we have tried to show that mathematics can be fun.

ACKNOWLEDG-MENTS

The writing of this book was a formidable challenge but also a tremendously rewarding experience. We wish to acknowledge, first of all, the many students in our course who shared our excitement for the subject. The many stories we heard of their having rediscovered the joy of doing and learning mathematics confirmed our conviction that this was a worthwhile project.

In the development of this book there were several people whose contributions we specifically wish to acknowledge. The following individuals reviewed various stages of the manuscript and shared many helpful ideas with us:

Carmen Q. Artino, *College of Saint Rose*
Donald Beaton, *Norwich University*
Terry L. Cleveland, *New Mexico Military Institute*
Leslie Cobar, *University of New Orleans*
William Hamilton, *Comm. College of Rhode Island*

Harold Jacobs, *East Stroudsburg University*

Matthew G. Pickard, *University of Puget Sound*

Lana Rhoads, *William Baptist College*

David E. Rush, *University of California at Riverside*

David Stacy, *Bellevue Community College.*

We would like to extend special thanks to Professor Benoit Mandelbrot of Yale University who read the manuscript for Chapter 12 and made several valuable suggestions.

We gratefully acknowledge Vahack Haroutunian, who class-tested some of the earliest versions of the material and Ronald Wagoner who helped with the development of many of the exercises.

And, last but not least, our heartfelt thanks to our editors, Valerie Zaborski and Priscilla McGeehon, whose advice and direction made the final product possible.

THE NEW YORK TIMES and **PRENTICE HALL** are sponsoring **A CONTEMPORARY VIEW:** a program designed to enhance student access to current information of relevance in the classroom.

Through this program, the core subject matter provided in the text is supplemented by a collection of time-sensitive articles from one of the world's most distinguished newspapers, **THE NEW YORK TIMES.** These articles demonstrate the vital, ongoing connection between what is learned in the classroom and what is happening in the world around us.

To enjoy the wealth of information of **THE NEW YORK TIMES** daily, a reduced subscription rate is available in deliverable areas. For information, call toll-free: 1-800-631-1222.

PRENTICE HALL and **THE NEW YORK TIMES** are proud to co-sponsor **A CONTEMPORARY VIEW.** We hope it will make the reading of both textbooks and newspapers a more dynamic, involving process.

1

Voting

The Paradoxes of Democracy

We are often reminded that the right to vote, one of the fundamental tenets of a democracy, is something that we should not take for granted. Voting is the vehicle by which a democratic society makes decisions, and the process by which the many and conflicting opinions of the individuals are consolidated into the single choice of the group is what **voting theory** is about. But why do we need a whole theory? It all sounds so simple. There surely must be a reasonable and easy way for a democratic society to make group decisions based on *numbers*, and we have every right to expect that mathematics with all its tools and power will give us the answer.

Mathematics does indeed provide the answer, but the answer is quite surprising and far from simple: There is *no* consistent method by which a democratic society can make a choice that is *always* fair when that choice must be made from among several (three or more) alternatives. This remarkable fact, discovered by the mathematical economist Kenneth Arrow in 1952, is known as **Arrow's impossibility theorem**. In 1972, Arrow received the Nobel Prize in Economics for his contributions to the mathematical theory of social decision making.

1

The purpose of this chapter is to clarify the meaning and significance of Arrow's impossibility theorem and at the same time to discuss several well-known methods of voting.

Our discussion for the entire chapter is centered around the following, deceptively simple example.

The MAC Election. An election is being held to choose the president of the Math Anxiety Club (MAC). There are four candidates: Alisha, Boris, Carmen, and Dave (*A*, *B*, *C*, and *D* for short). Each of the 37 members of the club is asked to submit a ballot indicating his or her first, second, third, and fourth choices (ties are not allowed in individual ballots). The 37 ballots submitted are shown in Fig. 1-1. Who should be president? Why?

	Ballot	Ballot	Ballot	Ballot	Ballot	Ballot	Ballot	Ballot	Ballot	Ballot
1st	A	B	A	C	B	C	A	D	A	A
2nd	B	D	B	B	D	B	B	C	B	B
3rd	C	C	C	D	C	D	C	B	C	C
4th	D	A	D	A	A	A	D	A	D	D

	Ballot	Ballot	Ballot	Ballot	Ballot	Ballot	Ballot	Ballot	Ballot	Ballot
1st	C	A	C	D	C	A	D	D	C	C
2nd	B	B	B	C	B	B	C	C	B	B
3rd	D	C	D	B	D	C	B	B	D	D
4th	A	D	A	A	A	D	A	A	A	A

	Ballot	Ballot	Ballot	Ballot	Ballot	Ballot	Ballot	Ballot	Ballot	Ballot
1st	A	B	C	C	D	A	D	C	A	D
2nd	B	D	B	B	C	B	C	B	B	C
3rd	C	C	D	D	B	C	B	D	C	B
4th	D	A	A	A	A	D	A	A	D	A

	Ballot	Ballot	Ballot	Ballot	Ballot	Ballot	Ballot
1st	B	A	C	A	A	D	A
2nd	D	B	D	B	B	C	B
3rd	C	C	B	C	C	B	C
4th	A	D	A	D	D	A	D

Figure 1-1

PREFERENCE BALLOTS AND PREFERENCE SCHEDULES

Ballots such as the ones shown in Fig. 1-1 in which the voter is asked to list all of the options in order of preference are called **preference ballots**. A closer inspection of the 37 preference ballots in Fig. 1-1 shows that in many instances individual voters voted exactly the same way. In fact, the outcome of the election can be summarized in Fig. 1-2 and more conveniently in the table that follows, which is called a **preference schedule** for the election.

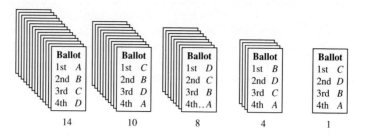

Figure 1-2

Number of voters	14	10	8	4	1
1st choice	A	C	D	B	C
2nd choice	B	B	C	D	D
3rd choice	C	D	B	C	B
4th choice	D	A	A	A	A

Some Basic Assumptions

Ballot
1st C
2nd B
3rd D
4th A

Figure 1-3

Before we take on the task of deciding who will be elected president of the Math Anxiety Club, we need to clarify certain assumptions that we will make about individual preferences. One of our basic assumptions is that individual preferences are **transitive**: If a voter prefers X to Y and Y to Z, then it is only reasonable to assume that the voter prefers X to Z. It follows therefore that a ballot such as the one shown in Fig. 1-3 tells us not only that this voter prefers C to B, B to D, and D to A, but also implicitly that the voter prefers C to D, C to A, and B to A.

A second important assumption is that relative preferences are not altered by the elimination of one or more candidates. In other words, suppose that a voter prefers X to Y, Y to Z, and Z to W. If, for whatever reason, Y drops out of the race, the other relative preferences remain unaffected: The voter still prefers X to Z and Z to W.

While there are occasional situations in which these assumptions do not hold,[1] they are by and large exceptions. In any voting system in which voters act *rationally*, it is reasonable to expect that the above assumptions will hold, and for the rest of this chapter we will pretty much take them for granted.

---

[1] See, for example, the discussion of how the 1996 Summer Olympics were awarded to Atlanta in Appendix 2 at the end of this chapter.

THE PLURALITY METHOD

A very well-known and seemingly reasonable way to decide an election is the **plurality method**: The candidate with the *most* first-place votes wins. For the MAC election, the results are

A: 14 first-place votes
B: 4 first-place votes
C: 11 first-place votes
D: 8 first-place votes.

Since Alisha has the most first-place votes, she is the winner!

Our fondness for the plurality method stems not only from its simplicity but also from the fact that it is a natural extension of the **majority rule**: In any democratic election between two candidates, the one with the majority (more than half) of the votes wins. Unfortunately, the majority rule cannot always be used when there are more than two candidates: In the MAC election with 37 voters, 19 first-place votes are needed for a majority, and none of the candidates received 19 first-place votes. Of course, if a candidate had received 19 or more first-place votes, that candidate would clearly have to be the winner under the plurality method since no other candidate could possibly receive as many. In the language of voting theory, we say that the plurality method *satisfies* the **majority criterion**.

> **Majority Criterion.** If there is a candidate or alternative that is the first choice of a majority of the voters, then that candidate or alternative should be the winner of the election.

On its surface, the majority criterion seems so logical and so much in accordance with our idea of democracy that we are inclined to take it for granted. Surprisingly, we will soon see that there are some well-known and widely used voting methods that actually violate this criterion.

■ Problems with the Plurality Method

In spite of its frequent usage, the plurality method has several flaws and is generally considered a very poor method for choosing the winner of an election between more than two candidates. The principal weakness of the plurality method is that it fails to take into account preferences other than first choice. In fact, when we look at the MAC election preference schedule, we notice that a majority of the voters (23 to be exact) were not terribly crazy about Alisha being club president and made her their last choice.

To underscore how the plurality method can sometimes give really bad results, consider the next example.

Example 1.   An election is held among the 100 members of the Tasmania State University marching band to decide in which of five different bowl games they will march. The alternatives are the Rose Bowl (R), the Hula Bowl (H), the Cotton Bowl (C), the Orange Bowl (O), and the Sugar Bowl (S). The results of the election are shown in the following preference schedule:

Number of voters	49	48	3
1st choice	R	H	C
2nd choice	H	S	H
3rd choice	C	O	S
4th choice	O	C	O
5th choice	S	R	R

When the plurality method is used, the choice is the Rose Bowl (R), in spite of the fact that it is the *last* choice of a majority (51) of the band members, while the Hula Bowl (H) is the first choice of 48 voters (only 1 less than R) and also has 52 second-place votes. Under any reasonable interpretation, the Hula Bowl is much more representative of the band's wishes than the Rose Bowl. In fact, we can make the following persuasive argument on behalf of the Hula Bowl: When compared on a *one-to-one* basis with any other bowl, the Hula Bowl would always get a majority of the votes. Take, for example, a comparison between the Hula Bowl and the Rose Bowl: The Hula Bowl would get 51 votes (48 from the second column plus the 3 votes in the last column) versus 49 votes for the Rose Bowl. Likewise, a comparison between the Hula Bowl and the Cotton Bowl would result in 97 votes for the Hula Bowl (first and second columns) and 3 votes for the Cotton Bowl. Finally, we see that the Hula Bowl is preferred to both the Orange Bowl and the Sugar Bowl by all 100 voters.

We can now summarize the problem as follows: Although H wins in a one-to-one comparison between it and any other alternative, the plurality method fails to choose H as the winner. In the language of voting theory, we say that the plurality method *violates* a basic principle of fairness called the **Condorcet**[2] **criterion.**   ■

[2] Named after Marie Jean Antoine Nicolas Caritat, Marquis de Condorcet (1743–1794). Condorcet was a French aristocrat, mathematician, philosopher, economist, and social scientist. As a member of a group of liberal thinkers (the *encyclopédistes*), his ideas were instrumental in leading the way to the French Revolution. Unfortunately, his independent ways eventually fell into disfavor and he died in prison.

> **Condorcet Criterion.** If there is a candidate or alternative that wins in a one-to-one comparison between it and any other alternative, then that candidate or alternative should be the winner of the election.

A candidate that is preferred by the voters in one-to-one comparisons with all other candidates is called a **Condorcet winner**, so what the Condorcet criterion says is that when there is a Condorcet winner, it should indeed be the winner of the election.

We will return to the idea of comparing candidates on a one-to-one basis shortly. In the meantime, we conclude this section on plurality by discussing one last weakness of the method known as **insincere** or **strategic voting**. Consider once again Example 1 and look at the last column of the preference schedule which represents the preference votes of three band members—let's say Hekyll, Jekyll, and Hyde. Let's suppose, moreover, that all three have a pretty good idea of how everybody is going to vote and know that under the plurality method they will be stuck with going to the Rose Bowl (which they really don't want). Since they know that their first choice (the Cotton Bowl) has no chance of winning the election, their strategy is clear: Make the Hula Bowl the first choice on their ballots (even though it really isn't), thereby keeping the Rose Bowl from being the winner.

The ability of voters to manipulate the results of an election by insincere voting is a common but undesirable feature of many voting methods, and the plurality method is particularly susceptible to it. In national elections it has had a profound effect on the viability of third-party candidates: Often a voter may prefer such a candidate but thinks his vote may be wasted, so instead he votes (insincerely) for his preference between the two major party candidates.

THE BORDA COUNT METHOD

In the **Borda count method**[3] each place on a ballot is assigned points. In an election with N candidates we give 1 point for last place, 2 points for second from last place, . . ., and N points for first place. The points are tallied for each candidate separately, and the candidate with the highest total wins.

Let's use the Borda count method to choose the winner of the MAC election. Putting the point values in the preference schedule, we have

---

[3] This method is named after its proposer, the Frenchman Jean-Charles de Borda (1733–1799). Borda was a military man—a cavalry officer and naval captain—who wrote papers on such diverse subjects as mathematics, physics, the design of scientific instruments, and voting theory.

Number of voters	14	10	8	4	1
1st choice: 4 points	*A*: 56 pts	*C*: 40 pts	*D*: 32 pts	*B*: 16 pts	*C*: 4 pts
2nd choice: 3 points	*B*: 42 pts	*B*: 30 pts	*C*: 24 pts	*D*: 12 pts	*D*: 3 pts
3rd choice: 2 points	*C*: 28 pts	*D*: 20 pts	*B*: 16 pts	*C*: 8 pts	*B*: 2 pts
4th choice: 1 point	*D*: 14 pts	*A*: 10 pts	*A*: 8 pts	*A*: 4 pts	*A*: 1 pt

Consequently,

A gets $56 + 10 + 8 + 4 + 1 = 79$ points.
B gets $42 + 30 + 16 + 16 + 2 = 106$ points.
C gets $28 + 40 + 24 + 8 + 4 = 104$ points.
D gets $14 + 20 + 32 + 12 + 3 = 81$ points.

The winner of the election under the Borda count method is Boris (B)! This is not a pleasant bit of news for Alisha (the winner under the plurality method).

MATH ANXIETY CLUB ELECTION RESULTS

Winner	Voting Method
Alisha	Plurality
Boris	Borda count

■ Problems with the Borda Count Method

The most serious weakness of the Borda count method is that it clashes with what we have already agreed is a reasonable expectation in a democratic election: the majority criterion. Example 2 shows how under the Borda count method, it is possible for a candidate to have a majority of the first-place votes and yet fail to win the election.

Example 2. Consider an election with 4 candidates (A, B, C, and D), 11 voters, and the preference schedule given in the following table.

Number of voters	6	2	3
1st: 4 points	*A*	*B*	*C*
2nd: 3 points	*B*	*C*	*D*
3rd: 2 points	*C*	*D*	*B*
4th: 1 point	*D*	*A*	*A*

Let's pick the winner using the Borda count method.

A gets $(6 \times 4) + (2 \times 1) + (3 \times 1) = 29$ points.
B gets $(6 \times 3) + (2 \times 4) + (3 \times 2) = 32$ points.
C gets $(6 \times 2) + (2 \times 3) + (3 \times 4) = 30$ points.
D gets $(6 \times 1) + (2 \times 2) + (3 \times 3) = 19$ points.

In this election B has the most points and wins under the Borda count method in spite of the fact that A has *the majority of first-place votes*. In the language of voting theory, the Borda count method violates the majority criterion. ∎

Another problem with the Borda count method is a direct consequence of the above: It also violates the Condorcet criterion. It is not hard to see that A, having a majority of first-place votes, would automatically win a one-to-one comparison against any other alternative (Exercise 39).

In spite of these weaknesses, the Borda count method is a widely used and generally satisfactory method for deciding an election. In contrast to the plurality method, it uses *all* the information provided by the voters and produces as a winner the best compromise candidate.

In practice, the Borda count method is often used by assigning the points as follows: 0 points for last place, 1 point for second from last, 2 points for third from last, etc. It turns out (see Exercise 35) that tabulating the points this way does not in any way change the results of the election.

Sometimes elections are decided using sophisticated variations of the Borda count method. In choosing the winner of the Heisman trophy (an

The most coveted individual award in college football: The Heisman Memorial Trophy. (AP/Wide World Photos)

annual award given to the "best collegiate football player" in the country), each of the voters (about 870 sportswriters and 50 former Heisman trophy winners) submits a ballot with a first, second, and third choice (3 points are awarded for first place, 2 points for second, and 1 point for third). The candidate with the highest total wins the award. In essence this is a variation of the Borda count method in which the point values given to the places are (from top to bottom) 3, 2, 1, 0, 0, 0, . . .

THE PLURALITY-WITH-ELIMINATION METHOD

This is a sophisticated variation of the plurality method and is carried out in rounds using the following procedure:

■ **Round 1.** Count the first-place votes for each candidate, just as you would in the plurality method. If a candidate has a majority of first-place votes, that candidate is automatically declared the winner. Otherwise, eliminate the candidate (or candidates if there is a tie) with the fewest number of first-place votes.

■ **Round 2.** Cross out the name(s) of the candidates eliminated from the preference schedule and recount the first-place votes. If a candidate has a majority of first-place votes, declare that candidate the winner. Otherwise, eliminate the candidate with the fewest number of first-place votes.

■ **Rounds 3, 4, etc.** Repeat the process, each time eliminating one or more candidates. Eventually, there will be a candidate with a majority of first-place votes. (In the worst of cases we might have to repeat the process until we end up with only two candidates.)

Let's apply the plurality-with-elimination method to the MAC election.

■ **Round 1.**

A: 14 first-place votes
B: 4 first-place votes
C: 11 first-place votes
D: 8 first-place votes.

B has the fewest number of first-place votes and is eliminated.

Number of voters	14	10	8	4	1
1st choice	A	C	D	~~B~~	C
2nd choice	~~B~~	~~B~~	C	D	D
3rd choice	C	D	~~B~~	C	~~B~~
4th choice	D	A	A	A	A

The preference schedule (with *B* removed) now becomes

Number of voters	14	10	8	4	1
1st choice	A	C	D	D	C
2nd choice	C	D	C	C	D
3rd choice	D	A	A	A	A

which can be simplified by combining like preference votes to get

Number of voters	14	11	12
1st choice	A	C	D
2nd choice	C	D	C
3rd choice	D	A	A

■ Round 2.

A: 14 first-place votes

C: 11 first-place votes

D: 12 first-place votes

C is eliminated.

Number of voters	14	11	12
1st choice	A	~~C~~	D
2nd choice	~~C~~	D	~~C~~
3rd choice	D	A	A

This preference schedule after simplification becomes

Number of voters	14	23
1st choice	A	D
2nd choice	D	A

■ Round 3.

A: 14 first-place votes

D: 23 first-place votes.

In this round Dave has a majority and is therefore the winner.

For the reader who likes straight answers to simple questions, it may be somewhat discomforting to realize that the question, Who is the real winner of the MAC election? may not make a lot of sense. The answer seems to depend on the voting method we choose.

> ### MATH ANXIETY CLUB ELECTION RESULTS
>
> | Winner | Voting Method |
> |--------|---------------|
> | Alisha | Plurality |
> | Boris | Borda count |
> | Dave | Plurality with elimination |

We illustrate the plurality-with-elimination method with another example. This time we take a shortcut, and instead of rewriting the preference schedule after each round, we just cross out the candidates as they are eliminated and note that the topmost candidate that is not crossed out in a column is the first-place candidate in that column.

Example 3. Consider an election with 5 candidates (A, B, C, D, and E), 24 voters, and the preference schedule

Number of voters	8	6	2	3	5
1st choice	A	B	C	D	E
2nd choice	B	D	A	E	A
3rd choice	C	E	E	A	D
4th choice	D	C	B	C	B
5th choice	E	A	D	B	C

Using the plurality-with-elimination method we obtain the following:

■ **Round 1.**

Candidates: $A\ B\ C\ D\ E$
First-place votes: 8 6 2 3 5

C is eliminated.

Number of voters	8	6	2	3	5
1st choice	A	B	~~C~~	D	E
2nd choice	B	D	A	E	A
3rd choice	~~C~~	E	E	A	D
4th choice	D	~~C~~	B	~~C~~	B
5th choice	E	A	D	B	~~C~~

■ **Round 2.**

Candidates: *A B D E*

First-place votes: 10 6 3 5

D is eliminated.

Number of voters	8	6	2	3	5
1st choice	*A*	*B*	~~*C*~~	~~*D*~~	*E*
2nd choice	*B*	~~*D*~~	*A*	*E*	*A*
3rd choice	~~*C*~~	*E*	*E*	*A*	~~*D*~~
4th choice	~~*D*~~	~~*C*~~	*B*	~~*C*~~	*B*
5th choice	*E*	*A*	~~*D*~~	*B*	~~*C*~~

■ **Round 3.**

Candidates: *A B E*

First-place votes: 10 6 8

B is eliminated.

Number of voters	8	6	2	3	5
1st choice	*A*	~~*B*~~	~~*C*~~	~~*D*~~	*E*
2nd choice	~~*B*~~	~~*D*~~	*A*	*E*	*A*
3rd choice	~~*C*~~	*E*	*E*	~~*A*~~	*D*
4th choice	~~*D*~~	~~*C*~~	*B*	~~*C*~~	~~*B*~~
5th choice	*E*	*A*	~~*D*~~	~~*B*~~	~~*C*~~

■ **Round 4.**

Candidates: *A E*

First-place votes: 10 14

The winner of this election is *E*.   ■

Plurality with elimination is a method commonly used in elections for local political offices (city councils, county boards of supervisors, school boards, etc.). A slight variation of the method is used by the International Olympic Committee to select the cities in which the Olympic Games are held (see Appendix 2). Just like the Borda count method, plurality with elimination uses all the information provided by the voters' preferences (not just first choice), but unlike the Borda count method, it uses it sequentially rather than all at once. In one important respect, however, plurality with elimination is superior to the Borda count method: It satisfies the majority criterion (Exercise 37).

■ **Problems with the Plurality-with-Elimination Method**

The major flaw in the plurality-with-elimination method is quite subtle and is illustrated by the next example.

Example 4.   Consider an election with 3 candidates (A, B, and C), 29 voters, and the preference schedule

Number of voters	7	8	10	4
1st choice	A	B	C	A
2nd choice	B	C	A	C
3rd choice	C	A	B	B

Under plurality with elimination, B is eliminated in round 1, leaving us with the preference schedule

Number of voters	11	18
1st choice	A	C
2nd choice	C	A

It follows that the winner of this election is C.

Suppose now that because of some irregularities in the original election, the election is declared null and void and a second election is held. In the second election all the voters vote the same way except for the four voters represented by the last column of the original preference schedule. They have a change of heart and switch their first and second preferences between A and C in favor of C (they heard that C is going to lower their taxes). Since C won the original election and the only changes in the second election were to give C even more first-place votes, we would expect that C is still the winner. Just to be safe we will check it out.

Switching A and C in the last column makes that column the same as the third column, so the preference schedule for the second election is

Number of voters	7	8	14
1st choice	A	B	C
2nd choice	B	C	A
3rd choice	C	A	B

Under plurality with elimination, A is eliminated in round 1, leaving us with the preference schedule

Number of voters	15	14
1st choice	B	C
2nd choice	C	B

Now *B* is the winner! To anyone with a sense of fairness the results of this second election would seem absurd, and one can imagine how candidate *C* would feel knowing that she had the original election won, picked up some extra support for the reelection, and thereby proceeded to lose it. ∎

In the language of voting theory what happened in Example 4 is known as a violation of the **monotonicity criterion**.

Monotonicity Criterion. If candidate or alternative *X* is the winner of an election and, in a reelection, all the voters who change their preferences do so in a way that is favorable only to *X*, then *X* should still be the winner of the election.

Example 4 demonstrates that plurality with elimination violates the monotonicity criterion. We leave it to the reader (Exercise 41) to verify that plurality with elimination also violates the Condorcet criterion.

THE METHOD OF PAIRWISE COMPARISONS

So far, all three voting methods we have discussed violate the Condorcet criterion. There are several voting methods that satisfy the Condorcet criterion (they are known generically as *Condorcet methods*), but we will discuss only one of the simplest ones: the method of pairwise comparisons.[4]

The **method of pairwise comparisons** is a little like a *round-robin tournament*; every candidate is matched on a one-to-one basis with every other candidate. Each of these one-to-one pairings is called a **pairwise comparison**. When pairing two candidates (say *X* and *Y*) one on one, each vote is assigned to either *X* or *Y* by the order of preference indicated by that voter. (*X* gets the votes of all voters ranking *X* higher than *Y*.) A pairwise comparison between *X* and *Y* can result in a win for *X* (1 point), a tie ($\frac{1}{2}$ point for each), or a loss for *X* (0 points). The candidate with the most points after *all* pairwise comparisons are tabulated is declared the winner.

For the MAC election, the method of pairwise comparisons works like this. First we compare *A* with *B*.

---

[4] This method is sometimes called *Copeland's method* and attributed to A. H. Copeland.

Number of voters	14	10	8	4	1
1st choice	Ⓐ	C	D	B̲	C
2nd choice	B̲	B̲	C	D	D
3rd choice	C	D	B̲	C	B̲
4th choice	D	Ⓐ	Ⓐ	Ⓐ	Ⓐ

The 14 voters represented in the first column prefer A to B, but all the rest of the voters (23) prefer B to A. Consequently, in the pairwise comparison between A and B, B is the winner. We summarize this result by

A versus B (14 to 23): B wins (B gets 1 point; A gets 0 points).

Our next pairwise comparison is between A and C.

Number of voters	14	10	8	4	1
1st choice	Ⓐ	C̲	D	B	C̲
2nd choice	B	B	C̲	D	D
3rd choice	C̲	D	B	C̲	B
4th choice	D	Ⓐ	Ⓐ	Ⓐ	Ⓐ

Once again we can summarize the result by

A versus C (14 to 23): C wins (C gets 1 point).

In a like manner we compare A with D, B with C, B with D, and C with D. We leave it to the reader to check the following results:

A versus D (14 to 23): D wins (D gets 1 point)
B versus C (18 to 19): C wins (C gets 1 point)
B versus D (28 to 9): B wins (B gets 1 point)
C versus D (25 to 12): C wins (C gets 1 point).

The final scoreboard reads: A, 0 points; B, 2 points; C, 3 points, and D, 1 point. Since C has the most points, she is declared the winner by the method of pairwise comparisons. Another method, another winner!

MATH ANXIETY CLUB ELECTION RESULTS	
Winner	Voting Method
Alisha	Plurality
Boris	Borda count
Dave	Plurality with elimination
Carmen	Pairwise comparisons

The fact that the method of pairwise comparisons satisfies the Condorcet criterion is quite obvious: If a candidate wins every pairwise comparison between itself and the other candidates, then that candidate will automatically have the most points and win the election. It is also easy to see that the method of pairwise comparisons satisfies the majority criterion (Exercise 38). Somewhat less obvious but still true is the fact that the method of pairwise comparisons satisfies the monotonicity criterion (Exercise 42).

■ Problems with the Method of Pairwise Comparisons

Is there anything wrong with the method of pairwise comparisons? The answer, unfortunately, is yes. We illustrate one of the main problems with the method of pairwise comparisons in the next example.

Example 5. Consider an election with 5 candidates (A, B, C, D, and E), 22 voters, and the preference schedule

Number of voters	5	3	5	3	2	4
1st choice	A	A	C	D	D	B
2nd choice	B	D	E	C	C	E
3rd choice	C	B	D	B	B	A
4th choice	D	C	A	E	A	C
5th choice	E	E	B	A	E	D

The reader should check that the results are as follows:

A versus B (13 to 9): A wins

A versus C (12 to 10): A wins

A versus D (12 to 10): A wins

A versus E (10 to 12): E wins

B versus C (12 to 10): B wins

B versus D (9 to 13): D wins

B versus E (17 to 5): B wins

C versus D (14 to 8): C wins

C versus E (18 to 4): C wins

D versus E (13 to 9): D wins.

Since A has 3 points, B has 2 points, C has 2 points, D has 2 points, and E has 1 point, A is the winner of this election under the method of pairwise comparisons.

Now suppose that because of some irregularity, the votes have to be recounted, but before they are, candidates B, C, and D become discouraged and drop out of the race, leaving only candidates A and E. Eliminating candidates B, C, and D gives us the preference schedule

Number of voters	5	3	5	3	2	4
1st choice	A	A	~~C~~	~~D~~	~~D~~	~~B~~
2nd choice	~~B~~	~~D~~	E	~~C~~	~~C~~	E
3rd choice	~~C~~	~~B~~	~~D~~	~~B~~	~~B~~	A
4th choice	~~D~~	~~C~~	A	E	A	~~C~~
5th choice	E	E	~~B~~	A	E	~~D~~

After simplification, this schedule becomes:

Number of voters	10	12
1st choice	A	E
2nd choice	E	A

The winner is now E. The winner under the method of pairwise comparisons was A, but by eliminating some of the other candidates and reapplying the same voting method, the winner is changed to E. ∎

In the language of voting theory we say that the method of pairwise comparisons violates the **independence of irrelevant alternatives criterion**.

Independence of Irrelevant Alternatives Criterion. If candidate or alternative X is the winner of an election, and one or more of the other candidates or alternatives are removed and the ballots recounted, then X should still be the winner of the election.

A second problem with the method of pairwise comparisons is that often it is unable to produce a winner. Consider the following simple preference schedule:

Number of voters	3	2	4
1st choice	A	B	C
2nd choice	B	C	A
3rd choice	C	A	B

Here A beats B (7 to 2), B beats C (5 to 4), and C beats A (6 to 3), producing a seemingly illogical situation: The preferences of the group are not transitive. It is one of the great paradoxes of voting that while individual preferences are always transitive, the collective preferences of the group may not be. When standoffs such as the one described above occur, the method of pairwise comparisons is unable to declare a winner.

■ The Number of Pairwise Comparisons

The final difficulty with the method of pairwise comparisons is a practical one: It takes a lot of work to check all possible pairwise comparisons. In fact, how many pairwise comparisons do we really have to check? We saw in the MAC election that with 4 candidates we needed to check 6 pairwise comparisons, and in Example 5 we had 5 candidates and we checked 10 pairwise comparisons. To make it interesting let's say that we have an election with 12 candidates. How many pairwise comparisons are possible? Let's try to systematically count the comparisons, making sure that we don't count any comparison twice.

- ■ The first candidate must be compared with each of the other 11 candidates—*11 pairwise comparisons.*

- ■ Compare the second candidate with each of the other candidates *except* the first one, since that comparison has already been made—*10 pairwise comparisons.*

- ■ Compare the third candidate with each of the other candidates *except* the first and second candidates, since those comparisons have already been made—*9 pairwise comparisons.*

  $\vdots$

- ■ Compare the eleventh candidate with each of the other candidates *except* the first 10 candidates, since those comparisons have already been made. In other words, compare the eleventh candidate with the twelfth candidate—*1 pairwise comparison.*

We see that the total number of pairwise comparisons is

$$1 + 2 + 3 + 4 + 5 + 6 + 7 + 8 + 9 + 10 + 11 = 66.$$

How many pairwise comparisons are needed if there are 100 candidates? Well, using an argument similar to the preceding one, the total number of comparisons needed is

$$1 + 2 + 3 + 4 + \cdots + 98 + 99.$$

Can we determine the number of pairwise comparisons needed without actually having to do the work of adding these 99 numbers together? Let's try a different way to count the comparisons. Suppose that before a pairwise comparison between two candidates is made, each candidate gives the other one his or her business card. Then clearly, each candidate would end up with the business card of every other candidate, and there would be a total of $99 \times 100 = 9900$ cards handed out (each of the 100 candidates would hand out 99 cards, 1 to each of the other candidates). But since each comparison resulted in 2 cards being handed out, the total number of comparisons must be half as many as the number of cards. Consequently,

$$1 + 2 + 3 + 4 + \cdots + 98 + 99 = \frac{99 \times 100}{2} = 4950.$$

Similar arguments show that if there are N candidates, the number of pairwise comparisons needed is

$$1 + 2 + 3 + \cdots + (N - 1) = \frac{(N - 1)N}{2}.$$

Because the number of comparisons grows quite fast in relation to the number of candidates, the method of pairwise comparisons is cumbersome and time-consuming and is seldom used in practice as a method for deciding elections. This is unfortunate, since in spite of its many flaws, and if we are willing to go through all the trouble, pairwise comparisons can be a fairly good voting method.

BREAKING TIES

All the examples given so far in this chapter have been carefully chosen to avoid tied winners, but of course in the real world ties are sometimes inevitable.

In this section we will discuss very briefly the problem of how to break ties when they occur. Tie-breaking methods can raise some fairly complex issues, and it is not our purpose here to study such methods in great detail but rather to make the reader aware of the problem and give some inkling as to possible ways to deal with it.

For starters, consider the election with two candidates (A and B) and the preference schedule

Number of voters	10	10
1st choice	A	B
2nd choice	B	A

It should be clear that no reasonable voting method could choose A as the winner over B or choose B as the winner over A. If A and B were interchanged, the preference schedule would be exactly the same. Neither candidate is favored in any way over the other. A tie is inevitable. We might call this kind of a tie an **essential tie**.

The election with three candidates (A, B, and C) and the preference schedule

Number of voters	10	10	10
1st choice	A	B	C
2nd choice	B	C	A
3rd choice	C	A	B

also represents an essential tie: No reasonable voting method can choose one candidate over the others. All three are interchangeable in the sense that any two candidates could switch names and we wouldn't be able to tell the difference (identical twins do it all the time!).

Essential ties cannot be broken using a rational tie-breaking procedure, and we must rely on some sort of outside intervention such as chance (flip a coin, draw straws, etc.), a third party (the president of the club, the chair of the committee, mom, etc.), or even some form of tradition (seniority, age, etc.).

Most ties are not essential ties, and they can often be broken in a more rational way: either by implementing some tie-breaking rule or by using a different voting method. We illustrate some of these ideas with the following example.

Example 6. Consider an election with 5 candidates (A, B, C, D, and E), 22 voters, and the preference schedule

Number of voters	5	3	5	3	2	4
1st choice	A	A	C	D	D	B
2nd choice	B	B	E	C	C	E
3rd choice	C	D	D	B	B	A
4th choice	D	C	A	E	A	C
5th choice	E	E	B	A	E	D

Using the method of pairwise comparisons, we have

A versus B (13 to 9): A wins

A versus C (12 to 10): A wins

A versus D (12 to 10): A wins

A versus E (10 to 12): E wins

B versus C (12 to 10): B wins

B versus D (12 to 10): B wins

B versus E (17 to 5): B wins

C versus D (14 to 8): C wins

C versus E (18 to 4): C wins

D versus E (13 to 9): D wins.

Since A has 3 points, B has 3 points, C has 2 points, D has 1 point, and E has 1 point, A and B tie for the winner.

We can break this tie in several different ways. For example,

1. Use the results of head-to-head competition between the winners. In the above example, since A beats B in a pairwise comparison, the tie would be broken in favor of A.

2. Use the *total point differentials*. For example, since A beats B 13 to 9, the point differential for A is $+4$, and since A lost to E 10 to 12, the point differential for A is -2. Computing the total point differentials for A gives $4 + 2 + 2 - 2 = 6$. Likewise, the total point differential for B is $2 + 2 + 12 - 4 = 12$. In this case the point differential favors B which would therefore be declared the winner.

3. Use first-place votes. In the example, A has 8 and B has 4. With this method, the winner would be A.

4. Use a Borda count between the winners. In the example, we have

   Borda count for A: $(5 \times 8) + (3 \times 4) + (2 \times 7) + (1 \times 3) = 69$.
   Borda count for B: $(5 \times 4) + (4 \times 8) + (3 \times 5) + (1 \times 5) = 72$.

   With this method, the tie would be broken in favor of B. ■

By now we should not be at all surprised that different tie-breaking methods produce different winners and that there is no single *right* method for breaking ties. In retrospect, flipping a coin might not be such a bad idea!

RANKINGS

Quite often it is important not only to know who wins the election but also to know who comes in second, third, etc. Let's consider for example, the MAC election but suppose now that instead of electing just a president we need to elect a board of directors consisting of a president, a vice-president, and a treasurer. The club's constitution states that the winner of the election should become the president, the candidate in second place should become the vice-president, and the candidate in third place should become the treasurer. Clearly, we need our voting method not only to tell us the winner but also who is second, third, and (by default) last—in other words, a **ranking** of the candidates.

■ Extended Ranking Methods

Each of the four voting methods we discussed at the beginning of this chapter can be extended to produce a ranking of the candidates in a very natural way.

Let's start with the plurality method and see how we might extend it to produce a ranking of the four candidates in the MAC election. As the reader may recall, the outcome of the election under plurality was that

A had 14 first-place votes, *B* had 4 first-place votes, *C* had 11 first-place votes, and *D* had 8 first-place votes.

The obvious thing to do here is to rank the candidates by the number of first-place votes. We will call this approach the *extended plurality method*. For the MAC election it gives the following ranking:

Winner (president): *A* (14 first-place votes)
2nd place (vice president): *C* (11 first-place votes)
3rd place (treasurer): *D* (8 first-place votes)
Last: *B* (4 first-place votes).

Ranking the candidates using the *extended Borda count method* is equally obvious. In the MAC election, for example, the point totals under the Borda count method were *A*, 79 points; *B*, 106 points; *C*, 104 points; and *D*, 81 points. The resulting ranking, based on point totals would be

Winner (president): *B* (106 points)
2nd place (vice-president): *C* (104 points)
3rd place (treasurer): *D* (81 points)
Last: *A* (79 points).

Ranking the candidates using *extended plurality with elimination* is only a bit more subtle: We rank them in reverse order of elimination (the first candidate eliminated is ranked last, the second candidate eliminated is ranked next to last, etc.).[5] If we applied this method to rank the candidates in the MAC election, the result would be

Winner (president): *D*
2nd place (vice-president): *A* (eliminated last)
3rd place (treasurer): *C* (eliminated second)
Last: *B* (eliminated first).

Last, we can rank the candidates using the *extended method of pairwise comparisons* according to the number of points (recall that we count a tie as $\frac{1}{2}$ point). In the case of the MAC election, *C* had 3 points, *B* had 2 points, *D* had 1 point, and *A* had 0 points, so the results are

Winner (president): *C* (3 points)
2nd place (vice-president): *B* (2 points)
3rd place (treasurer): *D* (1 point)
Last: *A* (0 points).

---

[5] In cases where a candidate gets a majority of first-place votes before the ranking of all the candidates is complete, we continue the process of elimination to rank the remaining candidates.

■ Recursive Ranking Methods

There is a second, somewhat more involved but equally reasonable strategy for producing rankings which we will call the **recursive strategy**. When using this strategy, the basic approach is the same regardless of which voting method we choose. Let's say we are going to use voting method X and a recursive strategy to get a ranking of the candidates after an election. We first use method X to find the winner of the election. So far, so good. We then cross out the name of the winner on the preference schedule and obtain a new, modified preference schedule with one less candidate on it. We apply method X once again to find the "winner" based on this new preference schedule and this candidate is ranked second. We repeat the process again (cross out the name of the last "winner", calculate the new preference schedule, and apply method X to find the next winner who is then placed next in line in the ranking) until we have ranked as many of the candidates as we want.

We will illustrate the basic idea of recursive ranking with an abbreviated example.

Example 7. Let's apply recursive plurality with elimination to rank the candidates in the MAC election.

■ **Step 1.** We apply the plurality-with-elimination method to the original preference schedule and get a winner: D. (We did all the busy work earlier.) We now cross out the winner D on the preference schedule.

Number of voters	14	10	8	4	1
1st choice	A	C	~~D~~	B	C
2nd choice	B	B	C	~~D~~	~~D~~
3rd choice	C	~~D~~	B	C	B
4th choice	~~D~~	A	A	A	A

This results in the revised schedule

Number of voters	14	10	8	4	1
1st choice	A	C	C	B	C
2nd choice	B	B	B	C	B
3rd choice	C	A	A	A	A

which is the same as

Number of voters	14	19	4
1st choice	A	C	B
2nd choice	B	B	C
3rd choice	C	A	A

■ **Step 2.** Once again, we apply the plurality-with-elimination method to the revised schedule. B is eliminated first, and A second (we leave it to the reader to verify the details), leaving C as the winner. This means we rank C second in the original election. We now cross out C on the last preference schedule, leaving the revised preference schedule

Number of voters	14	23
1st choice	A	B
2nd choice	B	A

■ **Step 3.** The winner of the above election under plurality with elimination is B. This means we rank B third (and of course it follows that A is last).

The final ranking under *recursive plurality with elimination* is

Winner (president): D
2nd place (vice president): C
3rd place (treasurer): B
Last: A.

Note that this results in a ranking different from the one produced by the extended plurality-with-elimination method.   ■

While the recursive approach to ranking is seldom used in practice, it is worth exploring nonetheless. We leave it to the reader to give it a try. (Exercises 13, 17, 20, 23, 24, 27 through 30.)

CONCLUSION: FAIRNESS AND ARROW'S IMPOSSIBILITY THEOREM

When is a voting method fair? Throughout this chapter we have introduced several *fairness criteria*. Let's review what they are.

■ **Majority Criterion.** If there is a candidate or alternative that is the first choice of a majority of the voters, then that candidate or alternative should be the winner of the election.

■ **Condorcet Criterion.** If there is a candidate or alternative that wins in a one-to-one comparison between it and any other alternative, then that candidate or alternative should be the winner of the election.

■ **Monotonicity Criterion.** If candidate or alternative X is the winner of an election and, in a reelection, all the voters who change their preferences do so in a way that is favorable only to X, then X should still be the winner of the election.

■ **Independence of Irrelevant Alternatives Criterion.** If candidate or alternative X is the winner of an election, and one or more of the other candidates or alternatives are removed and the ballots re-counted, then X should still be the winner of the election.

All the above fairness criteria (as well as others that we did not mention—see, for example, Exercise 44) seem reasonable enough that one would expect any good voting method to satisfy them. We have seen that, in fact, none of the four methods we introduced in this chapter satisfies *all* the criteria, and as the reader may have guessed, no one knows of a voting method that does, a perfect or ideal voting method, if you will. This does not seem like a particularly remarkable situation since there are, after all, all kinds of things that we don't know but hope that someday we will. In this case, however, things are quite different. In 1952, Kenneth Arrow proved the famous Arrow's impossibility theorem which essentially states that no matter how hard we try, *there can be no perfect or ideal voting system*—fairness and democracy are inherently incompatible.

KEY CONCEPTS

Arrow's impossibility theorem
Borda count method
Condorcet criterion
Condorcet winner
essential tie
extended rankings methods
independence of irrelevant
   alternatives criterion
insincere (strategic) voting
majority criterion

method of pairwise comparisons
monotonicity criterion
plurality method
plurality-with-elimination method
preference ballot
preference schedule
rankings
recursive ranking methods
transitive preferences

EXERCISES

■ Walking

Ballot		Ballot		Ballot		Ballot	
1st	A	1st	C	1st	B	1st	A
2nd	B	2nd	D	2nd	D	2nd	B
3rd	C	3rd	B	3rd	C	3rd	C
4th	D	4th	A	4th	A	4th	D

Ballot		Ballot		Ballot		Ballot	
1st	C	1st	C	1st	A	1st	C
2nd	B	2nd	B	2nd	B	2nd	D
3rd	D	3rd	D	3rd	C	3rd	B
4th	A	4th	A	4th	D	4th	A

Ballot		Ballot		Ballot		Ballot	
1st	A	1st	A	1st	C	1st	A
2nd	B	2nd	B	2nd	B	2nd	B
3rd	C	3rd	C	3rd	D	3rd	C
4th	D	4th	D	4th	A	4th	D

1. The management of the XYZ Corporation has decided to treat their office staff to dinner. The choice of restaurants is The Atrium (A), Blair's Kitchen (B), The Country Cookery (C), and Dino's Steak House (D). Each of the 12 staff members is asked to submit a preference ballot listing his or her first, second, third, and fourth choices among these restaurants. The resulting preference ballots are as follows:

   (a) Write out the preference schedule for this election.
   (b) Is there a majority winner?
   (c) Find the winner of the election under the plurality method.

2. The Latin Club is holding an election to choose its president. There are three candidates, Arsenio, Beatrice, and Carlos (*A*, *B*, and *C* for short). The other 11 members of the club (the candidates are not allowed to vote) vote as shown below.

	Sue	Bill	Tom	Pat	Tina	Mary	Alan	Chris	Paul	Kate	Ron
1st place	C	A	C	A	B	C	C	A	C	B	A
2nd place	A	C	B	C	A	B	A	C	A	A	B
3rd place	B	B	A	B	C	A	B	B	B	C	C

    (a) Write out the preference schedule for this election.
    (b) Is there a majority winner?
    (c) Find the winner of the election under the plurality method.

Exercises 3 through 6 refer to the following: A math class is asked by the instructor to vote among four possible times for the final exam—A (December 15, 8:00 A.M.), B (December 20, 9:00 P.M.), C (December 21, 7:00 A.M.), and D (December 23, 11:00 A.M.). The class preference schedule is given by the following table:

	5	11	6	3	5	9	5	2	5
1st choice	A	A	A	A	B	B	B	C	D
2nd choice	B	B	C	D	A	A	D	B	B
3rd choice	C	D	B	B	C	D	A	A	A
4th choice	D	C	D	C	D	C	C	D	C

    3. (a) How many students in the class voted?
       (b) Find the winner of the election using the plurality method.

    4. Find the winner of the election using the Borda count method.

    5. Find the winner of the election using the plurality-with-elimination method.

    6. Find the winner of the election using the method of pairwise comparisons.

Exercises 7 through 13 refer to the election discussed in Example 5 with 5 candidates (A, B, C, D, and E), 22 voters, and the preference schedule given below.

Number of voters	5	3	5	3	2	4
1st choice	A	A	C	D	D	B
2nd choice	B	D	E	C	C	E
3rd choice	C	B	D	B	B	A
4th choice	D	C	A	E	A	C
5th choice	E	E	B	A	E	D

    7. Find the winner of the election using the Borda count method.

8. Find the winner of the election using

    (a) the plurality method
    (b) the method of plurality with elimination.

9. Find the ranking of the candidates using the extended plurality-with-elimination method.

10. Find the ranking of the candidates using the extended plurality method.

11. Find the ranking of the candidates using the extended pairwise comparisons method.

12. Find the ranking of the candidates using the extended Borda count method.

13. Find the ranking of the candidates using the recursive plurality-with-elimination method.

Exercises 14 through 20 refer to the election discussed in Example 1 with 5 alternatives (R, H, C, O, and S), 100 voters, and the preference schedule given below.

Number of voters	49	48	3
1st choice	R	H	C
2nd choice	H	S	H
3rd choice	C	O	S
4th choice	O	C	O
5th choice	S	R	R

14. Find the winner of the election using the Borda count method.

15. Find the winner of the election using the method of pairwise comparisons.

16. Find the winner of the election using the plurality-with-elimination method.

17. Find the ranking of the alternatives using the recursive plurality method.

18. Find the ranking of the alternatives using the extended plurality-with-elimination method.

19. Find the ranking of the alternatives using the extended Borda count method.

20. Find the ranking of the alternatives using the recursive pairwise comparisons method.

Exercises 21 through 27 refer to the election discussed in Example 3 with 5 candidates (A, B, C, D and E), 24 voters, and the preference schedule given below.

Number of voters	8	6	2	3	5
1st choice	A	B	C	D	E
2nd choice	B	D	A	E	A
3rd choice	C	E	E	A	D
4th choice	D	C	B	C	B
5th choice	E	A	D	B	C

21. Find the winner of the election using the Borda count method.

22. Find the winner of the election using the method of pairwise comparisons.

23. Find the ranking of the candidates using the recursive plurality method.

24. Find the ranking of the candidates using the recursive plurality-with-elimination method.

25. Find the ranking of the candidates using the extended plurality method.

26. Find the ranking of the candidates using the extended pairwise comparisons method.

27. Find the ranking of the candidates using the recursive Borda count method.

Exercises 28 through 30 refer to the MAC election. The preference schedule is given below.

Number of voters	14	10	8	4	1
1st choice	A	C	D	B	C
2nd choice	B	B	C	D	D
3rd choice	C	D	B	C	B
4th choice	D	A	A	A	A

28. Find the ranking of the candidates using the recursive Borda count method.

29. Find the ranking of the candidates using the recursive pairwise comparisons method.

30. Find the ranking of the candidates using the recursive plurality method.

■ Jogging

31. (a) Suppose that 50 players sign up for a round-robin tennis tournament (everyone plays everyone else). How many tennis matches must be scheduled?
    (b) If there are 50 people in a room and everyone kisses everyone else (on the cheek of course), how many kisses take place?

32. Show that the four voting methods we discussed in this chapter give the same winner when there are only two candidates, and in fact that the winner is just determined by straight majority.

33. **The mystery election problem**. You are given the following information about an election. There are 5 candidates and 21 voters. The preference schedule for the election has been lost—the only thing you know is that there were only *two* columns in the schedule. (As you can imagine, this is the key piece of information in the problem.)

    (a) Explain why there must be a majority winner in this election.
    (b) Explain why the majority winner is also the winner under the method of pairwise comparisons.
    (c) If the number of candidates is changed from 5 to any other number larger than 2, and the number of voters is changed from 21 to some other odd number larger than 3, explain why the arguments given in parts (a) and (b) above can be repeated.

34. An election is held between 4 candidates (*A*, *B*, *C*, and *D*) using the Borda count method. There are 11 voters. Suppose that after the ballots are in and the points are tallied, *B* gets 32 points, *C* gets 29 points, and *D* gets 18 points. How many points does *A* get, and why?

35. An election involving 5 candidates and 21 voters is held, and the results of the election are to be determined using the Borda count method. Unfortunately, the elections committee completely botches up the addition of the points. After a complaint is lodged by one of the candidates, the results of the election are recomputed using 4 points for a first-place vote, 3 points for a second-place vote, 2 points for a third-place vote, 1 point for a fourth-place vote, and 0 points for a fifth-place vote. Explain why computing the results this way gives the same election outcome as using the traditional Borda count method.

36. Explain why the plurality method satisfies the monotonicity criterion.

37. Explain why the plurality-with-elimination method satisfies the majority criterion.

38. Explain why the method of pairwise comparisons satisfies the majority criterion.

39. Explain why any method that violates the majority criterion must also violate the Condorcet criterion.

40. Give an example of an election decided under the Borda count method in which the Condorcet criterion is violated but the majority criterion is not.

■ **Running**

41. Give an example (do not use one given in the book) of an election decided under the plurality-with-elimination method in which the Condorcet criterion is violated.

42. Show that the method of pairwise comparisons satisfies the monotonicity criterion.

43. Show that if in an election with an odd number of voters there is no Condorcet winner, then in any ranking of the candidates based on the extended pairwise comparison method must result in at least two candidates that end up tied in the rankings.

44. The Pareto Criterion. The following fairness criterion was proposed by the Italian economist Vilfredo Pareto (1848–1943): "If *every* voter prefers alternative *X* over alternative *Y*, then a voting method should not choose *Y* as the winner." Show that none of the four voting methods discussed in the chapter can violate the Pareto criterion. (A separate analysis is needed for each of the four methods.)

45. Suppose the following was proposed as a fairness criterion: If a *majority* of the voters prefer alternative *X* to alternative *Y*, then the voting method should rank *X* above *Y*. Give an example to show that all four of the extended voting methods discussed in the chapter can violate this criterion. (*Hint*: Consider an example with no Condorcet winner.)

Arrow's impossibility theorem tells us that the search for an ideal voting method is hopeless. We should not take this to mean more than that. In particular, we should not draw the false conclusion that the study of voting methods is therefore a fruitless activity. Short of perfection, voting methods can be anywhere from very bad to very good, and the careful analysis of both new and old methods is still an active field of research among social scientists, economists, and mathematicians.

A relatively recent development in voting theory is the idea that voters should not be asked to order candidates according to preference (first choice is . . .; second choice is . . .; etc.), but rather to evaluate the candidates in absolute terms. Voting methods based on this type of balloting are known as **nonpreferential** (as opposed to **preferential**) voting methods. We will briefly mention one of the best known methods of this type: **approval voting**.

In approval voting, each voter votes for as many candidates as he or she wants. Each of these votes is simply a yes vote for the candidate, and it means that the voter approves of that candidate. The voters are not asked to list the candidates in order of preference. The candidate with the most approval votes wins the election.

The following table summarizes the results of a hypothetical election using approval voting.

					Voters			
Candidates	Sue	Bill	Tito	Prince	Tina	Van	Devon	Ike
A	Yes		Yes	Yes	Yes		Yes	Yes
B			Yes		Yes	Yes		
C	Yes				Yes	Yes	Yes	

The results of this election are as follows: winner A (six approval votes); second place, C (four approval votes); last place, B (three approval votes). Note that a voter can cast anywhere from no approval votes at all (such as Bill did above) to approval votes for all the candidates (such as Tina did above). It is somewhat ironical that the effect of Tina's vote is exactly the same as that of Bill's.

In the last few years a strong case has been made suggesting that approval voting is a considerable improvement over any of the more traditional preferential voting methods, and while it is not without flaws, it does indeed have several things going for it. We will mention three that are relevant to our discussion and which are important in the context of national politics.

First, approval voting is easy to understand and simple to implement in practice. Second, the approval voting process is unaffected by the number of candidates. In particular, if new candidates throw their hats into the ring, the voters can approve or disapprove of them without having to reconsider their votes for the original candidates. Last but not least, approval voting encourages voter turnout. The reason for this is psychological: Voters are more likely to vote when they feel they can make intelligent decisions, and unquestionably it is easier for a voter to give an intelligent answer to the question, Do you approve of this candidate—yes or no? than it is to the question, Which candidate is your first choice, second choice, etc.? The latter requires a much deeper knowledge of the candidates, and in today's complex political world it is a knowledge that very few voters have.

**APPENDIX 2:
Realpolitik. Elections
in the Real World**

Olympic Venues.   Selecting the city in which the next Olympic Games are to be held is a decision entrusted to the members of the International Olympic Committee. It is a decision that has tremendous economic and political impact on the cities involved, and it goes without saying that it always generates a fair amount of controversy. The actual voting method used to select the winner is a slight variation of the plurality-with-elimination method which we studied in the chapter. In this case, instead of indicating their preferences all at once, the voters make their preferences known one round at a time. We illustrate the procedures with the actual details of how Atlanta was chosen to host the 1996 Summer Olympic Games.

On September 18, 1990, the 86 members of the International Olympic Committee met in Tokyo, Japan, to vote on selection of the site for the 1996 Summer Olympics. Six cities made bids: Athens (Greece), Atlanta (USA), Belgrade (Yugoslavia), Manchester (England), Melbourne (Australia), and Toronto (Canada). In each round, the delegates voted for just one city, and the city with the fewest votes was eliminated. The voting went as follows:

■ **Round 1.**

City	Athens	Atlanta	Belgrade	Manchester	Melbourne	Toronto
Votes	23	19	7	11	12	14

Belgrade is eliminated in round 1.

■ **Round 2.**

City	Athens	Atlanta	Manchester	Melbourne	Toronto
Votes	23	20	5	21	17

Manchester is eliminated in round 2.

■ **Round 3.**

City	Athens	Atlanta	Melbourne	Toronto
Votes	26	26	16	18

Melbourne is eliminated in round 3.

■ **Round 4.**

City	Athens	Atlanta	Toronto
Votes	34	30	22

Toronto is eliminated in round 4.

■ **Round 5.**

City	Athens	Atlanta
Votes	35	51

Athens is eliminated in round 5. Atlanta wins!

It is worth noting that while in theory this method is equivalent to plurality with elimination, the fact that the voters are allowed to indicate their preferences

Atlanta gets the 1996 Summer Olympics. (AP/ Wide World Photos)

in rounds rather than all at once allows for a certain irrational (or should we say fickle?) behavior on the part of the voters. After Belgrade was eliminated in round 1, six of the voters that voted for Manchester changed their votes. What connection is there, one wonders, between the elimination of Belgrade and Manchester's merit as an Olympic site? (Likewise, after Manchester was eliminated in round 2, five voters that had voted for Melbourne changed their votes.) One of the advantages of asking the voters to indicate their preferences all at once (by means of a preference ballot) is that this kind of irrational voting is eliminated.

There is one thing that is even more disturbing about the selection procedure: As we know, this voting method can violate the monotonicity criterion. A city may actually lose the election because of *too many* votes in an early round.

Consider, for example, the following hypothetical scenario. There are three cities in the running for the next Olympic site: Antioch, Babylon, and Carthage. Two days before the election Antioch has the support of 33 delegates, Babylon has the support of 24 delegates, and Carthage has the support of 29 delegates. Let's assume that the second choice of all Antioch supporters is Babylon, but that the second choice of all Babylon supporters is Carthage. If the election had been held two days earlier, it would have gone like this:

■ Round 1.

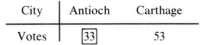

City	Antioch	Babylon	Carthage
Votes	33	24 ⟶	29

Babylon is eliminated, and all of Babylon's votes are shifted to Carthage.

■ Round 2.

City	Antioch	Carthage
Votes	33	53

Carthage wins the election.

Over the next two days, the Carthage representatives press their advantage and continue their lobbying efforts trying to win even more delegates. Things go so well for them that 12 of the 33 Antioch supporters are persuaded to vote for Carthage. When the election is finally held, this is what happens.

■ Round 1.

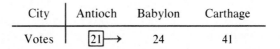

City	Antioch	Babylon	Carthage
Votes	21 ⟶	24	41

Antioch is eliminated, and all of Antioch's votes are shifted to Babylon.

■ **Round 2.**

City	Babylon	Carthage
Votes	45	41

They are dancing in the streets of Babylon.

The Associated Press College Football Poll.   In this election (which is held once a week during the college football season), the voters are a panel of approximately 60 national sportswriters, and the candidates are the top college football teams in the country. Each voter ranks their top 25 choices, with 25 points for first place, 24 for second place, . . ., and 1 point for twenty-fifth place. The team with the highest number of total points is ranked first in the poll, the one with the next highest number of points second, etc. Essentially this method of ranking is the extended Borda count method with one subtle variation: The list of candidates is open-ended (any college football team in the country can in principle be considered a candidate). As we know, a problem with this voting method is that it can violate the majority criterion: A football team could have a majority of first-place votes and yet not be ranked first. In practice, the probability of this happening is very small.   ■

The Academy Awards.   The Academy of Motion Picture Arts and Sciences gives its annual Academy Awards (''Oscars'') for various achievements in connection with motion pictures (best picture, best director, best actress, etc.). The winner in each category is chosen by means of an election held among the eligible members of the Academy. The election process varies slightly from award to award and is quite complicated. For the sake of brevity we will describe the election process for best picture. (The process is almost identical for each of the major awards.) The election takes place in two stages: (1) the nomination stage in which the five top pictures are nominated, and (2) the final balloting for the winner.

We describe the second stage first because it is so simple: Once the five top pictures are nominated, each eligible member of the Academy is asked to vote for one picture, and the winner is chosen by simple plurality. Because the number of voters is large (somewhere between 4000 and 5000), ties are not likely to occur, but if they do, they are not broken. Thus, it is possible for two candidates to share an award.

The process for selecting the five nominations is considerably more complicated and is based on a voting method called **single transferable voting**. Each eligible member ot the Academy is asked to submit a preference ballot with the names of their top five choices ranked from first to fifth. Based on the total number of valid ballots submitted, the minimum number of votes needed to get a nomination (called the **quota**) is established, and any picture with enough first-place votes to make the quota is automatically nominated.

The quota is always chosen to be a number that is over one-sixth (16.66%) but not more than one-fifth (20%) of the total number of valid ballots cast. (Setting the quota this way ensures that it is impossible for six or more pictures to get automatic nominations.) While in theory it is possible for five pictures to make the quota right off the bat and get an automatic nomination (in which case the nomination process is over), this has never happened in practice. In fact, what usually happens is that there are no pictures that make the quota automatically. Then, the picture with the fewest number of first-place votes (say X) is eliminated, and on all the ballots that originally had X as the first choice, X's name is crossed off the top and all the other pictures are moved up one spot. The ballots are then counted again. If there are still no pictures that make the quota, the process of elimination is repeated. Eventually, there will be one or more pictures that make the quota and are nominated.

The moment that one or more pictures are nominated there is a new twist: Nominated pictures ''give back'' to the other pictures still in the running (not nominated but not eliminated either) their ''surplus'' votes. This process of giving back votes (called a **transfer**) is best illustrated with an imaginary example. Suppose that the quota is 400 (a nice, round number) and at some point a picture (say Z) gets 500 first-place votes, enough to get itself nominated. The surplus for Z is $500 - 400 = 100$ votes, and these are votes that Z doesn't really need. For this reason the 100 surplus votes are taken away from Z and divided *fairly* among the second-place choices on the 500 ballots cast for Z. The way this is done may seem a little bizarre, but it makes perfectly good sense. Since there are 100 surplus votes to be divided into 500 equal shares, each second-place vote on the 500 ballots cast for Z is worth $\frac{100}{500} = \frac{1}{5}$ vote. While one-fifth of a vote may not seem like much, enough of these fractional votes can make a difference and help some other picture or pictures make the quota. If that's the case, then once again the surplus or surpluses are transferred back to the remaining pictures following the procedure described above; otherwise, the process of elimination is started up again. Eventually, after several possible cycles of eliminations, transfers, eliminations, transfers, . . ., five pictures get enough votes to make the quota and be nominated, and the process is over.

The method of single transferable voting is not unique to the Academy Awards. It is used to elect officers in various professional societies as well as the members of the Irish Senate. ■

**REFERENCES
AND FURTHER
READINGS**

1. Arrow, Kenneth J., *Social Choice and Individual Values*. New York: John Wiley & Sons, Inc., 1963.

2. Black, Duncan, *The Theory of Committees and Elections*. New York: Cambridge University Press, 1968.

3. Brams, Steven J., and Peter C. Fishburn, *Approval Voting*. Boston: Birkhäuser, 1982.

4. Fishburn, Peter C., *The Theory of Social Choice*. Princeton, N.J.: Princeton University Press, 1973.

5. Fishburn, Peter C., and Steven J. Brams, "Paradoxes of Preferential Voting," *Mathematics Magazine,* 56 (1983), 207–214.

6. Gardner, Martin, "Mathematical Games (From Counting Votes to Making Votes Count: The Mathematics of Elections)," *Scientific American,* 243 (October 1980), 16–26.

7. Niemi, Richard G., and William H. Riker, "The Choice of Voting Systems," *Scientific American,* 234 (June 1976), 21–27.

8. Straffin, Philip D., Jr., *Topics in the Theory of Voting,* UMAP Expository Monograph. Boston: Birkhäuser, 1980.

2 Weighted Voting Systems

The first principle of a civilized state is that power is legitimate only when it is under contract.

WALTER LIPPMANN*

The Power Game

In many voting situations, the *one person–one vote* principle is not justified. In a diverse society, it is in the very nature of things that individuals or groups are not always equal, and this needs to be recognized in the way that the voting is conducted by giving some voters more say than others. At a corporate shareholder's meeting, for example, each shareholder has as many votes as the number of shares he or she owns; in many committee situations the committee chairperson has tie-breaking powers not enjoyed by the other committee members; and even at home, when it's time to decide where to go on vacation, it always seems that mom and dad have more say than the kids.

In this chapter we will analyze voting systems in which the voters are not all equal. These are called **weighted voting systems** (the opinions of some voters have more weight than the opinions of others). To keep things simple we will assume that the voting will always be on two alternatives or candidates. Any vote involving only two alternatives can be thought of as a yes-no vote and is generally referred to as a **motion**.

We know from Chapter 1 (see Exercise 32) that when dealing with motions, we don't have to worry about the choice of voting method be-

* Walter Lippmann, *The Public Philosophy* (Boston: Little, Brown and Co., 1955).

cause all reasonable voting methods reduce to majority rule. The issue that will primarily concern us in this chapter is **power**: Who has it and how much of it do they really have? (Note that the issue of power is not very interesting in a one person–one vote situation. Clearly, if all voters are equal, then any voter has exactly the same amount of power as any other voter.) Understanding how power is distributed in a voting system is useful in several ways. In the first place it allows individuals to respond properly to a situation (if you are going to butter-up someone, you might as well butter-up those who have the most power). More importantly, it allows institutions to rationally analyze power so that abuses may be prevented (does the Constitution give the vice president of the United States too much power by allowing him to cast the tie-breaking vote in the Senate? We'll see about that later.)

WEIGHTED VOTING SYSTEMS

Before we pursue the trappings of power in more detail, we will introduce some terminology and discuss some examples of weighted voting systems.

■ Terminology

Every weighted voting system is characterized by three elements: the players, the weights of the players, and the quota. The **players** are just the voters themselves. (As much as possible we will adhere to the convention that the word ''voter'' refers to the one person–one vote situation, and the word ''player'' to the weighted voting situation.) We will use the symbols $P_1, P_2, \ldots, P_N$ throughout this chapter to represent the names of the players—it is a little less personal but a lot more convenient than referring to Reggie, Betty, Wally, and Veronica. Each player controls a certain number of votes, and this number is called the player's **weight**. We will use the symbols $w_1, w_2, \ldots, w_N$ to represent the weights of P_1, $P_2, \ldots, P_N$, respectively. Finally, there is the **quota**, the minimum number of votes needed to pass a motion. We will use the letter q to denote the quota.

It is important to note that the quota q need not be restricted to be exactly a strict majority of the votes. There are many voting situations in which a strict majority of the votes is not enough to pass a motion— the rules may stipulate a different definition of what is needed for passing. When the U.S. Senate is attempting to override a presidential veto, for example, the rules state that two-thirds of the votes are needed. In other situations the rules may stipulate that 75% of the votes are needed or 83% (why not?) or even 100% (unanimous consent). In fact, any number can be a reasonable choice for the quota q, as long as it is more than half of the total number of votes but not more than the total number of votes. To put it more bluntly,

$$\frac{w_1 + w_2 + \cdots + w_N}{2} < q \leq w_1 + w_2 + \cdots + w_N.$$

■ Notation and Examples

A convenient way to describe a weighted voting system is

$$[q: w_1, w_2, \ldots, w_N].$$

The quota is always given first, followed by a colon and then the respective weights of the individual players.

Example 1.   [25: 8, 6, 5, 3, 3, 3, 2, 2, 1, 1, 1, 1].
This is a weighted voting system with 12 players $(P_1, P_2, \ldots, P_{12})$. P_1 has 8 votes, P_2 has 6 votes, P_3 has 5 votes, etc. The total number of votes is 36. The quota is 25.   ■

Example 2.   [7: 5, 4, 4, 2].
This weighted voting system doesn't make any sense because the quota (7) is less than half of the total number of votes (15). If P_1 and P_4 voted yes and P_2 and P_3 voted no, both groups would win. This is the mathematical equivalent of anarchy, and we will not consider this a legal weighted voting system.   ■

Example 3.   [17: 5, 4, 4, 2].
Here the quota is too high. In this weighted voting system no motion could ever pass. We can't allow this to happen either.   ■

Example 4.   [11: 4, 4, 4, 4, 4].
In this weighted voting system all 5 players are equal. To pass a motion at least 3 out of the 5 players are needed. Note that if the quota ($q = 11$) were changed to 12, the situation would still remain the same—at least 3 out of the 5 players would be needed. What we really have here is a one person–one vote, strict majority situation, and we can just as well describe it with the voting system [3: 1, 1, 1, 1, 1].   ■

Example 5.   [15: 5, 4, 3, 2, 1].
Here we have 5 players with a total of 15 votes. Since the quota is 15, the only way a motion can pass is by unanimous consent of the players. How does this voting system differ from the voting system [5: 1, 1, 1, 1, 1]? Well, this one also has 5 players, and the only way a motion can pass is by unanimous consent of the players. So for all practical purposes these two voting systems represent the same situation.   ■

The surprising conclusion of Example 5 is that the weighted voting system [15: 5, 4, 3, 2, 1] represents a one person–one vote situation. This seems like a contradiction only if we think of a one person–one vote situation as implying that all players have an *equal number of votes rather than an equal say in the outcome of the election*. These two things are clearly not the same!

■ Power; More Terminology; More Examples

As Example 5 makes abundantly clear, weight alone cannot be used to measure power because a player's power is not directly proportional to the number of votes he or she holds. A frequent mistake is to interpret this to mean that there is *no connection* between weight and power. If player X and player Y have the same number of votes, then they will always have the same amount of power. However, if player X has four votes and player Y has two votes, we cannot conclude that player X has twice as much power as player Y. In fact just about anything is possible regarding the relative power of X and Y (of course, X can never be allowed to have *less* power than Y—that would be a gross violation of the idea of power!) (See Exercises 23 and 24.)

Before we formally describe the way in which power is defined mathematically (we will actually give two different definitions), let's consider a few more examples of weighted voting systems, now focusing in an intuitive way on the notion of power.

Example 6.   [11: 12, 5, 4].

This is a situation in which a single player (P_1) controls enough votes to pass any measure single-handedly. Such a player has all the power by himself, and not surprisingly, we call such a player a dictator.

Formally, we define a **dictator** as a player whose weight is bigger than or equal to the quota. Notice that whenever there is a dictator, all the other players have absolutely no power. A player without power is called a **dummy** (this is not a reflection on the player's intellect but rather on the fact that such a player has no say in the outcome of an election.) ■

Example 7.   [12: 11, 5, 4, 2].

Here we have a situation in which a player (P_1), while not a dictator, has enough votes to prevent any motion she doesn't like from passing. Even if the remaining players all band together, they can't force a motion against her will.

A player that is not a dictator but can single-handedly prevent any group of players from passing a motion is said to have **veto power.**   ■

Example 8.   [101: 99, 98, 3].

How is power distributed in this weighted voting system? At first glance it appears that P_1 and P_2 have lots of power while P_3 has very little power (if any). On closer inspection, however, we notice that *any two* of the three players can join forces to pass a motion so that in fact P_3, with his measly three votes has as much say as either of the other two players. This is, in fact, another hard-to-swallow but honest-to-goodness one person–one vote situation: All players have an equal say in this voting system.   ■

THE BANZHAF POWER INDEX

We are almost ready to formally introduce our first mathematical interpretation of power in a weighted voting system. This particular definition of power was suggested by John Banzhaf[1] in 1965.

Let's analyze the weighted voting system [101: 99, 98, 3] (Example 8) in a little more detail. Although this example itself is fairly simple, we will use it to introduce some important concepts.

Which groups of players could join forces to form a winning combination? Clearly, there are four such groups:

■ P_1 and P_2 (this group controls 197 votes)

■ P_1 and P_3 (this group controls 102 votes)

■ P_2 and P_3 (this group controls 101 votes, just enough to win)

■ P_1, P_2, and P_3 (this group controls all the votes).

From now on we will adhere to the standard language of voting theory and call any group of players that join forces to vote together a **coalition**. We use the word "coalition" in a rather generous way and will allow a single player to form a coalition all by himself. The total number of votes controlled by a coalition is called the **weight of the coalition**. Of course, some coalitions have enough votes to win and some don't. We will call the former **winning coalitions,** and the latter **losing coalitions**.

For this example there is a total of seven coalitions which we will list using set notation:

	Coalition	Coalition Weight	Winning or Losing?
a	$\{P_1\}$	99	Losing
b	$\{P_2\}$	98	Losing
c	$\{P_3\}$	3	Losing
d	$\{P_1, P_2\}$	197	Winning
e	$\{P_1, P_3\}$	102	Winning
f	$\{P_2, P_3\}$	101	Winning
g	$\{P_1, P_2, P_3\}$	200	Winning

[1] Banzhaf, who was a lawyer and not a mathematician, was mostly concerned with issues of equity and fairness in state and local systems of government.

If we now analyze the winning coalitions, we note the following: In coalitions d, e, and f both players are needed for the win (if either player were to desert the coalition, the rest of the coalition would lose); in coalition g no single player is essential (if any single player were to desert the coalition, the rest of the coalition would still win).

A player whose desertion turns a winning coalition into a losing coalition obviously holds a certain amount of power, and we will call such a player a **critical player** for the coalition. Notice that a winning coalition often has more than one critical player, and occasionally a winning coalition has no critical players. Losing coalitions never have critical players.

The critical player concept is the basis for the definition of the **Banzhaf power index**. Banzhaf's key idea is that a player's power is proportional to the number of times that player is critical, so that the more often the player is critical, the more power he or she holds.

In our example each player is critical twice, so they all have equal power. Since there are three players, we can say that each player holds one-third of the power.

We can now formalize our approach for finding the Banzhaf power index of a player (let's suppose it is John Doe, but let's nickname him P) in an arbitrary weighted voting system with N players. The general procedure would go as follows:

■ **Step 1.** Make a list of all possible coalitions.

■ **Step 2.** Determine which are winning coalitions.

■ **Step 3.** In each winning coalition, determine the critical players.

■ **Step 4.** Count the total number of times P is critical and call this number B.

■ **Step 5.** Count the total number of times all players are critical and call this number T.

The Banzhaf power index of P is then given by the ratio B/T.

Before we go on to the next example, we mention the following useful fact: In a weighted voting system with N players, the total number of possible coalitions is $2^N - 1$. (Exercise 38 outlines a proof of this fact.)

Example 9. Let's find the Banzhaf power index of each of the players in the weighted voting system [4: 3, 2, 1].

■ **Step 1.** There are three players, and therefore $2^3 - 1 = 7$ coalitions. They are

$$\{P_1\}, \{P_2\}, \{P_3\}, \{P_1, P_2\}, \{P_1, P_3\}, \{P_2, P_3\}, \{P_1, P_2, P_3\}.$$

■ **Step 2.** The winning coalitions are

$$\{P_1, P_2\}, \{P_1, P_3\}, \text{ and } \{P_1, P_2, P_3\}.$$

■ **Step 3.**

Winning Coalition	Critical Players
$\{P_1, P_2\}$	P_1 and P_2
$\{P_1, P_3\}$	P_1 and P_3
$\{P_1, P_2, P_3\}$	P_1 only

■ **Step 4.** P_1 is critical three times.
P_2 is critical one time.
P_3 is critical one time.

■ **Step 5.** $T = 5$.

The Banzhaf power index of each of the players is, respectively,

$$P_1: \tfrac{3}{5} \; ; P_2: \tfrac{1}{5} \; ; P_3: \tfrac{1}{5} \; .$$ ■

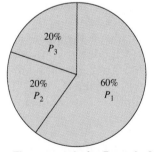

The power pie for Example 9.

We will refer to the complete listing of the Banzhaf power indexes as the **Banzhaf power distribution** of the weighted voting system.

Some people prefer to deal with percentages rather than fractions. Percentagewise, the Banzhaf power distribution of the weighted voting system in Example 9 is

$$P_1: 60\%; P_2: 20\%; P_3: 20\%.$$

Example 10. Let's consider the weighted voting system [6: 4, 3, 2, 1] and find its Banzhaf power distribution using the general procedure described earlier. Table 2-1 shows the situation after steps 1 through 3 have been carried out. (The reader is encouraged to fill in the details.) The critical players in each coalition have been underlined.

Coalition	Weight	Winning/Losing
$\{P_1\}$	4	L
$\{P_2\}$	3	L
$\{P_3\}$	2	L
$\{P_4\}$	1	L
$\{\underline{P_1},\underline{P_2}\}$	7	W
$\{\underline{P_1},\underline{P_3}\}$	6	W
$\{P_1,P_4\}$	5	L
$\{P_2,P_3\}$	5	L
$\{P_2,P_4\}$	4	L
$\{P_3,P_4\}$	3	L
$\{\underline{P_1},P_2,P_3\}$	9	W
$\{\underline{P_1},\underline{P_2},\underline{P_4}\}$	8	W
$\{\underline{P_1},\underline{P_3},\underline{P_4}\}$	7	W
$\{\underline{P_2},\underline{P_3},\underline{P_4}\}$	6	W
$\{P_1,P_2,P_3,P_4\}$	10	W

Table 2-1

The power pie for Example 10.

The Banzhaf power distribution is: P_1: $\frac{5}{12}$; P_2: $\frac{3}{12}$; P_3: $\frac{3}{12}$; P_4: $\frac{1}{12}$.

(Note that the power indexes add up to 1. This is not an accident! This fact provides a useful check on your calculations.) ∎

Example 11. Consider the weighted voting system [15: 9, 6, 3, 3, 3].

Again we want to apply the general procedure described earlier to this example to find the Banzhaf power index of each player. Rather than plow straight ahead, we are going to be lazy (as well as smart) and try to take some shortcuts. Instead of starting with a list of all the possible coalitions (there is a total of $2^5 - 1 = 31$ of them), why don't we try to figure out directly which are the winning coalitions. After all, they are the ones that count and there really are just a few of them. Table 2-2 shows just the winning coalitions with their critical players underlined.

The Banzhaf power distribution of the weighted voting system is

P_1: $\frac{11}{25}$ $= 44\%$; P_2: $\frac{5}{25}$ $= 20\%$; P_3: $\frac{3}{25}$ $= 12\%$; P_4: $\frac{3}{25}$ $= 12\%$; P_5: $\frac{3}{25}$ $= 12\%$. ∎

Example 12. A committee consists of four members, A, B, C, and D. Each committee member has one vote, and the quota is defined by strict majority except that in case of a 2-2 tie, the coalition containing the chairperson (A) wins. What is the Banzhaf power distribution in this committee?

Although we don't have the player's weights to work with, we have all the necessary information to play the game. The winning coalitions

Winning Coalitions

$\{\underline{P_1},\underline{P_2}\}$	Only possible winning two-player coalition.
$\{\underline{P_1},\underline{P_2},P_3\}$	
$\{\underline{P_1},\underline{P_2},P_4\}$	
$\{\underline{P_1},\underline{P_2},P_5\}$	
$\{\underline{P_1},\underline{P_3},\underline{P_4}\}$	Winning three-player coalitions must contain P_1.
$\{\underline{P_1},\underline{P_3},\underline{P_5}\}$	
$\{\underline{P_1},\underline{P_4},\underline{P_5}\}$	
$\{\underline{P_1},P_2,P_3,P_4\}$	
$\{\underline{P_1},P_2,P_3,P_5\}$	
$\{\underline{P_1},P_2,P_4,P_5\}$	Any four-player coalition wins.
$\{\underline{P_1},P_3,P_4,P_5\}$	
$\{P_2,P_3,\underline{P_4},P_5\}$	
$\{P_1,P_2,P_3,P_4,P_5\}$	The **grand coalition** (all players).

Table 2-2

are (1) any two-player coalition if it includes the chairperson, (2) any three-player coalition, and (3) the coalition containing all four players.

Table 2-3 shows the winning coalitions with the critical players underlined.

Winning Coalitions

$\{\underline{A},\underline{B}\}$
$\{\underline{A},\underline{C}\}$
$\{\underline{A},\underline{D}\}$
$\{\underline{A},B,C\}$
$\{\underline{A},B,D\}$
$\{\underline{A},C,D\}$
$\{\underline{B},\underline{C},\underline{D}\}$
$\{A,B,C,D\}$

Table 2-3

The Banzhaf power distribution is

A: $\frac{6}{12} = 50\%$; B: $\frac{2}{12} \approx 16.67\%$; C: $\frac{2}{12} \approx 16.67\%$; D: $\frac{2}{12} \approx 16.67\%$.

The seemingly harmless tie-breaking rule gives the chairperson three times as much power as that of any of the other committee members. Knowing this ahead of time might make a difference in how one goes about choosing the chairperson. ■

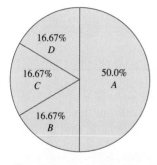

The power pie for Example 12.

Example 13. A committee consists of five members, the chairperson (Dr. K) and four other members of equal standing (B, C, D, and E). In this committee the quota is defined by strict majority, but the chairperson never votes except to break a tie. How is power distributed in this voting system?

Once more, let's write down the winning coalitions with the critical players underlined.

Winning coalitions without the chairman	$\{\underline{B}, \underline{C}, \underline{D}\}$ $\{\underline{B}, \underline{C}, \underline{E}\}$ $\{\underline{B}, \underline{D}, \underline{E}\}$ $\{\underline{C}, \underline{D}, \underline{E}\}$ $\{B, C, D, E\}$	Winning coalitions with the chairman	$\{\underline{K}, \underline{B}, \underline{C}\}$ $\{\underline{K}, \underline{B}, \underline{D}\}$ $\{\underline{K}, \underline{B}, \underline{E}\}$ $\{\underline{K}, \underline{C}, \underline{D}\}$ $\{\underline{K}, \underline{C}, \underline{E}\}$ $\{\underline{K}, \underline{D}, \underline{E}\}$

The Banzhaf power distribution is

$$K: \tfrac{6}{30} \; ; B: \tfrac{6}{30} \; ; C: \tfrac{6}{30} \; ; D: \tfrac{6}{30} \; ; E: \tfrac{6}{30} \; .$$

Surprise! All the members (including the chairperson) have the same amount of power. Clearly, this voting system represents a one person–one vote strict majority situation, and it can be formally described by [3: 1, 1, 1, 1, 1]. ■

The same situation described in Example 13 except on a larger scale exists in the U.S. Senate, where the vice-president of the United States (what's his name?) votes only to break a tie. An analysis similar to the one in Example 13 would show that, assuming all 100 senators are voting, he has exactly the same amount of power as any other member of the Senate.

■ Applications of the Banzhaf Power Index

The United Nations Security Council. The main body responsible for maintaining the international peace and security of nations is the Security Council of the United Nations. At present the composition of the Security Council is as follows: There are 5 **permanent members** of the Council (Britain, France, the People's Republic of China, the Soviet Union, and the United States), plus 10 additional **nonpermanent** slots filled by other countries on a rotating basis.[2] According to the voting rules of the Security Council, each of the permanent members has veto power, so that a res-

---

[2] When the original Security Council was set up by the League of Nations in 1945, the number of nonpermanent members was only 6. The number was increased to 10 in 1963.

olution cannot pass unless each of them votes yes. In addition, at least 4 of the 10 nonpermanent members must also vote yes. It is a challenging (but not unreasonable) exercise (Exercise 36) to show that under these rules the U.N. Security Council can be formally described as the weighted voting system [39: 7, 7, 7, 7, 7, 1, 1, 1, 1, 1, 1, 1, 1, 1, 1]. Once the Security Council is described this way, it is possible to compute the Banzhaf power index of each country. While the calculations are not difficult, they go beyond the scope of this book so we will omit them. The long and the short of it is that the Banzhaf power index of each permanent member is $\frac{848}{5080} \approx 16.7\%$, while the Banzhaf power index of each nonpermanent member is $\frac{84}{5080} \approx 1.65\%$. In retrospect, one could ask: When the U.N. Security Council was originally set up, was it really intended that a permanent member have more than 10 times as much power as a nonpermanent member? The answer is most likely no, and in an ideal world one would expect that the voting rules of the Security Council would be reconsidered. Needless to say, this is not about to happen soon, as the 5 permanent members are not likely to voluntarily give up their power. ■

The preceding example illustrates the most common way in which a measure of power such as the Banzhaf power index is used in practice—what one might call an after-the-fact approach. First, a weighted voting system is set up, and the rules by which the body will operate are "negotiated." Once these rules are in effect, the power of each of the players can be measured and then, looking back in retrospect, it can be determined if the power indexes of the players are in agreement with the original intent of the voting rules. More often than not the answer is no, and making appropriate changes in the voting rules may be the proper thing to do. Actually doing it often proves difficult or impossible.

The United Nations Security Council in session. (Yutaka Nagata/United Nations)

Our next example shows how a measure of power can be used a priori to actually define the rules by which a weighted voting system will operate. To be more specific, in this case the starting point is the desired power distribution and the mathematical problem is reversed: How many votes should each player get so that the power distribution is just right (or close to it)?

The Case of Cortland County, New York. Cortland County, New York, is made up of 19 different districts. Up until 1971 each district had exactly one vote in the county legislature regardless of its population. In 1971, in response to a suit brought against the county, the state court declared this arrangement unconstitutional. In essence, the court ordered the county to devise a voting system for the legislature based on two requirements: (1) the power of each district should be approximately proportional to its population, and (2) the Banzhaf power index should be used as the official measure of power to accomplish requirement 1. Based on this ruling, a new law took effect in 1974, setting up the Cortland County Legislature as a weighted voting system in which each district has one legislator but each legislator casts a different number of votes (ranging anywhere from a low of 19 to a high of 32). The actual number of votes assigned to each legislator is determined by a computer program.[3]

THE SHAPLEY-SHUBIK POWER INDEX

A second way of measuring power in weighted voting systems was given by Lloyd Shapley and Martin Shubik[4] in 1954. The key difference between their definition of power and Banzhaf's centers around the concept of a **sequential coalition**. According to Shapley and Shubik, coalitions are formed sequentially: Every coalition starts with a first player, she may then be joined by a second player, then a third, and so on. Thus, to an already complicated situation we are adding one more wrinkle—the question of the order in which the players joined the coalition.

Let's illustrate the difference with a simple example. According to Banzhaf, the coalition $\{P_1, P_2, P_3\}$ represents the fact that P_1, P_2, and P_3 have joined forces and will vote together. We don't care and don't even consider who joined the coalition first, second, or third. According to Shapley and Shubik the same three players can form *six* different coalitions: $\langle P_1, P_2, P_3 \rangle$ (this means P_1 started the coalition, P_2 joined in second, and P_3 third); $\langle P_1, P_3, P_2 \rangle$; $\langle P_2, P_1, P_3 \rangle$; $\langle P_2, P_3, P_1 \rangle$; $\langle P_3, P_1, P_2 \rangle$; $\langle P_3, P_2, P_1 \rangle$. Note the change in notation: From now on the notation $\langle \ \rangle$ will indicate that we are dealing with a sequential coalition; that is, we care about the order in which the players are listed.

[3] For specific details (and much more), the curious reader is referred to reference 3 at the end of the chapter.
[4] Lloyd S. Shapley was a mathematician at the Rand Corporation, and Martin Shubik an economist at Yale University.

■ Factorials

Before we continue with our discussion of the Shapley-Shubik approach to power, let's take a brief detour and consider the following mathematical question: For an arbitrary positive integer N, how many sequential coalitions containing N players are there? With two players (P_1 and P_2) there are two sequential coalitions, $\langle P_1, P_2 \rangle$ and $\langle P_2, P_1 \rangle$; we have already seen that with three players there are six sequential coalitions; with four players there are 24 (we'll let the reader write them all down); with five players there are 120; etc. So, what's the magic formula?

> The number of sequential coalitions with N players is
>
> $1 \times 2 \times 3 \times \cdots \times N$.

A good justification for the above statement will be given later in the book (see the multiplication principle in Chapter 15), but for now we will take its validity on faith. By the way, the number $1 \times 2 \times 3 \times \cdots \times N$ will show up several times in this book—in fact, it is a frequent flyer in the mathematical skies—and it is called the **factorial of N** and abbreviated $N!$. The factorial of 5, for example, is written $5!$ and equals 120 (because $1 \times 2 \times 3 \times 4 \times 5 = 120$), while $10! = 3,628,800$ (check it out!).

> For any positive integer N, the **factorial** of N is
>
> $N! = 1 \times 2 \times 3 \times \cdots \times N$.

■ Back to the Shapley-Shubik Power Index

Suppose that we have a weighted voting system with N players. We know from the preceding discussion that there is a total of $N!$ different sequential coalitions containing *all* the players. In each of these coalitions there is one player that tips the scales—the moment that he joins the coalition, the coalition changes from a losing to a winning coalition. We call such a player a **pivotal player** for the sequential coalition. The underlying principle of the Shapley-Shubik approach is that the pivotal player deserves special recognition. After all, the players who came before him did not have it, and the players who came after him are a bunch of Johnny-come-latelies. (Note that we can talk about ''before'' and ''after'' only because we are considering sequential coalitions.) According to Shapley and Shubik then, a player's power can be measured by the total number of times he is pivotal in relation to all other players.

The formal description of the procedure for finding the Shapley-Shubik power index of a player P is as follows:

■ **Step 1.**   Make a list of all sequential coalitions containing *all N* players. There are $N!$ of them.

■ **Step 2.**   In each sequential coalition determine the pivotal player. There is one in each sequential coalition.

■ **Step 3.**   Count the total number of times P is pivotal and call this number S.

The Shapley-Shubik power index of P is then given by the ratio $S/N!$. Listing the Shapley-Shubik power indexes of all the players gives the **Shapley-Shubik power distribution** of the weighted voting system.

Example 14.   Let's consider, once again, the weighted voting system [4: 3, 2, 1]. This is the same weighted voting system we discussed in Example 9, but this time we will find its Shapley-Shubik power distribution.

■ **Step 1.**   There are $3! = 6$ sequential coalitions of the three players. They are

$\langle P_1, P_2, P_3 \rangle$, $\langle P_1, P_3, P_2 \rangle$, $\langle P_2, P_1, P_3 \rangle$, $\langle P_2, P_3, P_1 \rangle$, $\langle P_3, P_1, P_2 \rangle$, $\langle P_3, P_2, P_1 \rangle$.

■ **Step 2.**

Sequential Coalition	Pivotal Player
$\langle P_1, P_2, P_3 \rangle$	P_2
$\langle P_1, P_3, P_2 \rangle$	P_3
$\langle P_2, P_1, P_3 \rangle$	P_1
$\langle P_2, P_3, P_1 \rangle$	P_1
$\langle P_3, P_1, P_2 \rangle$	P_1
$\langle P_3, P_2, P_1 \rangle$	P_1

■ **Step 3.** P_1 is pivotal four times.
P_2 is pivotal one time.
P_3 is pivotal one time.

The Shapley-Shubik power distribution is

P_1: $\frac{4}{6} \approx 66.7\%$; P_2: $\frac{1}{6} \approx 16.7\%$; P_3: $\frac{1}{6} \approx 16.7\%$.

Comparing with Example 9 we see that there is indeed a difference between the two methods. ■

Example 15. Consider the weighted voting system [6: 4, 3, 2, 1].

This is the same weighted voting system that was discussed in Example 10. We will now find the Shapley-Shubik power index of each player.

There are 24 different sequential coalitions involving the four players. They are listed in Table 2-4 with the pivotal player underlined.

$\langle P_1,\underline{P_2},P_3,P_4\rangle$	$\langle P_2,\underline{P_1},P_3,P_4\rangle$	$\langle P_3,\underline{P_1},P_2,P_4\rangle$	$\langle P_4,P_1\underline{P_2},P_3\rangle$
$\langle P_1,\underline{P_2},P_4,P_3\rangle$	$\langle P_2,\underline{P_1},P_4,P_3\rangle$	$\langle P_3,\underline{P_1},P_4,P_2\rangle$	$\langle P_4,P_1,\underline{P_3},P_2\rangle$
$\langle P_1,\underline{P_3},P_2,P_4\rangle$	$\langle P_2,P_3,\underline{P_1},P_4\rangle$	$\langle P_3,P_2,\underline{P_1},P_4\rangle$	$\langle P_4,P_2,\underline{P_1},P_3\rangle$
$\langle P_1,\underline{P_3},P_4,P_2\rangle$	$\langle P_2,P_3,\underline{P_4},P_1\rangle$	$\langle P_3,P_2,\underline{P_4},P_1\rangle$	$\langle P_4,P_2,\underline{P_3},P_1\rangle$
$\langle P_1,P_4,\underline{P_2},P_3\rangle$	$\langle P_2,P_4,\underline{P_1},P_3\rangle$	$\langle P_3,P_4,\underline{P_1},P_2\rangle$	$\langle P_4,P_3,\underline{P_1},P_2\rangle$
$\langle P_1,P_4,\underline{P_3},P_2\rangle$	$\langle P_2,P_4,\underline{P_3},P_1\rangle$	$\langle P_3,P_4,\underline{P_2},P_1\rangle$	$\langle P_4,P_3,\underline{P_2},P_1\rangle$

Table 2-4

The Shapley-Shubik power distribution is

$$P_1: \tfrac{10}{24}\ ;\ P_2: \tfrac{6}{24}\ ;\ P_3: \tfrac{6}{24}\ ;\ P_4: \tfrac{2}{24}\ .$$

Comparing with Example 10 we see that in this instance both power distributions are identical. Well, these things happen sometimes! ■

Example 16. A committee consists of four members, A, B, C, and D. Each member of the committee has one vote except for the chairperson (A) who has veto power. A measure needs a minimum of two votes to pass (one of which must be the chairperson's). What is the Shapley-Shubik power distribution in this committee?

We know that there are 4! = 24 different sequential coalitions involving the four players, but we are not going to write them all down. Instead, let's analyze how each player can be pivotal. The chairperson (A) is pivotal in every sequential coalition except those in which she is the first player. B can be pivotal only if he is the second player in the sequential coalition and the chairperson is the first player. There are only two such sequential coalitions: $\langle A, B, C, D\rangle$ and $\langle A, B, D, C\rangle$. Thus, the Shapley-Shubik power index of B is $\tfrac{2}{24} \approx 8.33\%$. Exactly the same argument applies to C and D, and so each of them has Shapley-Shubik power index $\tfrac{2}{24} \approx 8.33\%$. The balance of the power is $\tfrac{18}{24} = 75\%$ and belongs to A. Thus the power distribution for this committee is

$$A: \tfrac{18}{24} = 75\%;\ B: \tfrac{2}{24} \approx 8.33\%;\ C: \tfrac{2}{24} \approx 8.33\%;\ D: \tfrac{2}{24} \approx 8.33\%. \quad ■$$

As a final example involving the Shapley-Shubik power index, we choose one that is fairly complicated. Even those readers that have mastered this material shouldn't feel too bad if they have a harder time with this one.

Example 17. Let's find the Shapley-Shubik power distribution of the weighted voting system [15: 9, 6, 3, 3, 3]. This is the same weighted voting system we discussed in Example 11.

The number of sequential coalitions to consider is $5! = 120$. Maybe we should try to think of some shortcuts.

Let's start with player P_5 and try to figure out in how many sequential coalitions P_5 will be pivotal. If P_5 is the first player in the sequential coalition she can't be pivotal—that is clear. If P_5 is the second player in the coalition even in the best of cases (when preceded by P_1), P_5 can't be pivotal. When P_5 is the third player in the coalition, then P_5 can be pivotal only when preceded by P_1 and P_3 or P_1 and P_4. (Think about it— if P_1 is not already there, there won't be enough votes; if P_1 and P_2 are both there, there will be too many votes!) Schematically, we can draw the following diagram of this situation

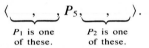

$$\langle \underbrace{\quad,\quad}_{\substack{P_1 \text{ is one} \\ \text{of these.}}}, P_5, \underbrace{\quad,\quad}_{\substack{P_2 \text{ is one} \\ \text{of these.}}} \rangle.$$

There is a total of eight such coalitions: $\langle P_1, P_4, P_5, P_2, P_3 \rangle$, $\langle P_4, P_1, P_5, P_2, P_3 \rangle$, . . . (we'll let the reader finish this list).

How can P_5 be pivotal when in the fourth position? A moment's reflection will show that this can happen only if P_1 does not precede P_5 in the coalition. In other words,

$$\langle \underbrace{\quad,\quad,\quad}_{P_2 \ P_3, \text{ and } P_4}, \ P_5, P_1 \rangle.$$

There is a total of six such coalitions. (It's not hard to list them.) Finally, P_5 can never be pivotal in the fifth (last) position.

We now know that P_5 is pivotal in a grand total of fourteen coalitions, so her Shapley-Shubik power index is $\frac{14}{120}$.

How about P_4 and P_3? That's easy. Anything that is true of P_5 is true of P_4 and P_3, so each of them is also pivotal in fourteen sequential coalitions. Let's do P_2 next, and let's try to let the pictures speak for themselves.

P_2 is pivotal in the second position:

$$\langle P_1, P_2, \underbrace{\quad,\quad,\quad}_{P_3, \ P_4, \text{ and } P_5} \rangle.$$

There is a total of six such coalitions. (There are 6 different ways in which P_3, P_4, and P_5 can be shuffled among themselves.)

P_2 is pivotal in the third position:

$$\langle \underbrace{\quad , \quad}_{\substack{P_1 \text{ is one} \\ \text{of these.}}} , P_2, \quad , \quad \rangle.$$

There is a total of twelve such coalitions. (Check it out!)

P_2 is pivotal in the fourth position:

$$\langle \underbrace{\quad , \quad , \quad}_{P_3, \ P_4, \text{ and } P_5} , P_2, P_1 \rangle.$$

There is a total of six such coalitions.

Since P_2 cannot be pivotal in either the first or last position, we are done. All in all, P_2 is pivotal in 24 coalitions, so that his Shapley-Shubik power index is $\frac{24}{120} = \frac{1}{5}$.

Finally there is P_1. By default, P_1 must be pivotal in each of the sequential coalitions that have not yet been counted. There are

$$120 - 14 - 14 - 14 - 24 = 54$$

of them. Thus, the Shapley-Shubik power index of P_1 is $\frac{54}{120}$.

Summarizing, the Shapley-Shubik power distribution is

P_1: $\frac{54}{120} = 45\%$; P_2: $\frac{24}{120} = 20\%$; P_3: $\frac{14}{120} \approx 11.66\%$; P_4: $\frac{14}{120} \approx 11.66\%$; P_5: $\frac{14}{120} \approx 11.66\%$.

Comparing with Example 11, we get a slightly different distribution of power. ∎

■ Applications of the Shapley-Shubik Power Index

The United Nations Security Council Revisited. As we pointed out earlier in the chapter, the U.N. Security Council as presently structured can be formally described as the weighted voting system [39: 7, 7, 7, 7, 7, 1, 1, 1, 1, 1, 1, 1, 1, 1, 1] and the Banzhaf power index of the five permanent members (the ones with seven votes) is about 10 times larger than that of the nonpermanent members (16.7% against 1.65%). What if we use the Shapley-Shubik power index instead? It turns out (the calculations are a bit too complicated to describe in detail) that the Shapley-Shubik power index of each of the five permanent members is 19.6%, while that of each of the nonpermanent members is about 0.2%. According to the Shapley-Shubik power index, any one of the five permanent members has almost 100 times more power than one of the nonpermanent members!

The Electoral College.   As we should all know, the president of the United States is chosen using an institution called the *electoral college*. In choosing the president, each state is allowed to cast a certain number of votes equal to the total number of members of congress (senators plus representatives) from that state. The votes are cast by individuals called *electors* who are chosen to represent the citizens of their respective states. The general rule is that all the electors from a particular state vote the same way (for the presidential candidate who wins a plurality of the votes in that state). While there have been challenges to the constitutionality of this rule (known as the *unit* rule or *winner-take-all* rule), and in a few instances the rule has been violated by individual electors, it is standard procedure in the electoral college. We can summarize by saying that while in theory the electoral college is not a weighted voting system, in practice, if we assume that the unit rule is respected by all electors and the election is between two candidates (which it almost always is), the electoral college can be thought of as a very real example of a weighted voting system. In this system the players are the states (actually the 50 states plus the District of Columbia), and the weight of a state is the number of senators plus representatives from that state (the weight of the District of Columbia is set at 3). The quota is defined by a strict majority of the electoral vote. Since the 1964 presidential election, the total number of electoral votes has been set at 538 and the quota at 270. (The appendix at the end of the chapter shows the electoral votes for each state based on the 1990 Census.)

The Shapley-Shubik power indexes of the various states in the electoral college have been computed. They range from approximately 10% for California to about 0.5% for the District of Columbia, Alaska, and each of the other states with three votes. An understanding of how the electoral college "power pie" is carved up among the states is a fundamental element in the strategy of a presidential campaign. As a general rule, candidates allocate their resources (money, time, and personnel) to states in accordance to their power indexes and not in accordance to population. The net effect of this is that candidates usually spend a disproportionate amount of their campaign resources in the states with the greatest weights. A presidential candidate can win by the barest majority of the popular vote in just 12 states (the 12 most populous ones) and have enough electoral votes to win the election, a fact that is not lost on campaign managers and political strategists.

CONCLUSION

In any society, no matter how democratic, some individuals and groups have more power than others. This is simply a consequence of the fact that individuals and groups are not all equal. Diversity is the inherent reason why the concept of power exists.

Power itself comes in many different forms. We often hear expressions such as "In strength lies power" and "Money is power," and in our new information age, "Knowledge is power." In this chapter we discussed the notion of power as it applies to formal voting situations (weighted voting systems) and saw how mathematical methods allow us to measure the power of an individual or group by means of a *power index*. In particular, we looked at two different kinds of power indexes: the *Banzhaf power index* and the *Shapley-Shubik power index*.

The Banzhaf power index and the Shapley-Shubik power index provide two different ways to measure power, and while they occasionally agree (Example 15), they are often significantly different. Of the two, which one is closer to reality?

Unfortunately, there is no simple answer. Both of them are useful, and in some sense the choice is subjective. Perhaps the best way to evaluate them is to think of them as being based on a slightly different set of assumptions. According to Banzhaf, players are free to come and go, negotiating their allegiance for power (somewhat like professional baseball players since the advent of free agency). Underlying the Shapley-Shubik interpretation of power is the assumption that when a player joins a coalition he or she is making a commitment to stay (as in an old-fashioned marriage). In the latter case a player's power is generated by his ability to be *in the right place at the right time*. In practice, the choice of which method to use in measuring power is based on which of the assumptions better fits the specifics of the situation. Mathematics, contrary to what we've often come to expect, does not give us the answer, just the tools that might help us make an informed decision.

KEY CONCEPTS

Banzhaf power distribution	**players**
Banzhaf power index	**quota**
coalition	**sequential coalition**
coalition weight	**Shapley-Shubik power**
critical player	**distribution**
dictator	**Shapley-Shubik power index**
dummy	**veto power**
factorial	**weighted voting system**
losing coalition	**weights**
motion	**winning coalition**
pivotal player	

EXERCISES

■ Walking

1. Consider the weighted voting system [10: 6, 5, 4].

    (a) How many players are there?

    (b) What is the quota?

    (c) What is the weight of the coalition formed by P_1 and P_3?

    (d) Write down all winning coalitions.

    (e) Which players are critical in the coalition $\{P_1, P_2, P_3\}$?

    (f) Find the Banzhaf power distribution of this weighted voting system.

2. Consider the weighted voting system [4: 3, 2, 1, 1].

    (a) How many players are there?

    (b) What is the quota?

    (c) What is the weight of the coalition formed by P_1 and P_3?

    (d) Which players are critical in the coalition $\{P_2, P_3, P_4\}$?

    (e) Which players are critical in the coalition $\{P_1, P_3, P_4\}$?

    (f) Write down all winning coalitions.

    (g) Find the Banzhaf power distribution of this weighted voting system.

3. **(a)** Find the Banzhaf power distribution of the weighted voting system [6: 5, 3, 2].

    (b) Find the Banzhaf power distribution of the weighted voting system [8: 5, 3, 2].

4. **(a)** Find the Banzhaf power distribution of the weighted voting system [8: 5, 4, 3, 2, 1]. (If possible do it without writing down all coalitions—just the winning ones.)

    (b) Find the Banzhaf power distribution of the weighted voting system [10: 5, 4, 3, 2, 1]. [*Hint*: Note that the only change from part (a) is in the quota and use this fact to your advantage.]

5. Find the Banzhaf power distribution of the weighted voting system [8: 5, 5, 3, 1, 1].

6. Find the Banzhaf power distribution of the weighted voting system [8: 5, 5, 2, 1, 1].

7. Consider the weighted voting system [10: 6, 5, 4].

    (a) Write down all the sequential coalitions involving all three players.

    (b) In each of the sequential coalitions in part (a), underline the pivotal player.

    (c) Find the Shapley-Shubik power distribution of this weighted voting system.

8. Consider the weighted voting system [6: 5, 4, 1].

    (a) Write down all the sequential coalitions involving all three players.

    (b) In each of the sequential coalitions in part (a), underline the pivotal player.

    (c) Find the Shapley-Shubik power distribution of this weighted voting system.

9. **(a)** Find the Shapley-Shubik power distribution of the weighted voting system [6: 5, 3, 2].

    (b) Find the Shapley-Shubik power distribution of the weighted voting system [8: 5, 3, 2].

10. Find the Shapley-Shubik power distribution of the weighted voting system [8: 4, 3, 2, 1].

11. In each of the following weighted voting systems, determine which players (if any) are dummies. Explain your reasons.

(a) [4: 2, 2, 1]

(b) [100: 100, 40, 30, 20].

12. In each of the following weighted voting systems, determine which players (if any) are dummies. Explain your reasons.

(a) [151: 100, 100, 100, 1]

(b) [10: 5, 5, 5, 2, 1, 1].

13. Consider the weighted voting system [q: 12, 8, 5, 4, 2].

(a) What is the smallest value that the quota q can take?

(b) What is the largest value that the quota q can take?

(c) How many coalitions are there for this weighted voting system?

(d) How many sequential coalitions are there involving all the players?

14. Consider the weighted voting system [q: 7, 3, 2, 1, 1].

(a) What is the smallest value that the quota q can take?

(b) What is the largest value that the quota q can take?

(c) How many coalitions are there for this weighted voting system?

(d) How many sequential coalitions are there involving all the players?

15. Consider the weighted voting system [q: 5, 3, 1]. Find the Shapley-Shubik power distribution of this weighted voting system when

(a) $q = 5$

(b) $q = 6$

(c) $q = 7$

(d) $q = 8$

(e) $q = 9$.

16. Consider the weighted voting system [q: 5, 3, 1]. Find the Banzhaf power distribution of this weighted voting system when

(a) $q = 5$

(b) $q = 6$

(c) $q = 7$

(d) $q = 8$

(e) $q = 9$.

17. This exercise is intended to help you develop a better understanding of factorials. If you have a fancy calculator with a factorial key, don't use it—use only the multiplication key!

(a) Calculate 8!

(b) Calculate 12!

(c) Calculate 13! (Think of some shortcut that you can take!)

(d) Suppose $25! = A$. Express 26! in terms of A.

(e) Suppose $14A = 14!$, find A.

18. This exercise should be done *without* a calculator. The amount of arithmetic is minimal. Remember that it is always better to do cancelations first and multiplications later.

    (a) Calculate $\frac{11!}{10!}$.

    (b) Calculate $\frac{100!}{98!}$.

    (c) Calculate $\frac{8!}{5!3!}$.

    (d) Calculate $\frac{25!}{21!4!}$.

19. A business firm is owned by four partners, A, B, C, D. When making group decisions, each partner has one vote and the majority rules except in the case of a 2-2 tie. Then, the coalition that contains D (the partner with the least seniority) *loses*. What is the Banzhaf power distribution in this partnership?

20. A business firm is owned by four partners, A, B, C, D. When making group decisions, each partner has one vote and the majority rules except in the case of a 2-2 tie. Then, the coalition that contains D (the partner with the least seniority) *loses*. What is the Shapley-Shubik power distribution in this partnership?

■ Jogging

21. Consider the weighted voting system [21: 6, 5, 4, 3, 2, 1]. (Note that here the quota is the sum of *all* the weights of the players.)

    (a) How many coalitions are there?

    (b) Write down the winning coalitions only and underline the critical players.

    (c) Find the Banzhaf power index of each player.

    (d) Explain why in any weighted voting system $[q: w_1, w_2, \ldots, w_N]$, if $q = w_1 + \cdots + w_N$, then the Banzhaf power index of each player is $1/N$.

22. Consider the weighted voting system [21: 6, 5, 4, 3, 2, 1] .

    (a) How many different sequential coalitions are there?

    (b) There is only one way in which a player can be pivotal in one of these sequential coalitions. Describe it.

    (c) In how many sequential coalitions is player 6 pivotal?

    (d) What is the Shapley-Shubik power index of player 6?

    (e) What are the Shapley-Shubik power indexes of the other players?

    (f) Explain why in any weighted voting system $[q: w_1, w_2, \ldots, w_N]$, if $q = w_1 + \cdots + w_N$, then the Shapley-Shubik power index of each player is $1/N$.

23. Give an example of a weighted voting system in which P_1 has twice as many votes as P_2 and

    (a) the Banzhaf power index of P_1 is greater than twice the Banzhaf power index of P_2.

    (b) the Banzhaf power index of P_1 is less than twice the Banzhaf power index of P_2.

    (c) the Banzhaf power index of P_1 is equal to twice the Banzhaf power index of P_2.

    (d) the Banzhaf power index of P_1 is equal to the Banzhaf power index of P_2.

24. Give an example of a weighted voting system in which P_1 has twice as many votes as P_2 and

   (a) the Shapley-Shubik power index of P_1 is greater than twice the Shapley-Shubik power index of P_2.
   (b) the Shapley-Shubik power index of P_1 is less than twice the Shapley-Shubik power index of P_2.
   (c) the Shapley-Shubik power index of P_1 is equal to twice the Shapley-Shubik power index of P_2.
   (d) the Shapley-Shubik power index of P_1 is equal to the Shapley-Shubik power index of P_2.

25. **(a)** Consider the weighted voting system [22: 10, 10, 10, 10, 1]. Are there any dummies? Explain your answer.
   (b) Without doing any work [but using your answer for part (a)], find the Banzhaf and Shapley-Shubik power distributions of this weighted voting system.
   (c) Consider the weighted voting system [q: 10, 10, 10, 10, 1]. Find all the possible values of q for which player 5 is not a dummy.

26. **(a)** Verify that the weighted voting systems [12: 7, 4, 3, 2] and [24: 14, 8, 6, 4] result in exactly the same Banzhaf power distribution. (If you need to make calculations, do them for both systems side by side and look for patterns.)
   (b) Based on your work in part (a), explain why the two proportional weighted voting systems [q: $w_1, w_2, ..., w_N$] and [cq: $cw_1, cw_2, . . ., cw_N$] always have the same Banzhaf power distribution.

27. **(a)** Verify that the weighted voting systems [12: 7, 4, 3, 2] and [24: 14, 8, 6, 4] result in exactly the same Shapley-Shubik power distribution. (If you need to make calculations, do them for both systems side by side and look for patterns.)
   (b) Based on your work in part (a), explain why the two proportional weighted voting systems [q: $w_1, w_2, ..., w_N$] and [cq: $cw_1, cw_2, . . ., cw_N$] always have the same Shapley-Shubik power distribution.

28. **A dummy is a dummy is a dummy.** . . . . This exercise shows that a player that is a dummy is a dummy regardless of which interpretation of power is used.

   (a) Show that if a player has a Banzhaf power index of 0 in a weighted voting system, then that player must also have a Shapley-Shubik power index of 0.
   (b) Show that if a player has a Shapley-Shubik power index of 0 in a weighted voting system, then that player must also have a Banzhaf power index of 0.

29. Consider the weighted voting system [q: 5, 4, 3, 2, 1].

   (a) For what values of q is there a dummy?
   (b) For what values of q do all players have the same power?

30. **(a)** Calculate $(4!)(4!)$.
   (b) Calculate $(5!)(3!)$.
   (c) Calculate $(6!)(2!)$.

(d) Compare each of your answers for parts (a) through (c) with 8!.

(e) Show that given any two positive integers M and N, $(M!)(N!) \leq (M + N)!$.

■ Running

31. (a) Give an example of a weighted voting system with four players and such that the Shapley-Shubik power index of player 1 is $\frac{3}{4}$.

(b) Show that in any weighted voting system with four players, a player cannot have a Shapley-Shubik power index of more than $\frac{3}{4}$ unless he or she is a dictator.

(c) Show that in any weighted voting system with N players, a player cannot have a Shapley-Shubik power index of more than $(N - 1)/N$ unless he or she is a dictator.

(d) Given an example of a weighted voting system with N players such that player 1 has a Shapley-Shubik power index of $(N - 1)/N$.

32. (a) Give an example of a weighted voting system with three players and such that the Shapley-Shubik power index of player 3 is $\frac{1}{6}$.

(b) Show that in any weighted voting system with three players, a player cannot have a Shapley-Shubik power index of less than $\frac{1}{6}$ unless he or she is a dummy.

33. (a) Give an example of a weighted voting system with four players and such that the Shapley-Shubik power index of player 4 is $\frac{1}{12}$.

(b) Show that in any weighted voting system with four players, a player cannot have a Shapley-Shubik power index of less than $\frac{1}{12}$ unless he or she is a dummy.

34. (a) Give an example of a weighted voting system with N players having a player with veto power who has a Banzhaf power index of $1/N$.

(b) Show that in any weighted voting system with N players, a player with veto power must have a Banzhaf power index of at least $1/N$.

35. (a) Give an example of a weighted voting system with N players having a player with veto power who has a Shapley-Shubik power index of $1/N$.

(b) Show that in any weighted voting system with N players, a player with veto power must have a Shapley-Shubik power index of at least $1/N$.

36. **The United Nations Security Council.** The U.N. Security Council is made up of 15 countries. There are 5 permanent members (China, France, the Soviet Union, the United Kingdom, and the United States) and 10 nonpermanent members. All 5 of the permanent members have veto power. A winning coalition must consist of the 5 permanent members plus at least 4 nonpermanent members. Explain why the Security Council can formally be described by the weighted voting system [39: 7, 7, 7, 7, 7, 1, 1, 1, 1, 1, 1, 1, 1, 1, 1].

37. **The original United Nations Security Council.** The original Security Council as set up in 1945 consisted of only 11 countries (5 permanent members plus 6 nonpermanent members). A winning coalition had to include all 5 permanent members plus at least 2 of the nonpermanent members. Give a formal description of the original Security Council as a weighted voting system.

38. The purpose of this exercise is to confirm the claim that in a weighted voting system with N players the total number of possible coalitions is $2^N - 1$.

(a) Write down the $2^3 - 1 = 7$ coalitions with players P_1, P_2, and P_3 and use them to show that there are $2^4 - 1 = 15$ coalitions with players P_1, P_2, P_3, and P_4. (*Hint*: Make two separate lists: those that don't include P_4 and those that include P_4. How many coalitions are there on each list?)

(b) Use the ideas from part (a) to calculate (without making any lists) how many coalitions with players P_1, P_2, P_3, P_4, and P_5 are possible. (*Hint*: How many don't include P_5? How many include P_5?)

(c) Show that $(2^N - 1) + 2^N = 2^{N+1} - 1$. What does this say about the number of coalitions with $N + 1$ players?

REFERENCES AND FURTHER READINGS

1. Brams, Steven J., William F. Lucas, and Philip D. Straffin, *Political and Related Models*. New York: Springer Verlag, 1983, chaps. 9 and 11.

2. Brams, Steven J., *Game Theory and Politics*. New York: Free Press, 1975, chap. 5.

3. Imrie, Robert W., "The Impact of the Weighted Vote on Representation in Municipal Governing Bodies of New York State," *Annals of the New York Academy of Sciences*, 219 (November 1973), 192–199.

4. Riker, William H., and Peter G. Ordeshook, *An Introduction to Positive Political Theory*. Englewood Cliffs, N.J.: Prentice-Hall, Inc., 1973, chap. 6.

5. Shubik, Martin, *Game Theory and Related Approaches to Social Behavior*. New York: John Wiley & Sons, Inc., 1964.

6. Straffin, Philip D., "The Power of Voting Blocs: An Example," *Mathematics Magazine*, 50 (1977), 22–24.

7. Straffin, Philip D., "Homogeneity, Independence and Power Indices," *Public Choice*, 30 (1977), 107–118.

8. Straffin, Philip D. Jr., *Topics in the Theory of Voting*, UMAP Expository Monograph. Boston: Birkhäuser, 1980, chap. 1.

State	Seats in the House of Representatives	Seats in the Senate	Electoral Votes
California	52	2	54
New York	31	2	33
Texas	30	2	32
Florida	23	2	25
Pennsylvania	21	2	23
Illinois	20	2	22
Ohio	19	2	21
Michigan	16	2	18
New Jersey	13	2	15
North Carolina	12	2	14
Georgia	11	2	13
Virginia	11	2	13
Indiana	10	2	12
Massachusetts	10	2	12
Missouri	9	2	11
Tennessee	9	2	11
Washington	9	2	11
Wisconsin	9	2	11
Maryland	8	2	10
Minnesota	8	2	10
Alabama	7	2	9
Louisiana	7	2	9
Arizona	6	2	8
Colorado	6	2	8
Connecticut	6	2	8
Kentucky	6	2	8

State	Seats in the House of Representatives	Seats in the Senate	Electoral Votes
Oklahoma	6	2	8
South Carolina	6	2	8
Iowa	5	2	7
Mississippi	5	2	7
Oregon	5	2	7
Arkansas	4	2	6
Kansas	4	2	6
Nebraska	3	2	5
New Mexico	3	2	5
Utah	3	2	5
West Virginia	3	2	5
Hawaii	2	2	4
Idaho	2	2	4
Maine	2	2	4
Nevada	2	2	4
New Hampshire	2	2	4
Rhode Island	2	2	4
Alaska	1	2	3
Delaware	1	2	3
Montana	1	2	3
North Dakota	1	2	3
South Dakota	1	2	3
Vermont	1	2	3
Wyoming	1	2	3
District of Columbia			3
Total			538

3

Fair Division

If you want to know the true character of a person, divide an inheritance with him.

BENJAMIN FRANKLIN

The Slice Is Right

If we have fifty pieces of candy, how do we divide them fairly among five children?

PROBLEM IN FOURTH GRADE SCHOOL BOOK

Sometime during our grammar school days, all of us have had to solve a problem like this—it is the classic application of whole-number division. It may come as a bit of a surprise, but we will resurrect the theme in this chapter. Why?

Implicit in the grammar school version of the problem is the assumption that the pieces of candy are all *identical*. But what if they aren't? What happens if we allow the pieces to be different (say 11 Baby Ruths, 9 Reeses Pieces, 13 mints, and 17 Lifesavers of assorted colors) and therefore have potentially different values. Moreover, what if, as it is usually the case, the children have different value systems (Joey would rather have one Baby Ruth than two Reeses Pieces, does not like mints, and prefers yellow Lifesavers to green ones, while Tanya likes Reeses Pieces a little more than Baby Ruths which she prefers to green Lifesavers but not to yellow ones, and on and on and on). Under these circumstances the complexity of the problem seems daunting, and the task of dividing the pieces fairly among the children can seem hopeless. Actually, under the right set of circumstances we can accomplish this using relatively simple procedures called *fair division schemes*. Moreover—and this is almost magical—we can often divide the candy in such a way that each child is satisfied that he or she has received a fair share and still end up with one or more pieces of candy left over.

In this chapter we will introduce several fair division schemes and discuss how they work and the assumptions under which they make sense. Before we present specific descriptions of the various fair division schemes, there is a critical question that must be addressed: Is this topic really that important? Who cares whether five little kids are happy with their share of candy? A moment's reflection will show, however, that children and candy are but a metaphor for an important, and frequently occurring, problem: How can an object or a set of objects to be shared by a set of participants be divided among them in a way that ensures that each is satisfied that he or she has received a fair share of the total.

The settlement of an estate among heirs, the division of common property in a divorce proceeding, the subdivision of a parcel of land, the allocation of airwaves to radio and television stations, and the apportionment of seats to states in the U.S. House of Representatives are all significant variations of a common theme: the problem of fair division.

FAIR DIVISION PROBLEMS AND FAIR DIVISION SCHEMES

In this section we will clarify the distinction between fair division problems and their possible methods of solution (fair division schemes).

The elements of every **fair division problem** are a set of players P_1, $P_2, \ldots, P_N$ and a set of goods which we will call S. The problem is to divide S into N shares $s_1, s_2, \ldots, s_N$ in such a way that each player gets a share that is in his or her own personal value system a fair share of the total set S.

What is a fair share? By a **fair share** we mean any share that *in the opinion of the player receiving it* has a value that is at least $1/N$ of the total value of the set of goods S. For example, if we have five players, then any share that, in the opinion of player X, is worth $\frac{1}{5}$ (20%) or more of the total can be considered a fair share for player X.

Implicit in our description of a fair division problem is the assumption that each player has the ability to judge whether a share is fair or not. In fact we will assume much more and give each player the ability to assign exact fractional values to each of the various shares (to me this piece is worth $\frac{1}{3}$ of the total, this other one is worth $\frac{3}{17}$ of the total, . . .). Although people often are hard pressed to precisely quantify their judgments about the worth of an object, it is a practice that is very much part of our culture. We exercise it every time we make a bid at an auction, make an offer on a used car, or give points for looks to a member of the opposite sex. Moreover, the ability to do this is an acquired skill. It is not unreasonable, therefore, that all our fair division schemes are based on the premise that each of the players has fine-tuned this skill.

By a **fair division scheme** we will mean any systematic procedure for solving a particular type of fair division problem (we will discuss various types of fair division problems soon). We will expect any fair division scheme to satisfy the following conditions.

- The procedure is **decisive**. This means that if the rules are followed, the procedure is guaranteed to result in a fair division of the set S.

- The procedure is **internal** to the players. This means that the procedure does not require the intervention of an outside authority such as a judge, an arbitrator, a parent, etc.

- The procedure assumes that the players have **no knowledge** about each others' value systems. This means that no player has information about the likes or dislikes of any of the other players.

- The procedure assumes that the players are **rational**. This means that the players' values and strategies are assumed to be based on logic and not emotion.

A final word about fair division schemes. A fair division scheme does not necessarily guarantee that each player will receive a fair share. It is possible for a player to misplay the game (the most common cause for this is greed) and end up with an unfair share. What a fair division scheme does guarantee is that it is impossible for the remaining players or bad luck to conspire to deny any player his or her fair share.

■ **Types of Fair Division Problems**

Depending on the nature of the set of goods S, fair division problems can be classified into three types: continuous, discrete, and mixed.

In a **continuous** fair division problem the set S is divisible in infinitely many ways, and shares can be increased or decreased by arbitrarily small amounts. Typical examples of continuous fair division problems are dividing a parcel of land, a cake, a pizza, ice cream, or a bottle of wine. A large enough sum of money, while not continuous in the theoretical sense (pennies cannot be divided), can be considered for all practical purposes continuous (nobody argues over pennies any more).

A fair division problem is **discrete** when the set S is made up of objects that are indivisible (houses, cars, boats, jewelry, etc.). Most candy, while continuous in the theoretical sense, is for all practical purposes discrete (hard pieces of candy cannot easily be broken up without crumbling; soft pieces are hard to cut up without making a big mess), and for convenience throughout this chapter we will think of candy as indivisible and therefore discrete.

A **mixed** fair division problem is one in which some of the components are continuous and some are discrete. Dividing an estate consisting of a car, a house, and a parcel of land is a mixed fair division problem.

Depending on whether a fair division problem is continuous, discrete, or mixed, a different strategy is needed for finding a solution. In the rest of this chapter we will present several different schemes for solving continuous fair division problems, and two very different schemes for solving discrete fair division problems. We will not discuss schemes for solving mixed fair division problems as they can usually be solved by dividing the continuous and discrete parts separately.

We start with, what is undoubtedly, the best known of all fair division schemes: the divider-chooser method.

THE DIVIDER-CHOOSER METHOD

This classic scheme can be used anytime there is a continuous fair division problem involving two players. Most of us have unwittingly used it at some time or another, and it is commonly known as the *you cut–I choose* method. As this name suggests, one player divides the cake (we will use the word "cake" as a metaphor for any continuous set *S*) into two pieces, and the other player picks the piece he or she wants, leaving the other piece to the divider. When played honestly, this method guarantees that each player will get a share that he or she believes to be worth *at least* one-half of the total. The divider can guarantee this for herself in the mere act of dividing, and the chooser because of the simple fact that when anything is divided into two parts, one of the parts must be worth at least one-half or more of the total.

Figure 3-1

Example 1.   On their first date, Bob and Rachel go to the county fair. With a $2.00 raffle ticket they win the chocolate-strawberry cake shown in Fig. 3-1. (Actually they had their hearts set on the red Corvette, but third prize is better than nothing.)

Since they bought the ticket jointly (each chipped in $1.00), they decide that the best way to split up the cake is to use the divider-chooser method. Bob volunteers to be the divider. Bob likes chocolate and strawberry equally well, so he makes a straight cut along a diameter of the cake as shown in Fig. 3-2, making sure that the cake is cut into two pieces of identical volume but otherwise not being concerned about the proportions of chocolate and strawberry on each piece (to him the value of chocolate and strawberry are the same).

Since it is their first date, Bob knows nothing about Rachel's likes and dislikes. In particular, he doesn't know that Rachel is allergic to chocolate and will not eat it at all. In her value system, the chocolate part has zero value and the entire value of the cake is contained in the strawberry part. Her choice is obvious—she will pick the bottom piece consisting of 75%

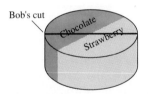

Figure 3-2

strawberry and 25% chocolate. Notice that while Bob gets a share that is (in his value system) worth exactly one-half of the total, Rachel ends up with a share that (in her value system) is worth three-fourths of the total. ∎

It is clear from Example 1 that although the divider-chooser method satisfies our requirement for fair division (each player gets a share that in his or her opinion is worth at least one-half of the total), there is a definite advantage in being the chooser. The simplest way to handle this problem is to randomly choose who gets to be the divider and who gets to be the chooser (toss a coin, draw straws, etc.).

The divider-chooser method for two players can be generalized for the case of more than two players in several ways. We will present three of them in this chapter: the lone divider method, the lone chooser method, and the last diminisher method. In the **lone divider method**, one of the players is the divider and all the rest are choosers; in the **lone chooser method**, one of the players is the chooser and all the rest are dividers; in the **last diminisher method**, each player has a chance to be both a divider and a chooser.

Before we describe these methods we must mention an important assumption needed for all of these methods to work: The value of the object S being divided does not diminish when the object is cut. Thus, when cutting our theoretical cakes, there will be no crumbs.

THE LONE DIVIDER METHOD

For the sake of simplicity, we first describe the method for the case of three players. Here we have one divider and two choosers. The fairest way to decide which player is the divider is by random selection (roll a die, draw straws, draw cards from a deck, etc.). Let's say Diva is the divider, and Chooch and Cher the choosers.

- **Move 1 (Division).**   The divider (Diva) cuts the cake into three slices (s_1, s_2, and s_3), each of which she judges to be a fair one-third of the cake.

- **Move 2 (Declarations).**   Each of the choosers declares independently (usually by writing it down on a slip of paper) which of the slices she believes to be a fair share (i.e., worth at least one-third) of the cake and therefore acceptable.

- **Move 3 (Distribution).**   Who gets what? Depending on the declarations, several things can happen.

   Case 1.   At least one of the choosers selects more than one slice as acceptable. In this case it is possible to distribute the three slices so that each player gets a slice that is acceptable.

Example 2. Chooch: $\{s_1, s_2\}$ (which means that Chooch declares that s_1 and s_2 are both acceptable); Cher: $\{s_1\}$. A fair division of the cake can be obtained by the distribution: Chooch gets s_2, Cher gets s_1, and Diva gets s_3. ∎

Example 3. Chooch: $\{s_1, s_3\}$; Cher: $\{s_1, s_3\}$. In this case we give s_2 to Diva and distribute s_1 and s_3 between the choosers. Again the fairest way to do it is to flip a coin and let the winning chooser be the first to choose. In any case everyone gets an acceptable slice. ∎

Case 2. Both choosers select just one slice as acceptable, but the slices are different. Once again we can distribute the slices so that each player gets a slice that is acceptable.

Example 4. Chooch: $\{s_3\}$; Cher: $\{s_2\}$. Here a fair division is given by giving s_1 to Diva, s_3 to Chooch, and s_2 to Cher. ∎

Case 3. Both choosers select just one slice as acceptable, and it is the same slice. We now have a standoff as both choosers covet the same slice. Our way out of the standoff is to get rid of the divider by giving her one of the other two slices (the fairest way to do it is to let the divider pick between the two). After the divider has chosen her slice the remaining two slices are recombined into a single piece and divided between the two choosers using the divider-chooser method for two players. It all sounds complicated but it really isn't, and you can clearly see how the method handles this case in the next example.

Example 5. Chooch: $\{s_1\}$; Cher: $\{s_1\}$. Here Diva gets to choose between s_2 and s_3. Let's say she chooses s_2. Now the two remaining slices s_1 and s_3 are combined into a single piece and divided between Chooch and Cher using the divider-chooser method for two players.

Why is this a fair division of the original cake? Clearly, Diva has no cause to complain—she gets s_2, a piece she believes to be a fair one-third of the cake. As far as Chooch and Cher are concerned, neither one of them considers s_2 to be worth one-third of the cake (otherwise they would have bid for it), which implies that both of them consider s_1 and s_3 combined to be worth more than two-thirds of the original cake. Dividing such a piece fairly between the two of them must result in shares that are acceptable to both. (In other words, if $x > \frac{2}{3}$ and $y \geq \frac{1}{2} x$, then we must have $y > \frac{1}{3}$.) ∎

The lone divider method as described above for three players can be generalized for any number of players (see Exercises 2 through 6 and 35, as well as Example 6 below).

We will use the following example to illustrate how the method works in a specific situation involving six players.

Example 6.   Al, Betty, Carla, Devon, Ellen, and Franco want to divide a cake using the lone divider method. They draw cards from a deck, and Devon gets to be the divider by virtue of having drawn the lowest card. Devon cuts the cake into six slices ($s_1, s_2, s_3, s_4, s_5, s_6$) which he considers to be equal in value and therefore fair shares of the original cake. Al, Betty, Carla, Franco, and Ellen make their independent declarations by each writing on a separate slip of paper the names of the slices they consider acceptable. The declarations are Al: $\{s_4, s_5\}$; Betty: $\{s_1, s_3\}$; Carla: $\{s_2, s_3, s_4, s_6\}$; Franco: $\{s_3\}$; and Ellen: $\{s_1\}$. Note that there is no possible way to distribute the six slices ($s_1, s_2, s_3, s_4, s_5, s_6$) to everyone's satisfaction. The standoff involves Betty, Franco, and Ellen—the three of them covet two of the six slices (s_1 and s_3). Our strategy then is to proceed with the rest of the group (Al, Carla, and Devon) and distribute three of the other four pieces (s_2, s_4, s_5, and s_6) to them. We first let Al and Carla flip a coin to see who gets to pick first (Devon, being the divider, gets to watch). Let's say Al wins and picks s_4. Now Carla gets to choose and picks s_6. Devon, to whom all pieces have equal value, randomly picks between s_2 and s_5. Let's say he picks s_2. At this point, Al, Carla, and Devon are presumably satisfied with their slices. Slices s_1, s_3, and s_5 are left over and recombined into a single piece ($s_1 + s_3 + s_5$) to be divided among Betty, Franco, and Ellen using the lone divider method for three players.

There is only one question that remains to be answered: Why should Betty, Franco, and Ellen feel that the piece ($s_1 + s_3 + s_5$) is good enough for them to share? The answer lies in a simple addition and subtraction of fractions: Since none of them thought that s_2 or s_4 or s_6 was worth one-sixth of the total (otherwise they would have bid for them), it follows that none of them think that ($s_2 + s_4 + s_6$) is worth $\frac{3}{6} = \frac{1}{2}$ of the total. Therefore it also follows that they all think that the rest ($s_1 + s_3 + s_5$) is worth more than one-half of the total—enough to divide fairly among the three of them and make each of them satisfied. So, Betty, Franco, and Ellen are happy; Al, Carla, and Devon have been happy for a while. All's well that ends well.   ■

THE LONE CHOOSER METHOD

Once again, we start with a description of the method for the case of three players. Here we have one chooser and two dividers. As usual, we decide who is what by random selection. Let's say that Chuck is the chooser, and Dave and Dirk the dividers.

■ **Move 1 (First Division).**   Dave and Dirk cut the cake (Fig. 3-3) into two slices using the divider-chooser method. Dave gets slice 1, and Dirk gets slice 2 (Fig. 3-4). Each considers his slice worth at least one-half of the total.

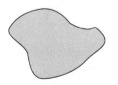

Figure 3-3 The original cake.

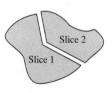

Figure 3-4 First division. The cake is divided between Dave (slice 1) and Dirk (slice 2) using the divider-chooser method.

Figure 3-5 Second division. Slice 1 is divided into three pieces of equal value by Dave; Slice 2 is divided into three pieces of equal value by Dirk.

■ **Move 2 (Second Division).**   Dave divides slice 1 into three pieces that in his opinion have equal value. Likewise, Dirk divides slice 2 into three pieces that he considers to be of equal value (Fig. 3-5).

■ **Move 3 (Selection).**   Chuck picks one of Dave's three pieces and one of Dirk's three pieces. These two pieces constitute Chuck's share. Dave gets to keep the remaining two pieces from slice 1, and Dirk gets to keep the remaining two pieces from slice 2. See Fig. 3-6.

Figure 3-6 Selection.

Why is this a fair division of the cake? Since the first division was a fair division of the cake between two people, Dave's assessment of slice 1 is that it is worth at least one-half of the total. Since he got to keep two thirds of slice 1, his share must be worth at least two-thirds of one-half of the total, thereby making it a fair share ($\frac{2}{3} \times \frac{1}{2} = \frac{1}{3}$). The same argument applies to Dirk. As far as Chuck is concerned, slice 1 is worth some fraction of the total. We do not know what that fraction is, so we will call it F. We can now say that the value of slice 2 to Chuck is $1 - F$, since between them, slice 1 and slice 2 make up the whole cake. Since Chuck is the chooser and presumably picks slice 1a because he likes it best among slice 1a, slice 1b, and slice 1c, the value of slice 1a to Chuck must be at least one third of F. Likewise, the value of slice 2c to Chuck must be at least one third of $1 - F$. Since

$$\tfrac{1}{3} F + \tfrac{1}{3} (1 - F) = \tfrac{1}{3},$$

Chuck gets at least one-third, and therefore a fair share of the cake.

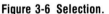

We leave it as an exercise to extend the lone chooser method to the general case of N players (see Exercise 21). The generalization is not nearly as complicated as the one for the lone divider method.

THE LAST DIMINISHER METHOD

We will describe the method for the case of five players. The method[1] can be generalized in a straightforward way to any number of players (see Exercise 22). Here there are no designated dividers and choosers, but before any action takes place the players are randomly assigned an order (P_1 first, P_2 second, . . ., P_5 last). The players will play in this order throughout the game, and the play takes place in rounds.

■ **Round 1.** P_1 cuts a slice from the cake that she believes to be a fair share (one-fifth) of the cake. This piece is P_1's staked claim (at least for the moment). P_1 must be careful to stake a claim that is neither too small (she may end up with that piece) nor too large (somebody else might end up with it). P_2 now has the right to pass or to play on P_1's claim. If he thinks that P_1's claim is a bad choice (worth less than one-fifth of the cake), he passes, remaining in contention for a fair share of the rest of the cake (worth in his opinion more than four-fifths of the cake). On the other hand if he thinks that P_1's claim is a good one (perhaps it appears to P_2 that P_1 might have been a little greedy), then he can make a claim on P_1's claim by staking out a subpiece of it. If P_2 opts to do this, then we call him a **diminisher**. Again, P_2 must be careful to make his claim just right—not too small (he might end up with it) and not too large (someone else might end up with it). Also, P_2 cannot claim part of P_1's claim and also remain in contention for the rest of the cake (he cannot have his cake and eat it too!). If P_2 becomes a diminisher, then the difference between P_1's claim and P_2's claim is added to the remainder of the cake, and P_1 returns to the group of players contending for that remainder. P_1 should be pleased, since she is now in contention for a fair one-fourth of a piece that in her opinion is worth more than four-fifths of the whole. It is now P_3's turn to play. P_3 has a choice to pass and contend for the

---

[1] To the best of our knowledge, this method was first proposed by the Polish mathematician Hugo Steinhaus (see reference 4 at the end of this chapter). With characteristic modesty, Steinhaus credits the discovery of this method to two other Polish mathematicians: S. Banach and B. Knaster.

remainder of the cake or become a diminisher on the current claim (whether it is P_1's or P_2's). Similarly, P_4 and P_5 in turn have a chance to become diminishers on the current claim, or pass and remain in contention for the remainder. After all the players have had a chance to play, the player whose claim is current (the **last diminisher**) gets that piece and departs.

■ **Round 2**. Reconstitute the cake, putting together all the slivers left over by the various cuts and subcuts. Start the process over with the remaining four players. At the end of the round the last diminisher gets his or her staked claim and departs.

■ **Round 3**. There are three players left. Repeat the process with the remainder of the cake. The last diminisher gets his or her staked claim and departs.

■ **Round 4**. At this point there are two players left and the leftover cake can be divided using the divider-chooser method.

Example 7. Five sailors are marooned on a deserted tropical island. Knowing how to make the best out of a good situation they decide to divide the island into fair shares and stay there forever. After some debate, they settle on the last diminisher method.

■ Round 1. (Fig. 3-7, below).

A map of the island.

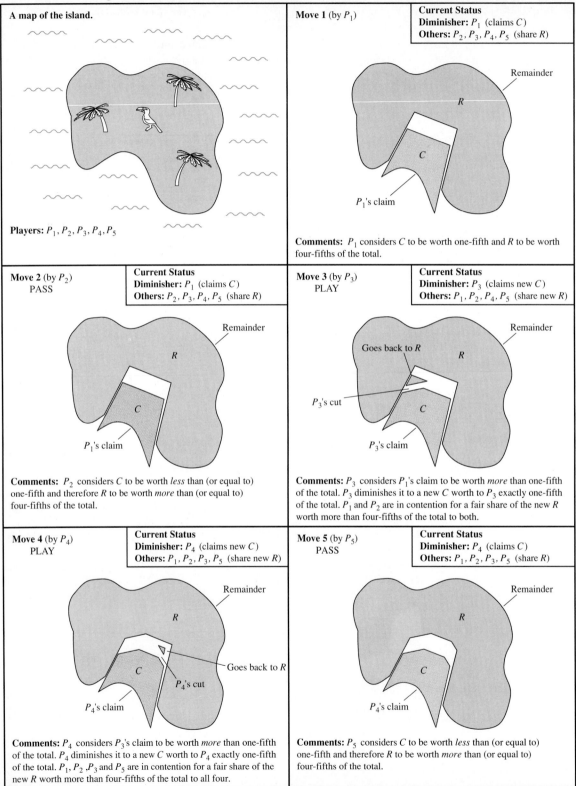

Players: P_1, P_2, P_3, P_4, P_5

Move 1 (by P_1)

Current Status
Diminisher: P_1 (claims C)
Others: P_2, P_3, P_4, P_5 (share R)

Remainder

R

C

P_1's claim

Comments: P_1 considers C to be worth one-fifth and R to be worth four-fifths of the total.

Move 2 (by P_2)
PASS

Current Status
Diminisher: P_1 (claims C)
Others: P_2, P_3, P_4, P_5 (share R)

Remainder

R

C

P_1's claim

Comments: P_2 considers C to be worth *less* than (or equal to) one-fifth and therefore R to be worth *more* than (or equal to) four-fifths of the total.

Move 3 (by P_3)
PLAY

Current Status
Diminisher: P_3 (claims new C)
Others: P_1, P_2, P_4, P_5 (share new R)

Remainder

Goes back to R

R

P_3's cut

C

P_3's claim

Comments: P_3 considers P_1's claim to be worth *more* than one-fifth of the total. P_3 diminishes it to a new C worth to P_3 exactly one-fifth of the total. P_1 and P_2 are in contention for a fair share of the new R worth more than four-fifths of the total to both.

Move 4 (by P_4)
PLAY

Current Status
Diminisher: P_4 (claims new C)
Others: P_1, P_2, P_3, P_5 (share new R)

Remainder

R

P_4's cut

Goes back to R

C

P_4's claim

Comments: P_4 considers P_3's claim to be worth *more* than one-fifth of the total. P_4 diminishes it to a new C worth to P_4 exactly one-fifth of the total. P_1, P_2, P_3 and P_5 are in contention for a fair share of the new R worth more than four-fifths of the total to all four.

Move 5 (by P_5)
PASS

Current Status
Diminisher: P_4 (claims C)
Others: P_1, P_2, P_3, P_5 (share R)

Remainder

R

C

P_4's claim

Comments: P_5 considers C to be worth *less* than (or equal to) one-fifth and therefore R to be worth *more* than (or equal to) four-fifths of the total.

Round 1 is now over (all players have had a chance to pass or play). The final outcome of round 1 is to give the last diminisher (P_4) his or her claim C; the remainder R will be divided fairly among the four remaining players (P_1, P_2, P_3, and P_5).

■ **Round 2.** (Fig. 3-8).

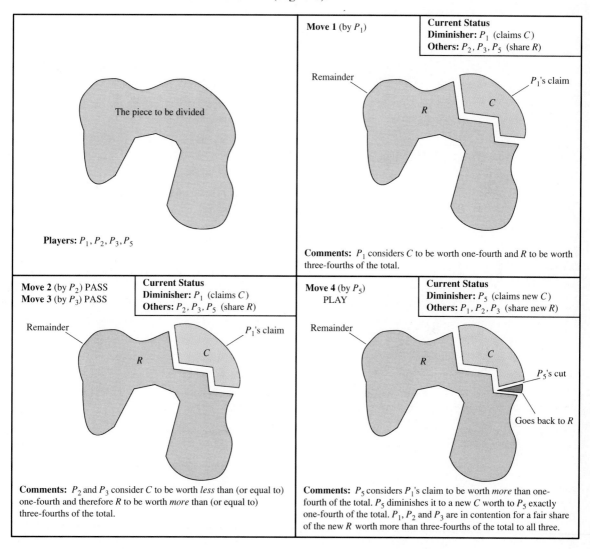

Move 1 (by P_1)

Current Status
Diminisher: P_1 (claims C)
Others: P_2, P_3, P_5 (share R)

The piece to be divided

Players: P_1, P_2, P_3, P_5

Remainder

R

C

P_1's claim

Comments: P_1 considers C to be worth one-fourth and R to be worth three-fourths of the total.

Move 2 (by P_2) PASS
Move 3 (by P_3) PASS

Current Status
Diminisher: P_1 (claims C)
Others: P_2, P_3, P_5 (share R)

Remainder

R

C

P_1's claim

Comments: P_2 and P_3 consider C to be worth *less* than (or equal to) one-fourth and therefore R to be worth *more* than (or equal to) three-fourths of the total.

Move 4 (by P_5)
PLAY

Current Status
Diminisher: P_5 (claims new C)
Others: P_1, P_2, P_3 (share new R)

Remainder

R

C

P_5's cut

Goes back to R

Comments: P_5 considers P_1's claim to be worth *more* than one-fourth of the total. P_5 diminishes it to a new C worth to P_5 exactly one-fourth of the total. P_1, P_2 and P_3 are in contention for a fair share of the new R worth more than three-fourths of the total to all three.

Figure 3-8

Round 2 is now over (all players have had a chance to pass or play). The final outcome of round 2 is to give the last diminisher (P_5) his or her claim C; the remainder R will be divided fairly among the remaining three players (P_1, P_2, and P_3).

■   **Round 3.**   (Fig. 3-9).

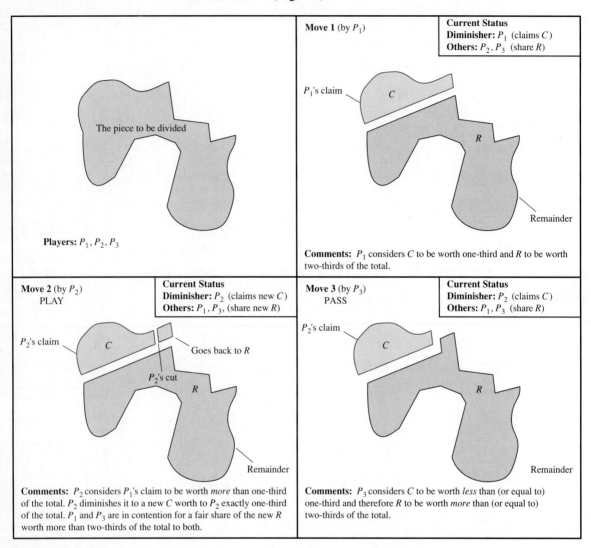

Move 1 (by P_1)

Current Status
Diminisher: P_1 (claims C)
Others: P_2, P_3 (share R)

P_1's claim

C

R

Remainder

Comments: P_1 considers C to be worth one-third and R to be worth two-thirds of the total.

Move 2 (by P_2)
PLAY

Current Status
Diminisher: P_2 (claims new C)
Others: P_1, P_3, (share new R)

P_2's claim

C

Goes back to R

P_2's cut

R

Remainder

Comments: P_2 considers P_1's claim to be worth *more* than one-third of the total. P_2 diminishes it to a new C worth to P_2 exactly one-third of the total. P_1 and P_3 are in contention for a fair share of the new R worth more than two-thirds of the total to both.

Move 3 (by P_3)
PASS

Current Status
Diminisher: P_2 (claims C)
Others: P_1, P_3 (share R)

P_2's claim

C

R

Remainder

Comments: P_3 considers C to be worth *less* than (or equal to) one-third and therefore R to be worth *more* than (or equal to) two-thirds of the total.

Players: P_1, P_2, P_3

The piece to be divided

Figure 3-9

Round 3 is now over (all players have had a chance to pass or play).

The final outcome of round 3 is to give the last diminisher (P_2) his or her claim C; the remainder R will be divided fairly among the two remaining players (P_1 and P_3).

■   **Round 4.**   (Fig. 3-10).

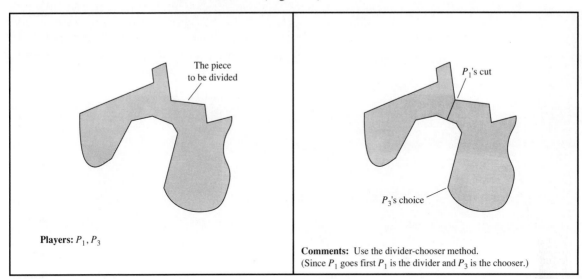

The piece to be divided

P_1's cut

P_3's choice

Players: P_1, P_3

Comments: Use the divider-chooser method.
(Since P_1 goes first P_1 is the divider and P_3 is the chooser.)

Figure 3-10

The final division of the island is shown in Fig. 3-11.

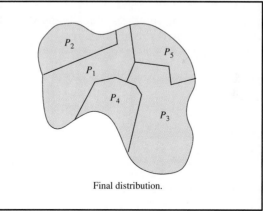

P_2

P_5

P_1

P_4

P_3

Figure 3-11

Final distribution.   ■

In the next two sections we will discuss discrete fair division schemes. In this case the set S consists of one or more objects that are indivisible (houses, cars, paintings, candy, etc.)

**THE METHOD
OF SEALED BIDS**

The best known of all discrete fair division schemes is the **method of sealed bids** originally proposed by the Polish mathematician Hugo Steinhaus. The easiest way to illustrate how this method works is by means of an example.

Example 8. In her last will and testament, Grandma plays a little joke on her four grandchildren (Art, Betty, Carla, and Dave) by leaving three indivisible items—a house, a Rolls Royce, and a Picasso painting—to be divided equally among the four of them. After some discussion someone

How can we divide this estate fairly among four heirs? A classic example of a discrete fair division problem. (Rolls Royce courtesy Aldo Mastrocola/Lightwave)

suggests that all three items be sold and the money divided equally among the four heirs. Unfortunately, each of the four heirs has a different opinion as to what the items are really worth (as shown in the following table), and they cannot therefore reach an agreement on what a reasonable selling price for each item ought to be. By the way, we are assuming that all four grandchildren are honest people and that the figures shown in the table represent their true and sincere assessment of the dollar value of the items.

	Art	Betty	Carla	Dave
House	220,000	250,000	211,000	198,000
Rolls Royce	40,000	30,000	47,000	52,000
Picasso	280,000	240,000	234,000	190,000

As a last recourse, they decide to consult a mathematician who suggests that it is possible to make everybody happy. His division of the estate (with explanation) goes as follows:

■ **Art.** By his own estimation, the total value of the estate is $540,000. (This figure is obtained by adding the values that Art has assigned to each of the three items.) Since he is entitled to one-fourth of the estate, his fair share is worth one-fourth of that amount, $135,000. Since he was the highest bidder for the Picasso painting, he will get the painting, but since the painting is worth much more than what he is entitled to he must make up the difference by paying it back to the estate. In conclusion, Art gets the Picasso painting but pays back to the estate $145,000 ($280,000 −

$135,000). Note that since Art was honest in his assessment of the value of each item, he is satisfied that this settlement is perfectly fair.

■ **Betty.** By her own estimation, the total value of the estate is $520,000, of which she is entitled to exactly one-fourth, which comes to $130,000. Being the highest bidder for the house, Betty gets it, but she must pay the estate the difference between her assessment of the value of the house ($250,000) and her fair share ($130,000). Thus, the settlement for Betty is that she gets the house but has to pay the estate $120,000.

■ **Carla.** Her assessment of the total value of the estate is $492,000, and her fair share is one-fourth of that or $123,000. Since Carla is not the highest bidder on any of the three items, she doesn't get any of them. She does, however, get her fair share of $123,000 in cash.

■ **Dave.** Dave's assessment of the total value of the estate is $440,000, and his fair share is therefore $110,000. Since he is the highest bidder for the Rolls, he gets it for $52,000. The balance of $58,000 (the difference between his fair share and the value of the Rolls Royce) is given to him in cash.

At this point each of the heirs ought to be satisfied that they have received a fair share, and the following table summarizes how things stand.

	Item Received	Cash
Art	Picasso painting	Pays $145,000 (+)
Betty	House	Pays $120,000 (+)
Carla	Nothing	Gets $123,000 (−)
Dave	Rolls Royce	Gets $58,000 (−)
		$84,000 surplus cash

If we add Art's and Betty's payments to the estate and subtract the payments made by the estate to Carla and Dave, we discover that something truly remarkable has happened. Even though each of the heirs is satisfied with his or her share and no one feels cheated, there is a surplus of $84,000 left over. Clearly, the only fair thing to do with this money is to offer it to the mathematician who came up with this brilliant scheme. Being a modest person, however, he declines and suggests instead that the money be divided equally among the four heirs. This is done, and each of the four heirs gets a bonus of $21,000. ■

The general procedure used in Example 8, which we will call the **method of sealed bids**, can be described as follows:

■ **Move 1 (Bidding).** Each player makes a sealed bid giving his or her honest assessment of the dollar value of each of the items in the estate.

■ **Move 2 (Original Allocation).** Each item goes to the highest bidder for that item. (In case of a tie, a predetermined tie-breaking procedure such as flipping a coin should be invoked.) Note that it is possible for a player to get more than one item—in fact even all of them. Each player's fair share is calculated by dividing the total of that player's bids by the number of players. Each player puts in or takes out (in cash) from a common pot (the estate) the difference between his or her fair share and the total value of the items allocated to that player.

■ **Move 3 (Dividing the Surplus).** After the original allocations are completed, there may be a surplus of cash left in the estate. This surplus is divided equally among the players.

The method of sealed bids is the mathematical equivalent of pulling rabbits out of a hat. For the method to work, however, certain conditions must be satisfied.

1. Each player must have enough money to play the game. If a player is going to make honest bids on the items, he must be prepared to take some or all of them, which means that he may have to pay the estate certain sums of money. If the player does not have this money available, he is at a definite disadvantage in playing the game.

2. Each player must accept money (if it is a sufficiently large amount) as a substitute for any item. This means that no player can consider any of the items priceless (I want Mom's diamond ring, and *no* amount of money in the world is going to make me change my mind! is not an attitude conducive to a good resolution of the problem.)

3. Players must have no useful information about each other's value systems prior to the bidding. In particular it is critical that the bids be sealed and that no player sees another player's bids ahead of time. To illustrate why this is critical for a fair division, consider the following example: Player *A* thinks that the real value of item *X* is $1500 but before writing down this bid sees that player *B* has bid $2000 for item *X*. She can now bid as much as $1999.99 for *X* with the certain knowledge that she will not be stuck with *X* at that price, while at the same time inflating the total value of her own fair share.

The method of sealed bids takes a particularly simple form in the case of two players and one item. Consider the following example:

Example 9. Al and Betty are getting a divorce. The only common property of value is their house. Since the divorce is amicable and they are not particularly keen on going to court or hiring an attorney, they decide to divide the house using the method of sealed bids. The bids are

Al	$130,000
Betty	$142,000

Betty, being the highest bidder, gets the house but must pay the estate $71,000 (she is entitled to only half of the value of the house). Al's fair share is half of his bid, namely, $65,000. The surplus of $6000 is divided equally between Al and Betty, and the bottom line is that Betty gets the house but pays Al $68,000. Notice that this result is equivalent to assessing the value of the house as the value halfway between the two bids ($136,000) and splitting this value equally between the two parties, with the house going to the highest bidder and the cash to the other party (Exercise 26). ∎

THE METHOD OF MARKERS

This is a discrete fair division scheme that does not require the players to put up any of their own money. In this sense it has a definite advantage over the method of sealed bids. On the other hand, unlike the method of sealed bids, this method cannot be used effectively unless there are many more items to be divided than there are players.

The basic idea in this method is that the items are lined up in a row (we call such a row an **array**) and that each player then breaks up the array of items into consecutive segments (as many segments as there are players). Each player must do this in such a way that the segments represent what in his or her opinion are fair (and therefore acceptable) shares of the entire set of items. Each player performs this task by laying down markers, each marker indicating the end of one segment and the beginning of another. The players lay down their markers independently and in such a way that no player can see the markers of any of the other players.

After all the players are through laying down their markers, the items are allocated among the players. We will describe how this is done with an example.

Example 10. Four players (P_1, P_2, P_3, and P_4) must divide the array of 20 items shown in Fig. 3-12. P_1 plays first and lays down her markers

Figure 3-12

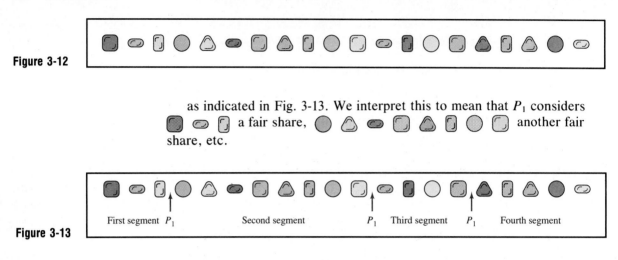

as indicated in Fig. 3-13. We interpret this to mean that P_1 considers a fair share, another fair share, etc.

Figure 3-13

Without seeing P_1's markers (P_1's markers can be recorded by a neutral person and then removed), P_2 now proceeds to play in a similar fashion. P_2 lays down his markers as shown in Fig. 3-14.

Figure 3-14

Similarly, P_3 and P_4 each get a turn to play by laying down their own markers, and they do so as shown in Figs. 3-15 and 3-16, respectively.

Figure 3-15

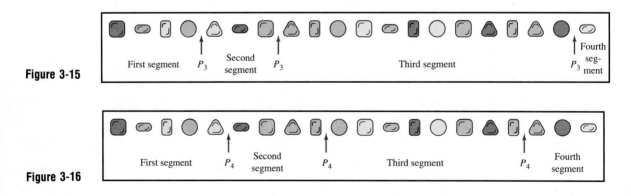

Figure 3-16

We are now ready to distribute the items among the players. To do so we need to look at the entire picture, showing all the markers of all the players, as in Fig. 3-17.

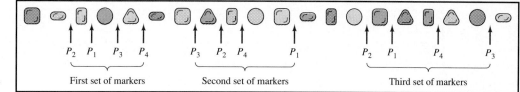

Figure 3-17

Moving from left to right, we assign the first segment to the player owning the first marker. That's P_2, who gets ▭ ▭. We know that these two items are in P_2's opinion a fair share. P_2 is now out of the game. All of P_2's markers can now be removed, and the new situation is described by Fig. 3-18.

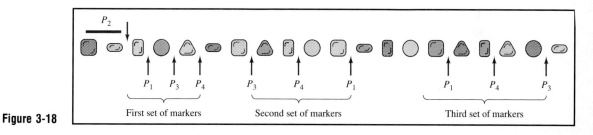

Figure 3-18

Once again we scan the picture from left to right until we find the first second marker. It belongs to P_3, and it is located to the right of the seventh item. We now look back and locate P_3's first marker (which is located to the right of the fourth item) and give P_3 ▲ ▭ ▭ as shown in Fig. 3-19.

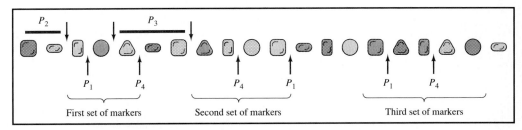

Figure 3-19

We repeat the process, locating the first third marker which belongs to P_1 (remember P_2 and P_3 are out of the running now) and scanning backward to find P_1's previous (second) marker. The items between these markers are given to P_1 (Fig. 3-20).

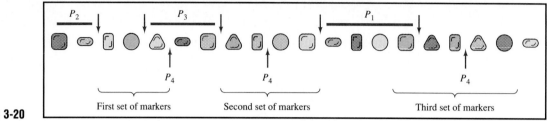

Figure 3-20

Finally, there is one last player left (P_4) who gets all the items from his last marker to the end of the array (Fig. 3-21).

Figure 3-21

Much to our surprise, we see that after each player has received his or her fair share, there are a few items left over (Fig. 3-22). These leftover items can be distributed in various ways—given to charity, allocated to

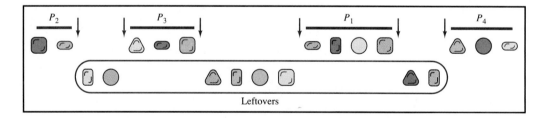

Figure 3-22

the players by drawing lots, or if there are enough of them, the method of markers can be used again.   ∎

The general description of the **method of markers** is as follows: There are N players (P_1, P_2, P_3, . . ., P_N) and M items arranged in an array. (Unlike the method of sealed bids, we must have $M > N$.)

∎ **Move 1 (Bidding).** Each player independently and secretly divides the linear array into N consecutive segments any one of which the player will accept as a fair share of the total set of items.

■ **Move 2 (Allocation).** Assign to the player owning the first marker (going from left to right) in the first set of markers that player's first segment. Remove that player's markers. Assign to the player owning the first marker in the second set of markers that player's second segment. Remove that player's markers. Continue this process until each player has received a fair share. The last player gets his or her last segment.

■ **Move 3 (Leftovers).** If there are leftover items, they can be allocated to the players by lottery, or if there are more items left than players, the method of markers can be used again.

In spite of its elegance and the nice touch added by the leftovers, the method of markers can be used only under some fairly restrictive conditions. In particular, the method assumes that every player is able to divide the array of items into segments in such a way that each of the segments has equal value (to that player). This precludes, for example, the existence of an item that has a disproportionately large value in relation to all the other items. Suppose that in Example 10, the 20 items consist of a gold watch and 19 pieces of candy (of assorted types). It would be impossible to distribute these items fairly among the four players, since no one except an insane candy lover could possibly divide them into four segments of equal value.

CONCLUSION

The problem of dividing an object or set of objects among the members of a group is a practical problem that comes up regularly in our daily lives. When the object is a pizza, a cake, or a bunch of candy, we don't always pay a great deal of attention to the issue of fairness, but when the object is an estate, a piece of land, or some other valuable asset, dividing things fairly becomes a critical issue.

On the surface, problems of fairness seem far removed from the realm of mathematics. We are more likely to think of ethics or law as being the proper fields for a discussion of this topic. It is surprising therefore, that when certain basic conditions are satisfied, mathematics can provide not only good but in fact remarkably elegant methods for guaranteeing fairness.

In this chapter we discussed several such methods, which we called *fair division schemes*. The choice of which is the best fair division scheme to use in a particular situation is not always clear, and in fact there are many situations in which a fair division is mathematically unattainable (we will discuss an important example of this in Chapter 4). At the same time, in a large number of everyday situations the fair division schemes we described in this chapter (or simple variations thereof) will work. Remember these methods the next time you must divide an inheritance, a piece of real estate, or even a fine old bottle of wine. They may serve you well.

KEY CONCEPTS

continuous fair division problem
discrete fair division problem
divider-chooser method
fair division problem
fair division scheme
fair share

last diminisher method
lone chooser method
lone divider method
method of markers
method of sealed bids

EXERCISES

■ Walking

1. Three players want to divide a cake fairly using the lone divider method. The divider cuts the cake into three slices (s_1, s_2, s_3).
   (a) If the chooser declarations are
      Chooser 1: $\{s_2, s_3\}$
      Chooser 2: $\{s_1, s_2\}$
      describe a possible fair division of the cake.
   (b) If the chooser declarations are
      Chooser 1: $\{s_1, s_2, s_3\}$
      Chooser 2: $\{s_3\}$
      describe a possible fair division of the cake.
   (c) If the chooser declarations are
      Chooser 1: $\{s_1\}$
      Chooser 2: $\{s_3\}$
      describe a possible fair division of the cake.
   (d) If the chooser declarations are
      Chooser 1: $\{s_2\}$
      Chooser 2: $\{s_2\}$
      describe how to proceed to obtain a possible fair division of the cake.

2. Suppose that four players want to divide a cake fairly using the lone divider method. The divider cuts the cake into four slices (s_1, s_2, s_3, s_4), and the choosers make the following declarations:
   Chooser 1: $\{s_2, s_3\}$
   Chooser 2: $\{s_1, s_3\}$
   Chooser 3: $\{s_1\}$.

(a) Describe a fair division of the cake.

(b) Explain why the answer in part (a) is the only possible fair division of the cake.

3. Suppose that four players want to divide a cake fairly using the lone divider method. The divider cuts the cake into four slices (s_1, s_2, s_3, s_4), and the choosers make the following declarations:
Chooser 1: $\{s_2, s_4\}$
Chooser 2: $\{s_1, s_4\}$
Chooser 3: $\{s_1, s_2\}$.
(a) Describe a fair division of the cake.
(b) Describe a fair division of the cake different from the one given in part (a).
(c) Is it possible to find a fair division of the cake such that the divider doesn't get s_3? Explain your answer.

4. Suppose that four players want to divide a cake fairly using the lone divider method. The divider cuts the cake into four slices (s_1, s_2, s_3, s_4) and the choosers make the following declarations:
Chooser 1: $\{s_3, s_4\}$
Chooser 2: $\{s_3, s_4\}$
Chooser 3: $\{s_3\}$.
Describe how to proceed to obtain a possible fair division of the cake.

5. Suppose that five players want to divide a cake fairly using the lone divider method. The divider cuts the cake into five slices (s_1, s_2, s_3, s_4, s_5), and the choosers make the following declarations:
Chooser 1: $\{s_3, s_4\}$
Chooser 2: $\{s_3, s_4\}$
Chooser 3: $\{s_2, s_3, s_4\}$
Chooser 4: $\{s_2, s_3, s_5\}$.
(a) Describe a fair division of the cake.
(b) Describe a fair division of the cake different from the one given in part (a).
(c) Is it possible to find a fair division of the cake such that the divider doesn't get s_1? Explain your answer.

6. Suppose that six players want to divide a cake fairly using the lone divider method. The divider cuts the cake into six slices (s_1, s_2, s_3, s_4, s_5, s_6), and the choosers make the following declarations:
Chooser 1: $\{s_1, s_3\}$
Chooser 2: $\{s_2\}$
Chooser 3: $\{s_4, s_5\}$
Chooser 4: $\{s_4, s_5\}$
Chooser 5: $\{s_2\}$.
Describe how to proceed in order to obtain a fair division of the cake.

7. A cake is to be divided among four people using the last diminisher method. The four people are randomly ordered and are called P_1, P_2, P_3, and P_4. In round 1, P_1 cuts a piece, P_2 and P_3 think it is fair and don't diminish it, and P_4 diminishes it.

(a) Is it possible for P_2 to end up with that piece or just a part of that piece?

(b) Who gets the piece at the end of round 1?

(c) Who cuts the piece at the beginning of round 2?

(d) Who is the last person who has an opportunity to diminish the piece in round 2?

(e) How many rounds are required to divide the cake among the four people?

8. A cake is to be divided among 15 people using the last diminisher method. The 15 people are randomly ordered and are called $P_1, P_2, P_3, \ldots, P_{15}$. In round 1, P_1 cuts a piece, and P_3, P_7, and P_{10} are the only diminishers. In round 2, the only diminisher is P_3, and in round 3 there are no diminishers.

(a) Who gets the piece at the end of round 1?

(b) Who cuts the piece at the beginning of round 2?

(c) Who is the last person who has an opportunity to diminish the piece in round 2?

(d) Who gets the piece at the end of round 2?

(e) Who cuts the piece at the beginning of round 3?

(f) Who is the last person who has an opportunity to diminish the piece in round 3?

(g) Who gets the piece at the end of round 3?

(h) How many rounds are required to divide the cake among the 15 people.

9. Three sisters (A, B, and C) wish to divide up four pieces of furniture they shared as children using the method of sealed bids. Their bids on each of the items are given in the following table.

	A	B	C
Dresser	$250	$310	$265
Desk	195	150	185
Vanity	175	215	235
Tapestry	510	490	475

Describe the outcome of this fair division problem.

10. Five heirs (A, B, C, D, and E) wish to divide up an estate consisting of six items using the method of sealed bids. The heirs bids on each of the items are given in the following table.

	A	B	C	D	E
Item 1	$352	$295	$400	$368	$324
Item 2	93	102	98	95	105
Item 3	461	449	510	501	476
Item 4	852	825	832	817	843
Item 5	512	501	505	515	491
Item 6	725	738	750	744	761

Describe the outcome of this fair division problem.

11. Three people (A, B, and C) wish to divide up four items using the method of sealed bids. Their bids on each of the items are given in the following table.

	A	B	C
Item 1	$ 20,000	$ 18,000	$ 16,000
Item 2	43,000	42,000	37,000
Item 3	4,000	2,000	1,000
Item 4	201,000	193,000	183,000

(a) What does A end up with? (Does he pay anything?)
(b) What does B end up with? (Does she pay anything?)
(c) What does C end up with? (Does she pay anything?)

12. Bob, Ann, and Jane wish to dissolve their partnership using the method of sealed bids. Bob bids $240,000 for the partnership, Ann bids $210,000, and Jane bids $270,000.
(a) Who gets the business and for how much?
(b) What do the other two people get?

13. Three players (P_1, P_2, and P_3) agree to divide the 13 items shown by lining them up in order and using the method of markers. The players' bids are as indicated.

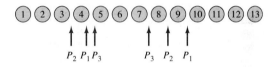

(a) Describe the allocation of items to each player.
(b) Which items are left over?

14. Two players (P_1 and P_2,) agree to divide the 12 items shown by lining them up in order and using the method of markers. The players' bids are as indicated.

(a) Describe the allocation of items to each player.
(b) Which items are left over?
(c) Illustrate how this situation might come up with a specific example.

15. Three players (P_1, P_2, and P_3) agree to divide the 12 items shown by lining them up in order and using the method of markers. The players' bids are as indicated.

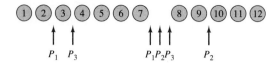

(a) Describe the allocation of items to each player.
(b) Which items are left over?

16. Three players (P_1, P_2, and P_3) agree to divide the 12 items shown by lining them up in order and using the method of markers. The players' bids are as indicated.

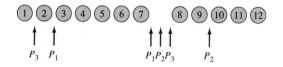

(a) Describe the allocation of items to each player.
(b) Which items are left over?

17. Five players (P_1, P_2, P_3, P_4, and P_5) agree to divide the 20 items shown by lining them up in order and using the method of markers. The players' bids are as indicated.

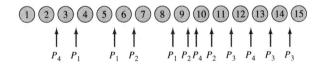

(a) Describe the allocation of items to each player.
(b) Which items are left over?

18. Four players (P_1, P_2, P_3, and P_4) agree to divide the 15 items shown by lining them up in order and using the method of markers. The players' bids are as indicated.

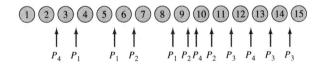

(a) Describe the allocation of items to each player.
(b) Which items are left over?

19. Two friends (David and Paul) decide to divide the pizza shown in the following figure using the divider-chooser method. David likes pepperoni, sausage, and mushrooms equally well but hates anchovies. Paul likes anchovies, sausage, and pepperoni equally well but hates mushrooms.

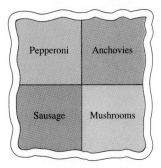

(a) Suppose that David is the divider. Which of the cuts (i) through (iv) show a division of the pizza into fair shares according to David?

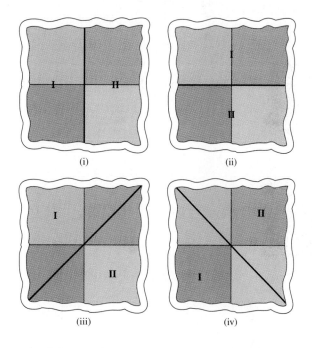

(b) For each of the cuts in part (a), which piece should Paul choose?

20. Raul and Trudy want to divide a chocolate-strawberry mousse cake. Raul values chocolate three times a much as he values strawberry ($3S = C$); Trudy values strawberry four times as much as she values chocolate.

(a) If Raul is the divider, which of the following cuts are consistent with Raul's value system?

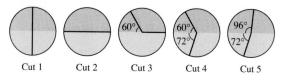

Cut 1   Cut 2   Cut 3   Cut 4   Cut 5

Hint: In Raul's eyes the cake looks (valuewise) like

(b) What does the cake look like (valuewise) in Trudy's eyes? [See the hint for part (a).]
(c) For each of the cuts in part (a), indicate which of the pieces is Trudy's best choice.
(d) Suppose Trudy is the divider. Draw three different cuts that are consistent with her value system.
(e) Explain why the following cut is consistent with any value system.

■ Jogging

21. (a) Describe how the lone chooser method would work in the case of four players.
    (b) Describe the lone chooser method for the general case of N players.

22. Describe the last diminisher method for the general case of N players.

23. Three players (P_1, P_2, and P_3) agree to divide the property shown using the last diminisher method. The order of the players is P_1, P_2, P_3. The first player to play, P_1, makes a claim C as shown in the figure.

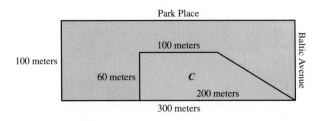

We know that both P_2's and P_3's value systems are the same and that they value the land uniformly.

(a) Give a geometric argument why P_2 and P_3 would both pass in round 1 and P_1 would end up with C.

(b) Describe a possible cut that the divider in round 2 might make.

(c) Suppose that after round 1 is over P_2 and P_3 discover that the city requires that the next cut be made parallel to Park Place. Describe a possible cut that the divider in round 2 might make in this case.

(d) Repeat part (c) for a cut that must be made parallel to Baltic Avenue.

24. This exercise is based on Example 8. Suppose that in her will, Grandma stipulates that the estate should be divided among the four heirs as follows: Art, 25%; Betty, 35%; Carla, 30%; and Dave, 10%. Describe a variation of the method of sealed bids that will accomplish this. (Feel free to use a calculator.)

25. Say that N players $(P_1, P_2, \ldots, P_N)$ are heirs to an estate. According to the will, P_1 is entitled to $r_1\%$ of the estate, P_2 is entitled to $r_2\%$, etc. $(r_1 + r_2 + \ldots + r_N = 100)$. Describe a general variation of the method of sealed bids that gives a fair division for this estate. (You should try this exercise only after you have finished Exercise 24.)

26. Two players (A and B) wish to dissolve their partnership using the method of sealed bids. A bids x and B bids y, where $x < y$.

(a) How much are A and B's original fair shares worth?

(b) How much is the surplus after the original allocations are made?

(c) When all is said and done, how much must B pay A for her half of the partnership?

27. Three players (P_1, P_2, and P_3) agree to divide some candy using the method of markers. The candy consists of three Reese's Pieces (R), six caramels (C), and six mints (M). The players' value systems are as follows:

P_1 loves Reese's Pieces but does not like caramels or mints at all.

P_2 loves caramels and Reese's Pieces equally well (i.e., 1 Reese's Piece = 1 caramel) but does not like mints at all.

P_3 loves caramels and mints equally well (i.e., 1 caramel = 1 mint) but is allergic to Reese's Pieces.

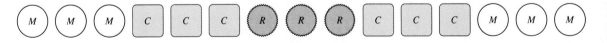

The candy is lined up as shown.

(a) What bid would P_1 make to ensure that she gets her fair share (according to her value system)?

(b) What bid would P_2 make to ensure that he gets his fair share (according to his value system)?

(c) What bid would P_3 make to ensure that she gets her fair share (according to her value system)?

(d) Describe the allocations to each player.

(e) What items are left over?

28. Repeat Exercise 27 except that the candy is lined up as follows.

$$\textcircled{M}\ \boxed{C}\ \textcircled{R}\ \boxed{C}\ \textcircled{R}\ \textcircled{M}\ \textcircled{R}\ \textcircled{M}\ \boxed{C}\ \textcircled{M}\ \boxed{C}\ \boxed{C}\ \textcircled{M}\ \textcircled{M}\ \boxed{C}$$

29. Suppose that two players (*A* and *B*) buy a chocolate-strawberry mousse cake with a caramel swirl and assorted frostings for $10. Since *A* contributes $7 and *B* only $3, they both agree that a fair division of the cake is one in which *A* gets a piece that is worth (in *A*'s opinion) at least 70% of the cake and *B* gets a piece that is worth (in *B*'s opinion) at least 30% of the cake. Describe a variation of the lone divider method that can be used in this situation. *Hint:* Think of this problem as an ordinary fair division problem with many players.

30. This problem is a variation of Exercise 29. Three players are involved (*A*, *B*, and *C*). *A* contributes $2.50, *B* contributes $3.50, and *C* contributes $4.00 toward the purchase of the cake. Describe a fair division scheme for this problem.

31. Consider the following variation of the divider-chooser method for two players: After the divider cuts the cake in two pieces, the chooser (who is unable to see either piece) picks his piece randomly by flipping a coin. The divider, of course, gets the other piece.
   (a) Is this a fair division scheme according to our definition? Explain your answer.
   (b) Who would you rather be—divider or chooser? Explain.

■ Running

32. **Alternative notion of a fair share.** To explain this exercise we will need some additional terminology. We will say that a player is *fairly happy* if she receives a share that is (in her opinion) worth at least 1/*N* of the total (what we have been calling a fair share). We will say that a player is *ecstatic* if she receives a share that is worth (in her opinion) at least as much as anyone else's—in other words, the player would not want to trade her share with anyone else. All the fair division schemes we presented in the chapter guarantee that all the players will be fairly happy but not that all the players will be ecstatic. Human nature being what it is, being fairly happy but not ecstatic might make some of the players unhappy. The purpose of this exercise is to discuss this alternative interpretation of fairness.
   (a) Explain why the divider-chooser method for two players guarantees that each player will be ecstatic.
   (b) Give an example of a fair division using the lone divider method with three players in which at least one of the players ends up with a piece that makes him happy but not ecstatic.
   (c) Give an example of a fair division using the lone divider method with three players in which all three players are ecstatic.

33. Using the terminology of Exercise 32, describe a continuous fair division method for three players that guarantees that all three players will be ecstatic. (See references 2 and 6.)

34. (a) Explain why after the original allocation is made in the method of sealed bids, the surplus produced must be either positive or zero.

(b) Under what condition is the surplus zero?

35. The purpose of this exercise is to extend the ideas of the lone divider method for three players to any number of players.

(a) Describe how the lone divider method would work in the case of four players. (*Hint*: Consider several cases following the format for the case of three players. Also look at Exercises 2 through 4.)

(b) Describe the lone divider method for the general case of *N* players.

**REFERENCES
AND FURTHER
READINGS**

1. Dubins, L. E., "Group Decision Devices," *American Mathematical Monthly,* 84 (1977), 350–356.

2. Gardner, Martin, *aha! Insight*. New York: W. H. Freeman, 1978, 124.

3. Kuhn, H. W., "On Games of Fair Division," in *Essays in Mathematical Economics,* ed. Martin Shubik. Princeton, N.J.: Princeton University Press, 1967, 29–37.

4. Steinhaus, Hugo, "The Problem of Fair Division," *Econometrica,* 16 (1948), 101–104.

5. Steinhaus, Hugo, *Mathematical Snapshots*. New York: Oxford University Press, 1960, 65–71.

6. Stromquist, Walter, "How to Cut a Cake Fairly," *American Mathematical Monthly,* 87 (October 1980), 640–644.

4 Apportionment

Making the Rounds

If we have fifty pieces of candy how do we divide them fairly among five children?

PROBLEM IN FOURTH GRADE SCHOOL BOOK

Hard as it is to believe, we have yet to say the last word on this question. In this chapter we will discuss one more twist to the remarkable story of fair division (after which we really promise to lay the issue to rest!): What happens when the pieces are all identical and indivisible but each child is entitled to a different percentage of the total? The following, all too familiar situation, illustrates what we mean.

A home lesson in capitalism. Fifty identical and indivisible pieces of candy (let's say for the sake of argument they are M&M's) are to be divided among the five children in a household. Instead of dividing the candy equally among the five kids, mother has an inspired idea: Why not divide the candy in proportion to the amount of time each child spends helping with the chores around the house? Stopwatch in hand, mother boldly proceeds to carry out this ingenious exercise in capitalism. When the work is done and the final tally made, this is how things stand:

	Alan	Betty	Connie	Doug	Ellie	Total
Minutes worked	150	78	173	204	295	900

It is now time to divide the candy. According to the ground rules, Alan is entitled to $16\frac{2}{3}\%$ of the 50 pieces of candy ($\frac{150}{900} = 16\frac{2}{3}\%$), namely, $8\frac{1}{3}$ pieces. Unfortunately, the individual pieces of candy are indivisible (ever try to cut an M&M in thirds?). This means that Alan can't get his exact fair share: He can get 8 pieces (and get shorted) or he can get 9 pieces (and someone else will get shorted), but he can't get exactly what he earned! Of course, what goes for Alan also applies to each of the other children: Betty's exact fair share would be $4\frac{1}{3}$ pieces, Connie's $9\frac{11}{18}$ pieces, Doug's $11\frac{1}{3}$ pieces, and Ellie's $16\frac{7}{18}$ pieces. Needless to say, none of these shares can be realized. Clearly, mother is going to need some help. ∎

APPORTIONMENT PROBLEMS

The problem that we described in the preceding example, belongs to an important class of fair division problems called **apportionment problems.** Solving apportionment problems is the subject of this chapter.

Given the circumstances described in the example, is there a fair way to distribute the candy? In one sense, the answer is clearly no. The very fact that the fair allotments are fractional while only whole-number allotments are possible implies that some children will get more than their fair share while others will get less. On the other hand, the fact that an element of unfairness is unavoidable should not preclude us from looking for methods of solution that minimize it (the unfairness). It is not unreasonable to expect that at the very least the decision as to which children get more than their share and which get less is made rationally and according to predetermined rules. Any systematic approach to solving an apportionment problem is called an **apportionment method**, and in this chapter we will describe some of the most commonly used apportionment methods.

Just as we did in Chapter 3, we must warn the reader not to be misled by the frivolous nature of our original example. Once again we're using kids and M&M's as a simple, but convenient metaphor. Starting with our next example, we will embark on a journey that will show why the apportionment problem represents one of the most important issues to confront a democracy: the issue of proportional representation. In fact, we will soon discover that the theme of this chapter is as much a part of American history as it is a mathematical problem.

Example 1 (Parador's Congress). Parador is a new republic located in Central America. It is made up of six states: Azucar, Bahia, Cafe, Diamante, Esmeralda, and Felicidad (*A*, *B*, *C*, *D*, *E*, and *F* for short).

According to the recently adopted constitution of Parador, the 250 seats in their legislature must be divided among the six states according to their respective populations. Based on the most recent census, the populations of each state are

State	A	B	C	D	E	F	Total
Population	1,646,000	6,936,000	154,000	2,091,000	685,000	988,000	12,500,000

How many seats should each state get?   ■

It is not hard to see that this problem and the problem of dividing the M&M's are variations on the same theme: A certain number of players (N) must divide fairly a certain number (M) of identical and indivisible objects. Here "fairly" means that each player is entitled to a certain **quota**[1] of the M objects which is proportional to some numerical criterion. In the M&M's example the players are the children ($N = 5$), the objects are pieces of candy ($M = 50$), and the quotas are based on hours of work around the house. In Example 1 the players are the states ($N = 6$), the objects are the seats in the legislature ($M = 250$), and the quotas are based on population.

Although apportionment problems occur in many types of applications, the classic apportionment problem (and the one from which the problem gets its name) is that of allocating seats in a legislative body based on populations as typified by Example 1. For the rest of the chapter our discussion will be centered around this version of the problem. The reader should keep in mind, however, that other interesting applications of the apportionment problem exist (see Exercises 6 through 12 and 18, as well as Example 10).

**STANDARD
DIVISORS AND
STANDARD
QUOTAS**

Let's go back to the apportionment problem for Parador's legislature (Example 1). For the reader's convenience we show the population data again.

State	A	B	C	D	E	F	Total
Population	1,646,000	6,936,000	154,000	2,091,000	685,000	988,000	12,500,000

Since the total population of Parador is 12,500,000 and the number of seats is $M = 250$, the ratio 12,500,000/250 = 50,000 represents the number

---

[1] Here the word "quota" has a completely different meaning than in Chapter 2.

of people per single seat in the legislature. In general, for any apportionment problem in which the total population of the country is P and the number of seats to be apportioned is M, the ratio P/M gives the number of people *per seat* on a national basis. We will call the number P/M the **standard divisor**. (Note that the standard divisor makes sense in any apportionment problem, regardless of the context. In our original M&M's example, the standard divisor is $P/M = 900/50 = 18$, and it represents the exact number of minutes of work needed to earn a single M&M.)

Using the standard divisor we can now calculate the fraction of the total number of seats each state would be entitled to if the seats were not indivisible (with apologies for the double negative!). This number, which we will call the state's **standard quota**, is obtained by dividing the state's population by the standard divisor. In other words,

$$\text{State's standard quota} = \frac{\text{state's population}}{\text{standard divisor}}$$

The standard quotas for Example 1 are as follows:

State	A	B	C	D	E	F	Total
Population	1,646,000	6,936,000	154,000	2,091,000	685,000	988,000	12,500,000
Standard quota	32.92	138.72	3.08	41.82	13.70	19.76	250

The fact that the standard quota for state A is 32.92 simply means that if the seats in the legislature could be chopped up into fractional parts, then state A would get exactly 32.92 seats. Unfortunately this can't be done. The same thing goes for the rest of the states.

Notice that the sum of all the standard quotas should be the total number of seats M. (If it's close but not exactly M, it is due to roundoff errors.) This is an important fact, and we will use it later.

Mathematically, an apportionment problem can now be thought of as a simple *rounding* problem—how do we round off the quotas (which are usually decimals) to whole numbers while maintaining the *same total* (M). A seemingly reasonable idea would be to round off in the conventional way: If the fractional part is more than or equal to 0.5 round upward, if it is less than 0.5 round downward. Unfortunately, this approach doesn't always work. Let's try it for Example 1.

State	A	B	C	D	E	F	Total
Population (in thousands)	1646	6936	154	2091	685	988	12,500
Quota	32.92	138.72	3.08	41.82	13.70	19.76	250
Rounded-off quotas	33	139	3	42	14	20	251

In this particular example, conventional rounding off results in a serious constitutional problem: By law there are 250 legislative seats to give out, but we would be giving out 251 (somewhat analogous to overbooking an airplane flight!).

Since the conventional approach to rounding off cannot be counted on to work as an apportionment method, somewhat more sophisticated ideas will be needed. Many such ideas have come from some surprising sources, names we do not associate with mathematical problems: Alexander Hamilton, Thomas Jefferson, John Quincy Adams, and Daniel Webster.

A LITTLE BIT OF U.S. HISTORY

Because it is difficult to separate the mathematics and the politics of the apportionment problem (and also because it is unusual to discover an important chapter in American history lurking in a mathematics book), we will take a brief historical detour.

> *Representatives and direct Taxes shall be apportioned among the several States which may be included within this Union, according to their respective Numbers. . . . The actual Enumeration shall be made within three Years after the first Meeting of the Congress of the United States, and within every subsequent Term of ten Years, in such Manner as they shall by Law direct. The Number of Representatives shall not exceed one for every thirty Thousand, but each State shall have at Least one Representative; . . .*

ARTICLE 1, SECTION 2, CONSTITUTION OF THE UNITED STATES

The relevant facts implied by Article 1, Section 2, and stated in the language of this chapter are

1. The seats in the House of Representatives shall be *apportioned*[2] to the states on the basis of their respective populations.

2. The populations of the states shall be determined every 10 years. (This is done by a national census—the most recent census was taken in 1990.)

3. The apportionment method (including the total number of seats M that will be apportioned) is also to be established every 10 years by congressional law (subject to the same rules and regulations as any other congressional bill).

4. The number of people per single seat (the standard divisor) should be at least 30,000 (in other words, $P/M \geq 30,000$).

5. Each state should get at least one seat.

It didn't take long after the drafting of the Constitution for the controversy over apportionment methods to begin. Following the census of 1790, two competing methods of apportionment were proposed in Congress—

[2] **ap·pōr′tion**: to divide and assign in proportion or according to some plan (Webster's New Twentieth Century Dictionary).

George Washington and his Cabinet: lithograph, 1876, by Currier and Ives. (The Granger Collection)

one was the brainchild of Alexander Hamilton, and the other was suggested by Thomas Jefferson. After a heated debate Congress approved (barely) a bill to apportion the House of Representatives using the method proposed by Hamilton based on a House of Representatives of size $M = 120$. After giving the issue considerable thought, President Washington vetoed the bill (this was the first exercise of presidential veto power in U.S. history!). Unable to override the presidential veto and facing a damaging political stalemate, Congress eventually adopted Jefferson's method based on a House of Representatives of size $M = 105$.

HAMILTON'S METHOD (THE METHOD OF LARGEST FRACTIONS)

While historically Hamilton's method did not come first, we will discuss it first because it is mathematically the simplest. By the way, Hamilton's method was adopted by Congress after the 1850 Census, and it was used from 1852 to 1900.

Hamilton's Method:

- **Step 0.** Calculate each state's standard quota. (Recall that the simplest way to do this is to first calculate the standard divisor P/M and then divide each state's population by the standard divisor.)

- **Step 1.** Round off each state's standard quota downward and give each state (for the time being) that number of seats.

- **Step 2.** If there are surplus seats (there usually are), give them (one at a time) to the states with the largest fractional parts until we run out of surplus seats.

Example 2. The following table illustrates how Hamilton's method would apportion the legislature of our favorite country, Parador (Example 1).

State	A	B	C	D	E	F	Total
Population (in thousands)	1646	6936	154	2091	685	988	12,500
Step 0: Calculate the standard quotas	32.92	138.72	3.08	41.82	13.70	19.76	250
Step 1: Round the quotas downward	32	138	3	41	13	19	246
Step 2: Distribute the four surplus seats	+1 (1st)	+1 (4th)		+1 (2nd)		+1 (3rd)	4
Final apportionment	33	139	3	42	13	20	250

■

Is Hamilton's method a fair and reasonable apportionment method? Obviously, Hamilton thought it was, but we can already see in Example 2 that all is not quite as fair as we would hope. Consider the sad fate of state E, next in line for a surplus seat with a hefty fractional part of 0.70 and yet getting none. Is state B (with a fractional part of 0.72) truly that much more deserving than state E in getting that last surplus seat? The answer is not quite clear. If we look at fractional parts in absolute terms, the answer is yes (0.72 is more than 0.70). However, when we look at the fractional part as a percentage of the entire state's population, state E's 0.70 represents a much larger proportion of its quota (13.70) than state B's 0.72 is of its quota (138.72). Perhaps the last surplus seat should have gone to state E.

LOWNDES' METHOD

The idea that relative rather than absolute fractional parts should determine the order in which the surplus seats are handed out is the basis for a variation of Hamilton's method called **Lowndes' method**.[3]

Lowndes' method starts out exactly the same as Hamilton's method—each state's standard quota gets rounded downward and each state is given (for the time being) that number of seats. The next step, however, requires that we calculate for each state's quota the **relative fractional part**. Specifically, this means that we divide the fractional part of the quota by the integer part of the quota.

Example 3.

(a) State's quota: 41.82
   Fractional part: 0.82
   Integer part: 41
   Relative fractional part: 0.82/41 = 0.02
(b) State's quota: 3.08
   Fractional part: 0.08
   Integer part: 3
   Relative fractional part: 0.08/3 = 0.027
(c) State's quota: 15 (= 15.0)
   Fractional part: 0 (= 0.0)
   Integer part: 15
   Relative fractional part: 0/15 = 0   ■

Examples 3(a) and 3(b) underscore the big difference between fractional parts and relative fractional parts. In Example 3(a) there is a large

---

[3] William Lowndes was a representative from the state of South Carolina who proposed this method in 1822. The method was discussed but never adopted by Congress.

fractional part (0.82); in Example 3(b) there is a very small fractional part (0.08). In Example 3(a) however, the relative fractional part is smaller than the one in Example 3(b). The difference is caused by the denominators (remember that smaller denominators make for bigger fractions than large denominators!).

Back to Lowndes' method. After the relative fractional parts for each state are calculated, we assign the surplus seats to the states with the largest relative fractional parts until we run out of surplus seats.

Example 4.   We will apply Lowndes' method to the apportionment problem for Parador's legislature so that we can compare the results with those produced by Hamilton's method (Example 2).

State	A	B	C	D	E	F	Total
Population (in thousands)	1646	6936	154	2091	685	988	12,500
Standard quota	32.92	138.72	3.08	41.82	13.70	19.76	250
Round the quotas downward	32	138	3	41	13	19	246
Relative fractional parts	0.029	0.0052	0.027	0.02	0.054	0.040	
Distribute the four surplus seats	+1 (3rd)		+1 (4th)		+1 (1st)	+1 (2nd)	4
Final apportionment	33	138	4	41	13	20	250

If we look closely at Example 4, particularly the fates of states C and D, we might have some questions about the fairness of this situation. State C has a tiny fractional part (0.08), but under Lowndes' method it is able to parlay it to get the last surplus seat ahead of state D with a fractional part about 10 times as large (0.82). The reason this happens is that State C is a very small state.

Comparing the results of Hamilton's method with those of Lowndes' method, it is not hard to see why large states always prefered Hamilton's method while small states were the ones pushing for (without much luck) the adoption of Lowndes' method. At the same time, we can see why it is difficult to say which of the two methods is fairer, so we conclude this section with some food for thought: If weightlifter A weighs 165 pounds and can lift 340 pounds, while weightlifter B weighs 242 pounds and can lift 461 pounds, who is the better weightlifter?

JEFFERSON'S METHOD

We start this section with a query in the form of a multiple-choice question:

Under both Hamilton's method and Lowndes' method:

    (a) all quotas are rounded downward.
    (b) all quotas are rounded upward.
    (c) some quotas are rounded downward and some are rounded upward.

If your answer was (a) you were wrong. Just look at the quotas and final apportionments in Examples 2 and 4 and you can see that while every quota is rounded downward at the beginning, by the time everything is said and done the result of giving a state a surplus seat is to round its quota upward. As a matter of fact, there is an important general rule lurking in the background here: *In any apportionment method based on standard quotas, some states' quotas will be rounded downward and other states' quotas will be rounded up.* This is an inevitable consequence of the fact that when we add up the standard quotas, the total must be equal to M. (Since the total is constant, one state's gain is another state's loss.) Because both Hamilton's and Lowndes' methods use the standard divisor and the standard quota as the basis for making the apportionment decisions, they are known as **quota methods**.

Standard divisors and standard quotas are not, however, the only way to go. Our experience with Hamilton's and Lowndes' methods warns us that having to choose which states get their quotas rounded upward and which get their quotas rounded downward makes it tough to be fair. Let's say, for the sake of argument, that we decide ahead of time that the fairest approach is to round off all the quotas the same way—downward. Can we set up the ground rules in such a way that this happens? From our preceding discussion we know that we can't use the standard quotas as a basis for making apportionment decisions. In fact we need to come up with new quotas that are *larger* than the standard quotas, so that when we add up their integer parts the total is M. The way to accomplish this is by using a divisor that is *smaller* than the standard divisor, as the next example illustrates.

Thomas Jefferson. (The White House Collection)

Example 5. Once again we follow up on Example 1.

State	A	B	C	D	E	F	Total
Population (in thousands)	1646	6936	154	2091	685	988	12,500
Standard quota (population/50,000)	32.92	138.72	3.08	41.82	13.70	19.76	250

Recall that the standard quotas are calculated using the standard divisor 50,000 (12,500,000 ÷ 250 = 50,000). Dividing each state's population by 50,000 results in the standard quotas shown in the preceding table.

What would happen if we divided each state's population by 49,500 instead of 50,000? (Let's not worry for the time being about where the 49,500 came from—let's just say it dropped out of the blue.) The results are shown in the following table.

State	A	B	C	D	E	F	Total
Population (in thousands)	1646	6936	154	2091	685	988	12,500
Modified quota (population/49,500)	33.25	140.12	3.11	42.24	13.84	19.96	252.52

The first thing we notice in the preceding table is that the modified quotas are all larger than the original standard quotas. This is no surprise because we are, after all, dividing by a smaller amount. When we add all the modified quotas together, we get 252.52, but this should be no cause for panic. If we round all the modified quotas downward and then add, we get 250—exactly what the doctor ordered!

State	A	B	C	D	E	F	Total
Population (in thousands)	1646	6936	154	2091	685	988	12,500
Modified quota (population/49,500)	33.25	140.12	3.11	42.24	13.84	19.96	252.52
Apportionment	33	140	3	42	13	19	250

Our preceding discussion is, in essence, a description of Jefferson's method[4] which we will now formalize.

Jefferson's Method:

■ **Step 1.** Find a number D such that when each state's modified quota (state's population ÷ D) is rounded *downward*, the total is M.

■ **Step 2.** Apportion to each state the *integer part* of its modified quota.

Before continuing with our discussion of Jefferson's method, we will make official some of the terminology we have already used: We call the number D used in step 1 the **divisor**, and the result of dividing the state's population by D, the state's **modified quota** (based on the divisor D).

A practical question that comes up in using Jefferson's method is How do we find a divisor D that works? Let's start with the fact that D should be less than the standard divisor (remember, we want the modified quotas to be bigger than the standard quotas). We now proceed by educated trial and error. We pick a good educated guess D_1, calculate the modified quotas based on D_1, round them downward and add. If the total is M, we have our divisor and we are finished. If the total is $M+1$ or more, we need to lower the modified quotas. We accomplish this by raising the divisor, so we choose $D_2 > D_1$ and try again. On the other hand, if the total is less than M, we reverse the process and choose $D_2 < D_1$. We keep trying—each time making a *better* guess than the last time until we hit a divisor that works (there is usually more than one). Needless to say, a good calculator is essential.

When presented with Jefferson's method, most people's first reaction is: Can we get away with it? Is it legal to use divisors other than the standard divisor? The answer is yes. There is nothing in the Constitution that requires use of the standard divisor as the basis for apportionment. Remember that the standard divisor is the ratio of people per seat in the House of Representatives for the nation as a whole. Using the 1990 United States census, for example, we had a total U.S. apportionment population of $P = 249{,}022{,}783$ (For apportionment purposes the population of the District of Columbia is not included.) and a House of Representatives of size $M = 435$, giving a standard divisor of approximately 572,466—give or take a limb or two. This means that for the *nation as a whole* there were about 572,466 people per seat in the House of Representatives. For individual states, however, the figures are different, because for each state

---

[4] Jefferson's method was the first apportionment method used in the United States, starting with its reluctant adoption by Congress in 1792. The method was used until 1840 but has not been used since.

the ratio of people per seat in the House of Representatives is based on the state's actual apportionment rather than the quota. Take, for example, California, with a population of 29,839,250 and 52 seats in the House of Representatives. If we divide these numbers out, we get 573,831 Californians for every seat in the House of Representatives—a larger ratio than the national figure.

We can see from all the above that there is no constitutional (or mathematical) requirement to use the standard divisor as a basis for apportionment. We should think of the standard divisor as a guideline rather than a requirement for apportionment.

■ The Quota Rule

A perceptive reader may have noticed that Jefferson's method as used in Example 5 produced a windfall for state *B*. State *B*, with a standard quota of 138.72, ended up with an apportionment of 140 seats. Thus far our apportionments have always been the integer immediately below or above the standard quota, and Jefferson's method has broken with tradition in this regard.

Our desire that a good apportionment method not stray too far from the standard quotas has a formal name, **the quota rule**, and in the language of apportionment we say that Jefferson's method *violates* (i.e., does not satisfy) the quota rule. In this case the violation of the quota rule will be called a violation of the **upper quota**. It is a case of giving a state an apportionment that is more than the integer immediately above the standard quota.

Apportionment methods that use divisors other than the standard divisor as a basis for apportionment are called **divisor methods**, and we have already seen an example of one such method—Jefferson's. There are several other divisor methods that are of both historical and mathematical interest, and we will mention two more in the next two sections (a third one is discussed in Exercises 33 through 35). Because these methods are very similar to Jefferson's method, we will only sketch out the details.

ADAMS' METHOD

This method was proposed by John Quincy Adams in 1830 but was never adopted by Congress.

Adams' Method:

■ **Step 1.** Find a divisor *D* such that when rounding off the modified quotas upward to the next higher integer the total is *M*.

■ **Step 2.** Apportion to each state its modified quota rounded upward to the next higher integer.

John Quincy Adams. (The Metropolitan Museum of Art)

Note that while the divisor D chosen for Jefferson's method must be smaller than (or equal to) the standard divisor, the divisor D chosen for Adams' method must be larger than (or equal to) the standard divisor.

Example 6.   Let's apply Adams' method to Example 1. We start by guessing a possible divisor D (larger than the standard divisor 50,000). Let us try $D = 50,500$.

State	A	B	C	D	E	F	Total
Population (in thousands)	1646	6936	154	2091	685	988	12,500
Modified quota ($D = 50,500$)	32.59	137.35	3.05	41.41	13.56	19.56	
Modified quotas rounded upward	33	138	4	42	14	20	251

Unfortunately the total of 251 is too large—we must lower the modified quotas a bit further by raising the divisor D. Using $D = 50,700$, we have the following:

State	A	B	C	D	E	F	Total
Population (in thousands)	1646	6936	154	2091	685	988	12,500
Modified quota ($D = 50,700$)	32.47	136.80	3.04	41.24	13.51	19.49	
Modified quotas rounded upward	33	137	4	42	14	20	250

We have found a divisor D that works! ■

Once again, we call the reader's attention to the final apportionment for state B: 137 seats (in spite of a standard quota of 138.72). This situation illustrates the fact that Adams' method can also violate the quota rule. Because in this case the violation occurs by giving a state an apportionment that is less than the integer immediately below the standard quota, it is called a violation of the **lower quota**.

WEBSTER'S METHOD

This method was proposed by Daniel Webster in 1830 as a natural compromise between Jefferson's and Adams' methods. It was adopted by Congress in the 1840s and again from 1900 to 1940.

Webster's Method:

■ **Step 1.** Find a divisor D such that when rounding off the modified quotas in the conventional way (to the nearest integer), the total is M.

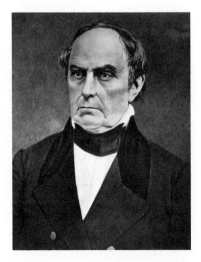

Daniel Webster. (U.S. Signal Corps photo)

■ **Step 2.** Apportion to each state its modified quota rounded off to the nearest integer.

Example 7. Let's use Webster's method to apportion Parador's congress. Our first decision is to make a guess at the divisor D. Should it be more than the standard divisor (50,000) or should it be less? Here we will use the standard quotas as a guideline.

If we round off the original quotas to the nearest integer (as we did at the beginning of this chapter), we get a total of 251. This tells us that we must find a divisor that lowers the total, so we must choose a divisor D slightly larger than 50,000. Let us try $D = 50,100$.

State	A	B	C	D	E	F	Total
Population (in thousands)	1646	6936	154	2091	685	988	12,500
Modified quota ($D = 50,100$)	32.85	138.44	3.07	41.74	13.67	19.72	
Apportionment	33	138	3	42	14	20	250

It worked! (Aren't we lucky?)   ■

Although Webster's method works in principle just like Jefferson's and Adams' methods, it is a little harder to use in practice. The divisor D we are looking for can be smaller, equal, or larger than the standard divisor. For guidelines as to how to go about making an educated guess, see Exercise 24.

PITFALLS AND PARADOXES

Of the various apportionment methods we have discussed in this chapter, which one is the best? Debate over this issue goes as far back as 1792, when Hamilton and Jefferson argued the merits of their respective methods with President Washington, and the question has come up in Congress time and time again. In fact it is likely that the debate over this question will flare up again soon.

Part of the problem in choosing among the various apportionment methods is that it is not even clear what it means for one apportionment method to be better than another. The aim of this section is to shed some light on these questions in a mathematical (as opposed to a political) vein. Our experience with Chapter 1 will serve us well here. There is in fact a surprising parallelism between voting methods and apportionment methods.

We'll start with the fact that none of the apportionment methods we have considered is perfect—they all have their flaws. In an effort to better understand these flaws, we will classify them into two kinds: pitfalls and paradoxes. **Pitfalls** are problems we can readily detect, whereas **paradoxes** are a more insidious kind of flaw—we have to dig deeper into the mathematics behind the method before we can detect them. A paradox, moreover, is a flaw that shocks because it defies intuition and common sense.

We have already encountered some pitfalls. In our discussion of Hamilton's method, we found that it seems to favor large states over smaller states and, conversely, we found that Lowndes' method seems to favor smaller states over larger ones.

We have also seen that both Jefferson's and Adams' methods can violate the quota rule. This was shown in the final apportionments for state *B* in Examples 5 and 6 respectively. There is, in fact, more than meets the eye here—both methods discriminate against large and small states in the way that the quota rule is violated. Under Jefferson's method, when the quota rule is violated, it is always a violation of the upper quota; under Adams' method quota rule violations are always violations of the lower quota. What about Webster's method? Although it doesn't happen in our example, Webster's method can also violate the quota rule. In this case both upper quota and lower quota violations can occur (but not in the same apportionment problem).

We will now graduate to a discussion of the more serious problems plaguing some of our apportionment methods. They are known as the **Alabama paradox**, the **population paradox**, and the **new states paradox**.

■ The Alabama Paradox

Immediately after the census of 1880, Congress started debating how to apportion the House of Representatives for the decade of the 1880s. Two of the several options discussed were

1. Use Hamilton's method with $M = 299$.

2. Use Hamilton's method with $M = 300$.

Options 1 and 2 were both based on the populations given by the census of 1880, so that the only difference was that under option 2 there was one more seat to give out. Under these conditions the logical thing one expects is that every state's apportionment under option 2 remain the same as it was under option 1 except that one lucky state would get the extra seat. Surprisingly this is not what happened. There were, in fact, three states affected: Illinois, with an apportionment of 18 seats under option 1, would have had 19 seats under option 2; Texas, with 9 seats under option 1, would have had 10 seats under option 2; and Alabama, *with 8 seats under option 1, would get 7 seats under option 2*. How is it possible that as the

House of Representatives got bigger (and nothing else changed) Alabama lost a seat?

We can illustrate the mathematical basis for the Alabama paradox by means of the following simple example.

Example 8. This example is based on the preceding true story. We have changed the numbers to simplify the computations. A country has three states: TX, IL, and AL. The following table shows the latest census figures.

State	TX	IL	AL	Total
Population	10,030	9030	940	20,000

Suppose the house of representatives has size $M = 200$. Then the standard divisor would be 100, and Hamilton's method would give

State	TX	IL	AL	Total
Population	10,030	9030	940	20,000
Standard quota	100.3	90.3	9.4	200
Step 1	100	90	9	199
Step 2			+1	+1
Final apportionment	100	90	10	200

Now let's change the size of the house of representatives to $M = 201$. The standard divisor would now be 99.5, and Hamilton's method would give

State	TX	IL	AL	Total
Population	10,030	9030	940	20,000
Standard quota	100.8	90.75	9.45	201
Step 1	100	90	9	199
Step 2	+1	+1		+2
Final apportionment	101	91	9	201

■

Example 8 shows clearly how the Alabama paradox can occur: The increase in the value of M changes the quotas in such a way that the order of fractional parts gets scrambled. In this example, AL goes from being first in line to being last, and in spite of the fact that there are more seats to be handed out, AL loses a seat!

■ **The Population Paradox**

To best understand this paradox we will first introduce a smaller paradox (kind of a warmup paradox, if you will). Based on the 1900 U.S. census, the population of Virginia was 1,854,184. Its standard quota (based on a House of size $M = 386$) was 9.599. By 1901, Virginia's population had grown to 1,874,024 (a 1.07% growth rate). Based on its 1901 population and the same-sized House ($M = 386$), Virginia's standard quota would have been 9.509 (had anyone taken the trouble to compute it). Why did Virginia's quota go down even though its population went up? The answer is simple: The nation as a whole grew at a faster rate than Virginia. Virginia's situation is analogous to that of a horse fading in a horserace—moving forward but at the same time falling further and further behind. That this can happen is not a big surprise, just a matter of simple arithmetic.

At about the same time that Virginia was growing at a rate of 1.07%, Maine was growing at a rate of 0.67%, a growth rate much smaller than Virginia's. The effect that this had on Maine's standard quota was that it dropped from 3.595 to 3.548. The relevant figures are shown in the following table:

	1900 Quota	Growth Rate 1900–1901	1901 Quota
Virginia	9.599	1.07%	9.509
Maine	3.595	0.67%	3.548

The real surprise is the way that Hamilton's method would have dealt with these changes. In 1900 Virginia would have been the last state to get a surplus seat, but in 1901 Maine would have moved ahead of Virginia in the fractional line. The following table shows the apportionments under Hamilton's method as they would have occurred in 1900 and 1901, respectively. The growth rates are repeated for emphasis.

	1900 Apportionment	Growth Rate 1900–1901	1901 Apportionment
Virginia	10	1.07%	9
Maine	3	0.67%	4

The irony of the above situation is that Virginia grew at a faster rate than Maine and at the same time lost a seat to Maine in the House of Representatives. This paradox is called the **population paradox**, and any apportionment method that forces one state to give up seats to another state that has become proportionally smaller is said to suffer from this paradox.

The following example gives a complete illustration of how the population paradox can occur when Hamilton's method is used. The example is based on the opening example for the chapter (remember mom, the 50 M&M's, and the five kids doing housework?)

Example 9.   The following table shows the original facts given at the beginning of the chapter and how the 50 M&M's would have been apportioned using Hamilton's method.

	Alan	Betty	Connie	Doug	Ellie	Total
Minutes of work	150	78	173	204	295	900
Standard quota (Standard divisor $= 18$)	$8\frac{1}{3}$	$4\frac{1}{3}$	$9\frac{11}{18}$	$11\frac{1}{3}$	$16\frac{7}{18}$	50
Step 1: Round quotas downward	8	4	9	11	16	48
Step 2: Distribute surplus			+1		+1	2
Final apportionment	8	4	10	11	17	50

Let's suppose now that, right before the candy is about to be handed out (in accordance with the last line of the preceding table), mom discovers that the dishwasher needs to be emptied (oops!)—an additional 9 minutes of work. The decision (hers) is to ask for volunteers to carry out the job, and based on the new work totals to reapportion the M&M's using Hamilton's method again. Only Connie (8 minutes) and Ellie (1 minute) help out—Alan, Betty, and Doug have had enough! The following table shows the way that Hamilton's method would apportion the M&M's now.

	Alan	Betty	Connie	Doug	Ellie	Total
Old work	150	78	173	204	295	900
Extra work	0	0	8	0	1	9
Total work	150	78	181	204	296	909
Standard quota (Standard divisor $= 18.18$)	8.25	4.29	9.96	11.22	16.28	50
Step 1: Round quotas downward	8	4	9	11	16	48
Step 2: Distribute surplus		+1	+1			2
Final apportionment	8	⑤	10	11	⑯	50

If we now compare the final apportionments before and after the extra work was added, the paradox can be seen clearly: Mom must tell Ellie that Betty earned one of her M&M's by virtue of having done no extra work.   ■

■ The New States Paradox

The third paradox lurking behind Hamilton's method came up after Oklahoma became a state in 1907. Before 1907 the apportionments for each of the already existing states were based on the 1900 U.S. Census (and a House of Representatives of size $M = 386$). Based on the same standard divisor used for the other states, Oklahoma was allocated 5 seats, and these 5 seats were added to the House of Representatives so that after Oklahoma joined the Union the size of the House of Representatives grew to 391. The obvious intent of adding the 5 seats to the House was to leave the old apportionments unchanged. After all, it seems reasonable that no state's apportionment should be affected by the addition of a new state.

Suppose that after Oklahoma became a state a new apportionment was carried out using exactly the same state population figures (Oklahoma included), the same number of seats ($M = 391$), and the same method (Hamilton's). The obvious thing to expect is that Oklahoma would get the 5 new seats and all other apportionments would remain the same. Much to our surprise we find that Maine's apportionment increased from 3 to 4 seats and New York's apportionment decreased from 38 to 37 seats. The mere addition of Oklahoma (with its fair share of seats) to the Union would have forced New York to give up a seat to Maine! The perplexing fact that the addition of a new state with its fair share of seats can by itself affect the apportionments of other states is called the **new states paradox**.

The following example gives a simple illustration of the new states paradox.

Example 10. Central School District has two high schools: North High has an enrollment of 1045 students; South High has an enrollment of 8955. The school district is allocated a counseling staff of 100 counselors, and these counselors are to be apportioned between the two schools using Hamilton's method. This results in an apportionment of 10 counselors to North High and 90 counselors to South High. The computation is summarized in the following table:

	North High	South High	Total
Enrollment	1045	8955	10,000
Standard quota (Standard divisor = 100)	10.45	89.55	100
Apportionment	10	90	100

Suppose now that a new high school (New High) is added to the district. New High has an enrollment of 525 students, so the district (using the same standard divisor of 100 students per counselor) decides to hire 5 new counselors and assign them to New High. After this is done, someone

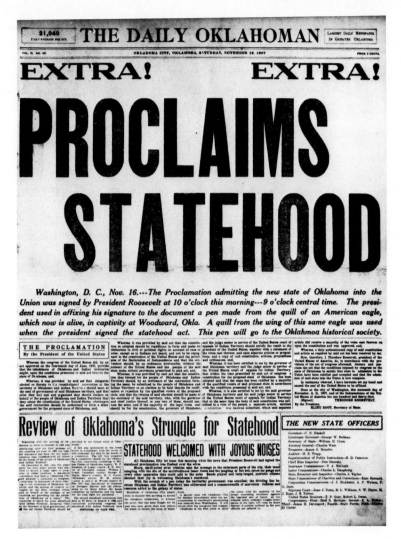

November 16, 1907:
Oklahoma becomes the 46th
state.

has the bright idea of recalculating the entire apportionment (still using Hamilton's method). The surprising result is shown in the following table:

	North High	South High	New High	Total
Enrollment	1045	8955	525	10,525
Standard quota (Standard divisor = 100.238)	10.4	89.34	5.24	105
Apportionment	11	89	5	105

The Alabama, population, and new states paradoxes are serious flaws shared by both Hamilton's and Lowndes' methods. It is a mathematically proven fact that none of the divisor methods (such as Jefferson's, Adams', or Webster's) can suffer from these flaws, so at least in terms of these paradoxes it is safe to say that divisor methods are far superior to quota methods. At the same time, neither Hamilton's nor Lowndes' method can violate the quota rule (Exercise 30), and as we have already discovered, all three of the divisor methods can.

CONCLUSION: BALINSKI AND YOUNG'S IMPOSSIBILITY THEOREM

In this chapter we introduced five different *apportionment methods*. The following table summarizes the results of apportioning Parador's congress (Example 1) under each of the five methods. It is worthwhile to note that each of the five methods produced a different apportionment.

State	A	B	C	D	E	F	Total
Population (in thousands)	1646	6936	154	2091	685	988	12,500
Quota	32.92	138.72	3.08	41.82	13.70	19.76	250
Hamilton	33	139	3	42	13	20	250
Lowndes	33	138	4	41	14	20	250
Jefferson	33	140	3	42	13	19	250
Adams	33	137	4	42	14	20	250
Webster	33	138	3	42	14	20	250

Of the five methods we discussed, two are *quota methods* (Hamilton's and Lowndes') and the other three are *divisor methods* (Jefferson's, Adams' and Webster's). The following table summarizes the characteristics of the five methods:

	Hamilton	Lowndes	Jefferson	Adams	Webster
Violates quota rule	No	No	Yes	Yes	Yes
Alabama paradox	Yes	Yes	No	No	No
Population paradox	Yes	Yes	No	No	No
New states paradox	Yes	Yes	No	No	No
Favoritism toward	Large states	Small states	Large states	Small states	Neutral

While some of the methods are better than others, none of them is perfect: All five methods have some flaw or another. For many years, the ultimate hope held by thoughtful students of the apportionment problem, both inside and outside Congress, was that mathematicians would eventually come up with an *ideal* apportionment method—one that never violates the quota rule, does not produce any paradoxes, and treats large and small states without favoritism. Even as far back as 1929, Representative Ernest Gibson of Vermont echoed these sentiments in Congress. "The apportionment of Representatives to the population is a mathematical problem. Then why not use a method that will stand the test [of fairness] under a correct mathematical formula?"[5]

Indeed, why not? The surprising answer was provided in 1980 by mathematicians Michel L. Balinski of the State University of New York at Stony Brook and H. Peyton Young of the University of Maryland who collaborated to produce a remarkable negative discovery known as **Balinski and Young's impossibility theorem**: *It is mathematically impossible for an apportionment method to be flawless. Any apportionment method that does not violate the quota rule must produce paradoxes and any apportionment method that does not produce paradoxes must violate the quota rule.*

As with other situations involving impossibility theorems (remember Arrow's impossibility theorem in chapter 1?) the knowledge that there is no ideal apportionment method forces us to look a little closer at the methods we already have. Even Daniel Webster intuitively understood this when, speaking about the apportionment problem, he said " . . . that which cannot be done perfectly must be done in a manner as near perfection as can be."[6]

Of the various apportionment methods known, which one is the best? Unfortunately, neither mathematicians nor politicians have been able to reach an agreement on this issue. The Huntington-Hill method currently used in Congress (See Exercises 33 through 35) is believed by some mathematicians to be the best, but recent discoveries have convinced many mathematicians that Webster's method is the best and should be reinstituted as the official apportionment method for Congress.

In 1882, Representative John A. Anderson of Kansas, speaking in Congress in favor of Webster's method, said:

Since the world began there has been but one way to proportioning numbers, namely, by using a common divisor, by running the "remainders" into decimals, by taking fractions above .5, and dropping those below .5; nor can there be any other method. This process is purely arithmetical . . . If a hundred men were being torn limb from limb, or a thousand babes were being crushed, this process would have no more feeling in the matter than would an iceberg; because the science of mathematics has no more bowels of mercy than has a cast-iron dog.

It is ironical, that a hundred years of debate and a major mathematical discovery later, these may still be the best words ever said on the subject.

[5] *Congressional Record*, 70th Congress, 2nd Session, 70 (1929), p. 1500.

[6] Daniel Webster, *The Writings and Speeches of Daniel Webster*, VI, National Edition, Boston, Little, Brown and Company, 1903, p. 108.

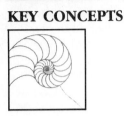

KEY CONCEPTS

Adams' method
Alabama paradox
apportionment method
apportionment problem
Balinski and Young's
  impossibility theorem
divisor methods
Hamilton's method
Jefferson's method

Lowndes' method
modified quota
new states paradox
population paradox
quota methods
quota rule
standard divisor
standard quota
Webster's method

EXERCISES

■ Walking

Exercises 1 through 5 refer to a small country consisting of four states. There are 160 seats in the legislature and the populations of the states are shown as follows.

State	A	B	C	D
Population (in millions)	1.33	2.67	0.71	3.29

1. **(a)** Find the standard divisor.
   (b) Find each state's standard quota.
   (c) Find each state's apportionment under Hamilton's method.

2. Find each state's apportionment under Lowndes' method.

3. **(a)** Using the divisor 49,400, find each state's modified quota.
   (b) Find each state's apportionment under Jefferson's method.

4. **(a)** Using the divisor 50,700, find each state's modified quota.
   (b) Find each state's apportionment under Adams' method.
   (c) Using the divisor 50,650, find each state's modified quota.
   (d) Using the divisor 50,600, find each state's modified quota.
   (e) Explain why the divisor 50,650 works for Adams' method but the divisor 50,600 doesn't.
   (f) Find a divisor different from 50,700 and 50,650 that will also work for Adams' method.

5. **(a)** Find a divisor that works for Webster's method. (*Hint:* You should be able to do this without using a calculator.)
   (b) Find the apportionment for each state under Webster's method.

Exercises 6 through 10 refer to a bus company that operates six bus routes (A, B, C, D, E, and F) and 150 buses. The buses are apportioned among the routes on the basis of average number of daily passengers per route which is given in the following table:

Route	A	B	C	D	E	F
Daily average number of passengers	12,550	38,623	19,781	31,112	33,280	14,654

6. (a) Find the standard divisor.
   (b) Find the standard quota for each bus route.
   (c) Apportion the buses among the routes using Hamilton's method.

7. Apportion the buses among the routes using Lowndes' method.

8. Apportion the buses among the routes using Jefferson's method.

9. Apportion the buses among the routes using Adams' method.

10. Apportion the buses among the routes using Webster's method.

Exercises 11 and 12 refer to a hospital with a nursing staff consisting of 225 nurses working in four shifts: A (7:00 A.M. to 1:00 P.M.); B (1:00 P.M. to 7:00 P.M.); C (7:00 P.M. to 1:00 A.M.); and D (1:00 A.M. to 7:00 A.M.). The number of nurses apportioned to each shift is based on the average number of patients per shift, as given in the following table:

Shift	A	B	C	D
Average number of patients	823	659	382	277

11. (a) Find the standard divisor. Explain what the standard divisor represents in this problem.
   (b) Find the standard quota for each shift.
   (c) Apportion the nurses to the shifts using Hamilton's method.

12. Apportion the nurses to the shifts using Lowndes' method.

Exercises 13 through 17 refer to a small country consisting of five states. The total population of the country is 24.8 million. The standard quotas of each state are given in the following table:

State	A	B	C	D	E
Quota	25.26	18.32	2.58	37.16	40.68

13. (a) Find the number of seats in the legislature.
    (b) Find the standard divisor.
    (c) Find the population of each state.
    (d) Find each state's apportionment under Lowndes' method.

14. Find each state's apportionment under Hamilton's method.

15. Find each state's apportionment using Jefferson's method.

16. Find each state's apportionment using Adams' method.

17. Find each state's apportionment using Webster's method.

18. Tasmania State University is made up of five different schools: Agriculture, Business, Education, Humanities, and Science. The number of faculty positions allocated to each school is based on the schools' respective enrollments as shown.

School	Agriculture	Business	Education	Humanities	Science	Total
Enrollment	2500	1300	2890	3400	4910	15,000

Given that there is a total of 500 faculty positions to be allocated:
    (a) Find the standard divisor. What does the standard divisor represent in this problem?
    (b) Find each school's standard quota.
    (c) Find the number of faculty members apportioned to each school under Lowndes' method.

19. A mother wishes to distribute 10 pieces of candy among her three children based on the number of minutes each child spends studying. She decides to use Hamilton's method with the information as shown in the table.

Child	Bob	Peter	Ron
Minutes studied	703	243	54

(a) Find each child's apportionment.
(b) Suppose that just prior to actually handing over the candy, she finds another piece of candy and includes it in the distribution. Find each child's apportionment using Hamilton's method and 11 pieces of candy.
(c) Did anything surprising occur?

20. A mother wishes to distribute 11 pieces of candy among her three children based on the number of minutes each child spends studying. She decides to use Hamilton's method with the information as shown in the table.

Child	Bob	Peter	Ron
Minutes studied	703	243	54

(a) Find each child's apportionment.

(b) Suppose that before she has time to sit down and do the actual calculations, the children decide to do a little more studying. Say, Ron studies an additional 10 minutes, Peter an additional 20 minutes, and Bob an additional 70 minutes. Find each child's apportionment based on the new total time studied.

(c) Did anything surprising occur?

■ Jogging

21. Make up an apportionment problem in which Hamilton's and Lowndes' methods result in exactly the same apportionment for each state.

22. Make up an apportionment problem in which Hamilton's and Webster's methods result in exactly the same apportionment for each state.

23. Make up an apportionment problem in which Lowndes' and Adams' methods result in exactly the same apportionment for each state.

24. (a) Consider the following situation:

State	A	B	C	D	E
Quota	11.23	24.39	7.92	36.18	20.28

You want to use Webster's method as a method of apportionment. Explain why you should look for a divisor that is *smaller* than the standard divisor.

(b) Consider the following situation:

State	A	B	C	D	E
Quota	11.73	24.89	7.92	35.68	19.78

You want to use Webster's method as a method of apportionment. Explain why you should look for a divisor that is *bigger* than the standard divisor.

(c) Under what conditions can you be assured that the standard divisor will work in Webster's method?

25. Consider an apportionment problem with only two states, A and B. Suppose that state A has quota q_1 and state B has quota q_2, neither of which is a whole number. (Of course $q_1 + q_2 = M$ must be a whole number.) Let f_1 represent the fractional part of q_1 and f_2 the fractional part of q_2.

(a) Explain why one of the fractional parts is bigger than or equal to 0.5, and why the other is smaller than or equal to 0.5.

(b) Assuming neither fractional part is equal to 0.5, explain why Hamilton's and Webster's methods must result in the same apportionment. [Refer to Exercise 24(c) for a clue.]

(c) Explain why in the above situation (two states only) Hamilton's method can never suffer from the Alabama paradox or the population paradox.

(d) Explain why in the above situation Webster's method can never violate the quota rule.

26. Consider an apportionment problem with only two states, A and B. Suppose that state A has quota q_1 and state B has quota q_2, neither of which is a whole number. (Of course $q_1 + q_2 = M$ must be a whole number.) Let f_1 represent the fractional part of q_1 and f_2 the fractional part of q_2.
    (a) Find values q_1 and q_2 such that Lowndes' and Hamilton's methods result in the same apportionment.
    (b) Find values q_1 and q_2 such that Lowndes' and Hamilton's methods result in different apportionments.
    (c) Write an inequality involving q_1, q_2, f_1, and f_2 that guarantees that Lowndes' and Hamilton's methods result in different apportionments.

27. The purpose of this example is to show that under rare circumstances a divisor method may not work. A small country consists of four states with populations as follows:

State	A	B	C	D
Population	500	1000	1500	2000

There are $M = 51$ seats in the house of representatives.
    (a) Find each state's apportionment using Jefferson's method.
    (b) Attempt to apportion the seats using Adams' method with a modified divisor $D = 100$. What happens if you take $D < 100$? What happens if you take $D > 100$?
    (c) Explain why Adams' method will not work for this example.

28. Following the 1800 U.S. Census, the states' populations were as given in the following table (U.S. Census Bureau data). The size of the House of Representatives was 141.

State	Population
Connecticut	250,622
Delaware	61,812
Georgia	138,807
Kentucky	204,822
Maryland	306,610
Massachusetts	574,564
New Hampshire	183,855
New Jersey	206,181
New York	577,805
North Carolina	424,785
Pennsylvania	601,863
Rhode Island	68,970
South Carolina	287,131
Tennessee	100,169
Vermont	154,465
Virginia	747,362

(a) Find each state's apportionment using Hamilton's method.

(b) Find each state's apportionment using Jefferson's method.

29. (a) Explain why when Jefferson's method is used, any violations of the quota rule must be upper quota violations.

(b) Explain why when Adams' method is used, any violations of the quota rule must be lower quota violations.

(c) Use parts (a) and (b) to justify why in the case of an apportionment problem with just two states, neither Jefferson's nor Adams' method can possibly violate the quota rule.

30. Explain why neither Hamilton's nor Lowndes' method can ever violate the quota rule.

■ Running

31. Make up an apportionment problem in which all five methods (Hamilton's, Lowndes', Jefferson's, Adams', and Webster's) result in exactly the same apportionment for each state.

32. (This exercise may seem out of place in this chapter, but it is necessary background for Exercises 33 through 35.)

The geometric mean. The geometric mean of two positive numbers is the square root of their product (i.e., the geometric mean of x and y is $\sqrt{xy}$).

(a) Using a calculator, complete the following table giving geometric means for consecutive whole numbers.

x	y	Geometric Mean
1	2	1.41
2	3	2.45
3	4	3.46
4	5	
5	6	
6	7	6.4807
7	8	
8	9	
9	10	
11	12	
12	13	
13	14	
14	15	
15	16	
16	17	
17	18	
18	19	
19	20	

(b) Explain why the geometric mean of two consecutive whole numbers falls between the two numbers.

(c) Explain why the fractional part of the geometric mean of two consecutive positive integers is always less than 0.5.

33. **The Huntington-Hill method.** This method of apportionment is a variation of Webster's method and often gives the same apportionment as Webster's method. (By the way, this is the method currently being used to apportion the U.S. House of Representatives.) The decision as to which way to round off the modified quota (upward or downward) is based on whether the fractional part of the modified quota is less than or more than the geometric mean (refer to Exercise 32) of the whole numbers immediately to the left and to the right of it. For example, suppose that the modified quota is 6.49. We must decide if we should apportion 6 or 7 seats. The geometric mean of 6 and 7 is 6.4807. . . . Since 0.49 is more than 0.4807 . . . , we round upward to 7. (Note that a modified quota of 6.49 would be rounded downward to 6 under Webster's method.) Similarly, a modified quota of 3.47 would be rounded upward to 4 because the geometric mean of 3 and 4 is 3.46.

Use Huntington-Hill's method to find the apportionments of each state for a small country that consists of five states. The total population of the country is 24.8 million. The standard quotas of each state are as follows:

State	A	B	C	D	E
Standard quota	25.26	18.32	2.58	37.16	40.68

34. This exercise refers to the Huntington-Hill method as described in Exercise 33. A country consists of six states with populations as follows:

State	A	B	C	D	E	F	Total
Population	344,970	408,700	219,200	587,210	154,920	285,000	2,000,000

There are 200 seats in the legislature.
(a) Find the apportionment under Webster's method.
(b) Find the apportionment under the Huntington-Hill method.
(c) Compare the divisors used in parts (a) and (b).
(d) Compare the apportionments found in parts (a) and (b).

35. This exercise refers to the Huntington-Hill method as described in Exercise 33. A country consists of six states with populations as follows:

State	A	B	C	D	E	F	Total
Population	344,970	204,950	515,100	84,860	154,960	695,160	2,000,000

There are 200 seats in the legislature.
(a) Find the apportionment under Webster's method.
(b) Find the apportionment under the Huntington-Hill method.
(c) Compare the divisors used in parts (a) and (b).
(d) Compare the apportionments found in parts (a) and (b).

**HISTORICAL
APPENDIX:
A Brief History
of Apportionment
in the United States**

1787

■ The Constitutional Convention drafts the Constitution of the United States.

■ Small states -versus- large-states controversy dominates the Convention.

■ Article I, Section 2, gives Congress the authority to determine the method of apportionment for the House of Representatives.

1791

■ Following the census of 1790 two methods of apportionment are proposed. Hamilton's method (a quota method) is supported by the Federalists; Jefferson's method (a divisor method) is supported by the Republicans.

■ After considerable controversy and debate Congress approves a bill to apportion the House of Representatives using Hamilton's method with $M = 120$.

■ President Washington vetoes the bill (the first exercise of a presidential veto in U.S. history).

■ Jefferson's method is adopted using $M = 105$ seats and a divisor $D = 33,000$. (Jefferson's method will remain in use until 1840.)

1792

■ Representative William Lowndes (South Carolina) proposes what we now call Lowndes' method. The proposal dies in Congress.

■ Jefferson's method is readopted using $M = 213$.

1832

■ The controversies between North and South, large states and small states, quota methods and divisor methods heat up.

■ John Quincy Adams (former President and at this time a congressman from Massachusetts) proposes what we now call Adams' method. The proposal fails.

■ Senator Daniel Webster (Massachusetts) proposes what we now call Webster's method. His proposal also fails.

■ Jefferson's method prevails once again (with $M = 240$).

1842

■ Webster's method is adopted with $M = 223$. This is the only time in U.S. history that M goes down (politicians are not inclined to legislate themselves out of work!).

1852

■ Representative Samuel Vinton (Ohio) proposes a bill adopting Hamilton's method as the permanent method of apportionment with $M = 233$ seats.

■ Congress approves the bill with the change $M = 234$. (For this particular value of M, Hamilton's method gives the same apportionment as Webster's method.)

1872

■ $M = 283$ seats is chosen so that Hamilton's and Webster's methods result in the same apportionment.

■ For unexplained political reasons 9 additional seats are added to the House.

■ The final apportionment does not agree with either Hamilton's or Webster's method. The whole thing is completely botched up by politics!

1876

■ Rutherford B. Hayes becomes President of the United States based on the botched apportionment of 1872.

■ The electoral college vote is 185 votes for Hayes and 184 votes for Samuel J. Tilden. However, based on the correct apportionment required by law, Tilden should have won the election.

The election of 1876. The botched apportionment of 1872 resulted in the election of Rutherford B. Hayes over Samuel Tilden. (Ohio Historical Society Library and the Granger Collection)

1880

■ The Alabama paradox surfaces as a serious flaw in Hamilton's method.

1882

■ Unable to understand that the Alabama paradox is a logical flaw rather than a political trick, some congressmen vent their frustration by blaming mathematics. (So, what else is new?)

I thought that mathematics was a divine science. I thought that mathematics was the only science that spoke to inspiration and was infallible in its utterances. I have been taught always that it demonstrated the truth. I have been told . . . that mathematics, like the voice of Revelation, said when it spoke "thus saith the Lord." But here is a new system of mathematics [Hamilton's method] that demonstrates the truth to be false.

TEXAS CONGRESSMAN ROGER MILLS

- Despite serious concerns about Hamilton's method, an apportionment bill with $M = 325$ seats is approved. (This number is chosen so that Hamilton's and Webster's methods result in the same apportionment.)

1901
- The Bureau of the Census submits to Congress tables showing apportionments based on Hamilton's method for all House sizes between $M = 350$ and $M = 400$.

- For all values of M between 350 and 400 except one ($M = 357$), Colorado would get an apportionment of 3 seats. For $M = 357$ Colorado gets only 2 seats (the Alabama paradox again!) The chairman of the House Census Committee (Representative Albert J. Hopkins of Illinois) does not care for the populist leanings of western states. The bill submitted by the Hopkins committee provides for (guess what?) $M = 357$ seats.

- Congress is in an uproar! Hopkins' bill is defeated, and Hamilton's method is finally abandoned for good.

- Webster's method is used with $M = 386$.

1907
- Oklahoma joins the Union. The new states paradox is discovered.

1911
- Webster's method is adopted with $M = 433$. (A provision is made for Arizona and New Mexico to get one seat each if admitted into the Union.)

- Joseph Hill (chief statistician of the Bureau of the Census) proposes a new method now known as the Huntington-Hill method. It is a slight variation of Webster's method (see Exercises 33 through 35).

1921
- No reapportionment is done after the 1920 U.S. Census (in direct violation of the Constitution). Between 1921 and 1928, 42 different reapportionment bills are presented in Congress. All are defeated.

1931
- Webster's method is used with $M = 435$.

1941

■ The Huntington-Hill method is adopted and the size of the House of Representatives is fixed by law at $M = 435$. It remains the permanent method of apportionment (until Congress votes to change it).

REFERENCES AND FURTHER READINGS

1. Balinski, Michel L., and H. Peyton Young, *Fair Representation: Meeting the Ideal of One Man, One Vote*. New Haven, Conn.: Yale University Press, 1982.

2. Balinski, Michel L., and H. Peyton Young, "The Quota Method of Apportionment," *American Mathematical Monthly*, 82 (August–September 1975), 701–730.

3. Hoffman, Paul, *Archimedes Revenge: The Joys and Perils of Mathematics*. New York: W. W. Norton & Co., 1988, chap. 13.

4. Huntington, E. V., "The Apportionment of Representatives in Congress," *Transactions of the American Mathematical Society*, 30 (1928), 85–110.

5. Saari, D. G., "Apportionment Methods and the House of Representatives," *American Mathematical Monthly*, 85 (1978), 792–802.

6. Steen, Lynn A., "The Arithmetic of Apportionment," *Science News*, 121 (May 8, 1982), 317–318.

7. Young, H. Peyton, *Fair Allocation*. Providence, R.I.: American Mathematical Society, 1985.

5

Euler Circuits

The Circuit Comes to Town

The map in Fig. 5-1 (on the next page) shows the complete street plan for the picturesque city of Cleansburg, California (population 9471). Cleansburg's primary source of revenue is tourism, and the city fathers are committed to keeping the streets of the city "spic and span." To accomplish this goal, the city has entered into an agreement with a new high-tech garbage company, Algorithms for Clean Living (ACL), which has contracted to do the garbage collection for the city. The agreement is that ACL will collect the garbage daily throughout the *entire* city—7 days a week, 365 days a year.

ACL is a garbage collection company owned and operated by an enterprising young couple (Cleo and Tony) who are determined to make a go of it even though they own only one garbage truck. They are convinced that just the two of them, with hard work and efficient management, can carry out their contractual obligations, keeping the streets of Cleansburg shining clean and the tourist dollars rolling in.

**ASSORTED
ROUTING
PROBLEMS**

While the above scenario is totally fictitious (and some would say even corny), it will serve us well, for it is a typical example of an extremely important category of problems in management science called **routing problems**. To put it in the most general way, routing problems are con-

131

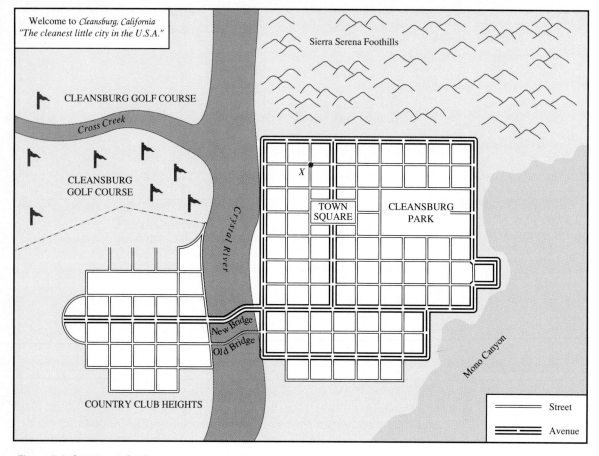

Figure 5-1 Street map for the city of Cleansburg.

cerned with finding efficient ways to route the delivery of goods or services to an assortment of destinations. The goods could be packages, mail, computer files, etc.; the services could be snow removal, curb sweeping, garbage collection, sewer, water, or telephone lines, etc.; and the destinations could be cities in a country, houses along city blocks, sprinklerheads in a garden, computer terminals located across a university campus, etc.

While routing problems often sound very similar, subtle details can make a great difference in both the method and the nature of the solution. In this chapter we will discuss one particular class of routing problems called **Euler circuit problems**. Our discussion will be motivated by our original example dealing with the garbage collection problem for the city of Cleansburg.

The Routing of a Garbage Truck (Part I)

Let's briefly review our original example. We have a small town (Cleansburg), a young couple (Cleo and Tony) with a garbage truck, and a job to be done (collecting the garbage along the streets of the city on a daily basis). The fact that almost all the streets in Cleansburg are residential (which means they are neither very wide nor overly busy with traffic), coupled with the fact that Tony and Cleo own only one garbage truck, dictates the following strategy: They will both ride the truck and pick up garbage on both sides of the street at the same time. (Cleo will pick up garbage on the right side of the street, and Tony on the left side.)

The above is an important observation—it means that the *garbage along an entire block can be picked up in a single trip along the block.* The conclusion that we draw from this is that Tony and Cleo will have to drive through every block of Cleansburg at least once (the garbage has to be collected) and possibly more than once (when there is a need to travel again through some streets just to reach a different area where garbage has not yet been collected or perhaps to get back home at the end of the day).

Because they will be traveling through the same streets 365 times a year, Cleo and Tony are particularly keen on finding the most efficient route for collecting the garbage. Specifically, we will interpret this to mean the following: If it is possible to find a route that starts at a specific location X, goes along *every* block of *every* street of Cleansburg *exactly once*, and finally returns to X, then Tony and Cleo would certainly like to use such a route for collecting the garbage. If there is no such route, Tony and Cleo would like to find the next best thing—a route that passes through each block of the city at least once and has the fewest possible number of deadhead blocks. In the jargon of the garbage collection business, a **deadhead block** is a block that one travels for purposes other than collecting the garbage. Typical reasons for deadhead travel are to get to another part of the city where garbage still needs to be collected, to return home (i.e., to point X) at the end of the day, and sometimes to get to the deli to pick up a sandwich and a soda. (We will not consider the last reason an acceptable one!)

To put it in a nutshell then, Tony and Cleo are looking for a route that crisscrosses the entire city and has either zero deadhead travel or as few deadhead blocks as possible. We will devote this entire chapter to analyzing and eventually solving this problem. In addition to garbage collection, routing problems similar to the kind we have just described have applications in other areas of urban planning such as snow removal, meter reading, street sweeping, and mail delivery.

The Königsberg Bridge Problem

There is a town in Soviet Lithuania called Kaliningrad (this one is for real!). In the eighteenth century it was a part of Prussia and was called Königsberg. The town (as shown in Fig. 5-2) is divided by a river (the

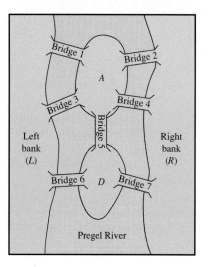

Figure 5-2

Pregel), there are two islands in the middle of the river (*A* and *D*) and there are seven bridges (1 through 7) joining the two banks of the river to the islands and the islands to each other.

For some unfathomable reason it became a great challenge to the locals to find a route that would allow them to walk around the town crossing each of the seven bridges once, never recrossing any of the bridges, and returning to the original starting point—a simple routing problem if there ever was one! While no one had been able to find such a route, people were not totally persuaded that the whole thing was impossible until one of the great mathematicians of all time, Leonhard Euler,[1] proved mathematically that such a route could not exist. The real significance of Euler's proof was that it set the foundations of an important branch of mathematics called **graph theory**. We will discuss graph theory and Euler's solution[2] to the Königsberg bridge problem in the next few sections. But before we do, let's bring up a ''different'' problem.

■ A Children's Game

Can we trace the picture in Fig. 5-3(a), starting and ending at the same place, without lifting our pencil or retracing any lines?

Many of us played such games in our childhood (those were the good old days before video games). The pictures may have been different from Fig. 5-3(a), but the name of the game was the same. At this point we don't

[1] Leonhard Euler (1701–1783) was one of the most brilliant and prolific mathematicians in history. It is estimated that, when finally compiled, his collected memoirs will fill nearly 100 volumes. In addition to mathematics, Euler loved children and had 13 of his own. In the words of one biographer, ''Euler was the Shakespeare of mathematics—universal, richly detailed and inexhaustible.''

[2] Euler's solution was that there is no solution—a perfectly legitimate solution.

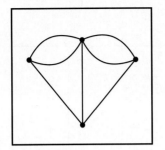

Figure 5-3(a)

care too much about the answer (the willing reader is encouraged to play with it for a few minutes), but rather about the observation that this is the Königsberg bridge problem all over again. Although this is a critical observation, we will not explain it. Instead, we encourage each reader to personally reason out the connection. In the spirit of compromise we will give a hint without words in the form of Fig. 5-3(b).

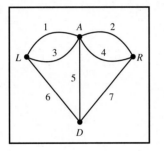

Figure 5-3(b)

GRAPHS

The paradox is now fully established that the utmost abstractions are the true weapons with which to control our thoughts of concrete fact.

ALFRED NORTH WHITEHEAD

A graph is an abstraction invented and used as a servant of practical reality. A graph consists of two types of objects: **vertices** and **edges**. It is customary (but not mandatory) to represent the vertices as dots and the edges as lines (not necessarily straight). The edges must always be connections between vertices, but an edge is allowed to connect a vertex with itself (we call such edges **loops**). We illustrate some of these concepts in Fig. 5-4.

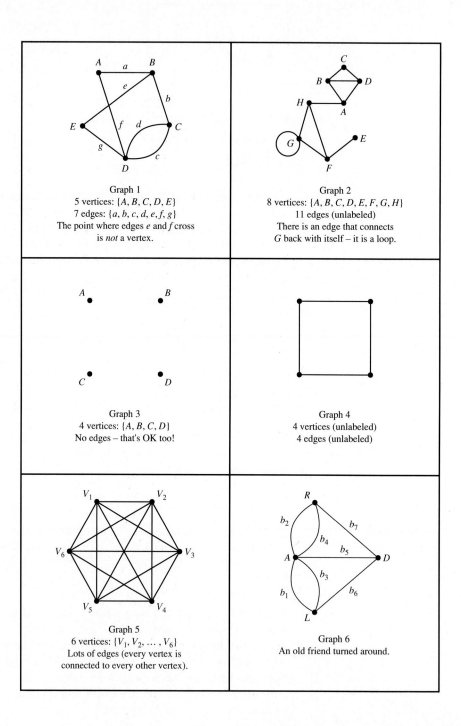

Graph 1
5 vertices: $\{A, B, C, D, E\}$
7 edges: $\{a, b, c, d, e, f, g\}$
The point where edges e and f cross
is *not* a vertex.

Graph 2
8 vertices: $\{A, B, C, D, E, F, G, H\}$
11 edges (unlabeled)
There is an edge that connects
G back with itself – it is a loop.

Graph 3
4 vertices: $\{A, B, C, D\}$
No edges – that's OK too!

Graph 4
4 vertices (unlabeled)
4 edges (unlabeled)

Graph 5
6 vertices: $\{V_1, V_2, \dots, V_6\}$
Lots of edges (every vertex is
connected to every other vertex).

Graph 6
An old friend turned around.

Figure 5-4

The reader has no doubt noticed in graphs 1 through 6 in Fig. 5-4 that we have taken some liberties with the labeling. In some of the graphs both vertices and edges are labeled; in others only the vertices are; in graph 4 neither vertices nor edges are. For labels we can use different letters or the same letter with subscripts, and the letters can be anything we want. As a convenience, we will agree to respect one rule: Use capital letters for vertices and lowercase letters for edges. Sometimes it is convenient to describe an edge by listing the two vertices connected by that edge. For example, in graph 1 in Fig. 5-4 edge f can also be described as edge AD.

Having said all this, we still haven't defined a graph. In the abstract, a graph is a relational structure: It tells us that we have a bunch of objects (the vertices) and that these objects are related among each other. The story of which objects are related to which other objects is given by the edges. That is the entire information carried by a graph—no more and no less.

Let's analyze some of the graphs in Fig. 5-4. This is what graph 1 tells us: There are five "things" (no one has told us what they are, and in fact we don't care), and they are represented in the abstract by the vertices A, B, C, D, and E. There is also a relationship among these five things (it could be love, hate, kinship, or whatever—again, we don't care). We know that A is related to B (edge a) and also to D (edge f); B is related to A, to C, and to E; C is related to B and to D twice (double whammy); etc.

The story of graph 3 is simple: We have four objects (A, B, C, D) and none are related. In graph 5, on the other hand, we have six objects (V_1, V_2, V_3, V_4, V_5, V_6), and every object is related to every other object.

Because a graph only describes relations between objects, there are infinitely many different ways to draw the same graph. In fact, almost all the things that matter in traditional geometry (distance, position, shape, etc.) are irrelevant in a graph.

Example 1.

Even though geometrically they look very different, Figs. 5-5(a) and (b) are two pictures of exactly the same graph. We leave it to the reader to verify (Exercise 8) that in both pictures we have the same five objects related in exactly the same way. While Fig. 5-5(a) might be a cleaner picture than Fig. 5-5(b), the information contained in both graphs is identical. ■

Example 2. Sue, Mary, Veronica, Kevin, Tom, and Paul go to a dance. Sue dances with Kevin and Paul, Mary dances with all the boys, and Veronica dances with no one. All the information about who danced with whom is shown by the graph in Fig. 5-6. ■

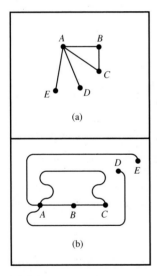

Figure 5-5

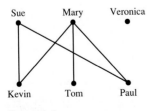

Figure 5-6

Example 3. For the last week of the baseball season (that's when the pennant race is hottest), the schedule for the National League East is

- Monday: Pittsburgh versus Montreal, New York versus Philadelphia, Chicago versus St. Louis

- Tuesday: Pittsburgh versus Montreal

- Wednesday: New York versus St. Louis, Philadelphia versus Chicago

- Thursday: Pittsburgh versus St. Louis, New York versus Montreal, Philadelphia versus Chicago

- Friday: Philadelphia versus Montreal, Chicago versus Pittsburgh

- Saturday: Philadelphia versus Pittsburgh, New York versus Chicago, Montreal versus St. Louis.

All the information about who plays whom is relational (each game "relates" two different teams) and can be conveniently represented by the graph shown in Fig. 5-7.

Note the following relevant observations about Fig. 5-7:

Figure 5-7

- The position of the vertices has nothing to do with the geographic location of the cities. In describing who plays whom, the actual placement of the vertices is irrelevant.

- The graph is in many ways a much more practical way to describe the schedule than the original listing. Say one wants to know if New York plays Pittsburgh the last week of the season. The answer (no) is obvious if one looks at the graph, but much less so if one has to check through the listed schedule of games. ■

■ Graph Concepts and Terminology

Every branch of mathematics has its own peculiar jargon, and the theory of graphs has more than its share. In this section we will introduce a few of the essential concepts and terms that we will need in the chapter.

1. Two vertices are said to be **adjacent** if there is an edge joining them. In the graph shown in Fig. 5-8, vertices A and B are adjacent, but C and D aren't. Vertex E is adjacent to itself because there is a loop at E.

2. The **degree** of a vertex is the total number of edges at that vertex. (A loop contributes 2 toward the degree.) In the graph shown in Fig. 5-8, vertex A has degree 4 [we can write this as deg(A) = 4], vertex

Figure 5-8

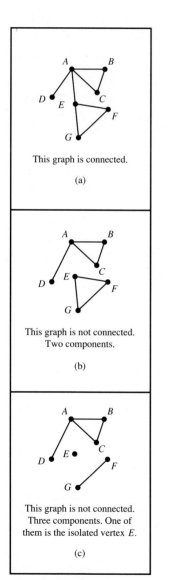

This graph is connected.

(a)

This graph is not connected.
Two components.

(b)

This graph is not connected.
Three components. One of
them is the isolated vertex E.

(c)

Figure 5-9

B has degree 3 [deg(*B*) = 3], deg(*C*) = 3, deg(*D*) = 2, deg(*E*) = 4 (because of the loop), and deg(*F*) = 0. (A vertex with degree 0 is called an **isolated** vertex.)

3. A **path** in a graph is a string of vertices, each one adjacent to the next one. Whereas a vertex can appear on the path several times, an edge can be part of a path *only once*. Here are some examples of paths taken from the graph in Fig. 5-8: *A, B, C* is a path; *A, B, C, A, E, D* is a longer path; *D, E, E, A, C, B, C* is also a path (*E* is adjacent to *E* because of the loop there, *C* and *B* are adjacent twice because of the double edge between them). *A, C, D, E* is *not* a path because *C* and *D* are not adjacent vertices; *A, B, C, A, D, E, A, C, B* is *not* a path, because the edge joining *A* and *C* appears twice in the list.

4. If a path starts and ends at the same vertex, it is called a **circuit**. In the graph in Fig. 5-8, *A, B, C, A* is a circuit; so are *D, E, A, C, B, A, D* and *B, C, B* and *E, E*. On the other hand, *A, B, C, B, A* is not a circuit, since the same edge (*AB*) is used twice.

5. A graph is **connected** if any two of its vertices can be joined by a path. This essentially means that it is possible to travel from any vertex to any other vertex along consecutive edges of the graph. If a graph is not connected, it is said to be **disconnected**. A graph that is disconnected is made up of pieces that are by themselves connected. Such pieces are called the **components** of the graph. The graph in Fig. 5-9(a) is connected. The graphs in Figs. 5-9(b) and (c) are disconnected. The one in Fig. 5-9(b) has two components; the one in Fig. 5-9(c) has three.

6. Sometimes in a connected graph there is an edge such that if we were to erase it, the graph would become disconnected. For obvious reasons such an edge is called a **bridge** (burn a bridge behind you, and you'll never be able to get back to where you were). In Fig. 5-9(a), the edge *AE* is a bridge because when we remove it, the graph becomes disconnected [as in Fig. 5-9(b)]. There is another bridge in Fig. 5-9(a). We leave it to the reader to find it.

7. When a path contains each and every edge of the graph it is called an **Euler path**. In order to have an Euler path a graph must be connected, but not every connected graph has an Euler path. For the graph in Fig. 5-9(a), the path *D, A, B, C, A, E, F, G, E* is an Euler path of the graph. The reader may verify this by tracing the path out and checking that each edge is traced exactly once. When an Euler path starts and ends at the same vertex it is called an **Euler circuit**.

The remainder of this chapter is devoted to answering problems concerning Euler circuits and Euler paths in graphs, in particular:

1. How can we tell if a graph has an Euler circuit (Euler path)?

2. If it does, how do we find it?

3. If it doesn't, how can we doctor up the graph so that it does?

**EULER'S
THEOREMS**

Remember the Königsberg bridge problem? Finding a walk through the city that crosses each of the bridges once and returns to the starting place [Fig. 5-10(a)] is exactly the same problem as finding an Euler circuit in the graph shown in Fig. 5-10(b). Euler's answer to this problem was that it can't be done!

Why is such a walk impossible? Let's say for the sake of argument that the walk starts on the left bank L (it actually makes no difference where we pick the starting point). At least once along the way the walker

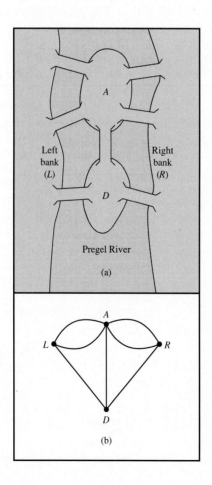

Figure 5-10

must pass through the island *A*, and we suspect that the walker will have to visit *A* more than once. Let's count exactly how many times. The first visit to *A* will use up two bridges (one getting there and a different one getting out); the second visit to *A* will use up two other bridges; and the third visit to *A* will use up two more. Oops! There are only five bridges to get in and out of *A*, so two visits to *A* won't do it (there would be an unused bridge) and three visits are too many (we would have to recross one of the bridges). It follows that the walk is impossible! It's the odd number of bridges at *A* (or anywhere else) that causes the problem. The argument can be extended and made general in a very natural way. We present it without any further ado.[3]

Euler's Theorem 1.

■ If a graph has vertices of odd degree, then it cannot have an Euler circuit.

■ If a graph is connected and every vertex has even degree, then it has at least one Euler circuit (usually more).

Note that a graph can have every vertex of even degree but be disconnected (as in Fig. 5-11), in which case it cannot have an Euler circuit.

What can we say about graphs with Euler paths? The same arguments apply for all the vertices (they must have even degree) except for the starting and ending vertices of the path which are different. The starting vertex requires one edge to get out and two more for each visit through that vertex, so it has odd degree. Likewise the ending vertex has odd degree (two edges for every visit plus one more to get to the vertex on the last leg of the trip). Thus, we have

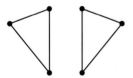

Figure 5-11

Euler's Theorem 2.

■ If a graph has more than two vertices of odd degree, then it cannot have an Euler path.

■ If a graph is connected and has exactly two vertices of odd degree, then it has at least one Euler path (usually more). Any such path must start at one of the odd-degree vertices and end at the other one.

---

[3] A translation of Euler's own account of how he developed these ideas is given in reference 4.

Example 4 (Fig. 5-12).

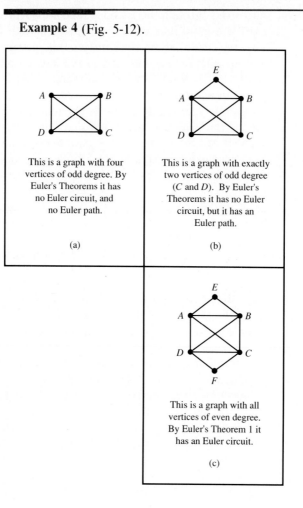

This is a graph with four vertices of odd degree. By Euler's Theorems it has no Euler circuit, and no Euler path.

(a)

This is a graph with exactly two vertices of odd degree (C and D). By Euler's Theorems it has no Euler circuit, but it has an Euler path.

(b)

This is a graph with all vertices of even degree. By Euler's Theorem 1 it has an Euler circuit.

(c)

Figure 5-12

FLEURY'S ALGORITHM

Euler's theorems are great because they give us an easy way to determine if a graph has an Euler circuit or an Euler path just by looking at the degrees of the vertices. Unfortunately they are of no help in finding the actual Euler circuit or path if there is one. Of course for graphs such as the ones shown in Figs. 5-12(b) and (c), we can find an Euler path (circuit) by simple trial and error. In most real life applications, however, graphs can have hundreds and even thousands of vertices and edges, and for such graphs simple trial and error won't do.

Our next major accomplishment will be to learn a method for finding an actual Euler circuit (or Euler path) when there is one. The method comes packaged in the form of an **algorithm,** a set of mechanical rules that when followed are guaranteed to produce an answer. The fact that the rules that make up an algorithm are mechanical means that there is no thinking involved—that's why mindless but efficient things like computers are ideally suited to carrying out algorithms. For human beings, the difficulties in carrying out algorithms (once the rules are understood) are not intellectual but rather procedural. Accuracy and fastidious attention to detail are the important virtues when carrying out the instructions in an algorithm and being a brilliant mathematical thinker in and of itself is of little value here. We mention this as a piece of friendly advice because most of the practical things we will do in this part of the book will be based on the ability to correctly carry out algorithms—an ability that is acquired primarily through practice.

If every vertex of a connected graph has even degree, then Euler's Theorem 1 assures us that the graph has an Euler circuit. We can find such an Euler circuit by using an algorithm known as **Fleury's algorithm**. Because this is our first encounter with a graph algorithm and because the algorithm is far from obvious, we will describe it informally first, then work out a couple of examples, and give the formal description of the algorithm last.

Our strategy will be to build the Euler circuit in steps (one edge at a time) and to mark the circuit as it is being built right on top of the original graph. To do so we will need a pen or marker of a bright color that contrasts with the color of the original graph. Since our graphs are in black, we will choose red to mark, as we travel through the graph, the edges of the Euler circuit. As we do so, we will also label the red edges $e_1, e_2, e_3, \ldots$ in the order in which we choose them.

At a typical stage in the algorithm, our original graph will be divided into two parts (as shown in Fig. 5-13): the red part (the piece of the Euler

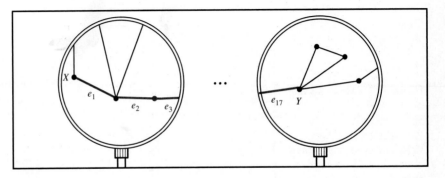

Figure 5-13 A typical stage in Fleury's algorithm.

circuit that we have built so far) and the black part (the rest of the graph). So here we are, standing at some vertex Y in the middle of this trip. Where do we go next? (In other words, what is the next edge we should color red?) There are two possible alternatives

- **Case 1.** There is only one way to go (see Fig. 5-14). In this case we ask no questions and proceed along that edge, color the edge red, and label it appropriately. Since we will never be coming back to Y (we have just used up the last edge out of Y) *we also color the vertex Y red*! (This instruction is very important.)

Figure 5-14 Case 1: Only one black edge is left at Y. Action: Mark the edge and vertex Y red and label the edge appropriately.

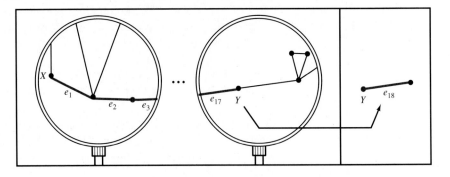

- **Case 2.** There are several edges to choose from. Pick an edge (any edge) that is not a bridge for the black part of the graph (see Fig. 5-15). Color the edge (only the edge) red and label it appropriately. (Remember that we defined a bridge as an edge whose removal disconnects the graph.) To put it in simple terms, we do not want to mark an edge red when doing so will break up the black part into disconnected pieces. The reason we don't want to do this is pretty clear: If we do so we will never be able to get to one of those two black pieces except by means of red (i.e., already traveled) edges.

Figure 5-15 Case 2: Several black edges are left at Y; some are not bridges for the black graph. Action: Color any edge that is *not* a bridge red and label; do not color the vertex Y red!

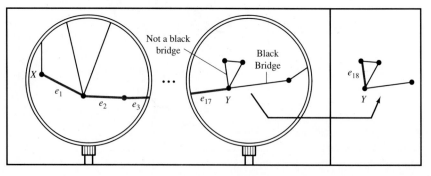

1a

1b

Imagine, if you will, having to visit each of the 48 state capitals in the continental United States, one after another, starting and ending in the same capital. In what order should you visit these cities so that the total length of the trip is as short as possible? The solution is shown on the top plate. The total length of the trip is approximately 12,000 air miles. This *optimal solution* was found using a sophisticated computer program which checked billions of possible combinations—an expensive proposition even under the best of circumstances. No algorithm is presently known that can efficiently solve this problem (commonly known as the *Traveling Salesman Problem*) when the number of cities becomes very large (see chapter 6). The bottom plate shows an *approximate solution* obtained using an extremely simple-minded procedure called the *nearest-neighbor algorithm*. The bad news is that the length of this trip is about 20% longer than the optimal solution (approx. 14,500 miles). The good news is that it can be found by anyone in a matter of minutes with just a pencil and a ruler.

2a

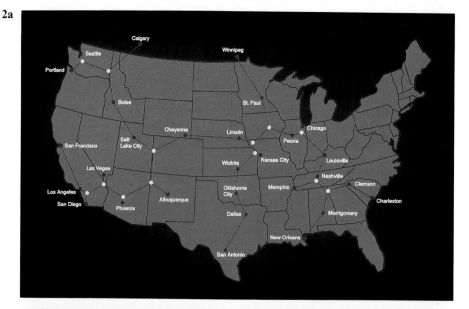

2b

A telephone company wants to connect the 29 cities shown on the map using the least possible amount of underground cable. The solution depends on whether junction points outside of the 29 cities are allowed. The top plate shows the shortest of all possible ways to connect the cities. The yellow dots are special junction points called *Steiner points*. This *optimal network,* which requires approximately 7400 miles of cable was found using a sophisticated computer algorithm developed by E. J. Cockayne and D. E. Hewgill of the University of Victoria. No algorithm is presently known that can efficiently solve this problem when the number of cities becomes large. The bottom plate shows the shortest network when no junction points other than the original cities are allowed. Although this network is about 3% longer than the optimal network (it requires approximately 7600 miles of cable) it can be found by hand in a matter of minutes using any one of several extremely simple algorithms (see chapter 7).

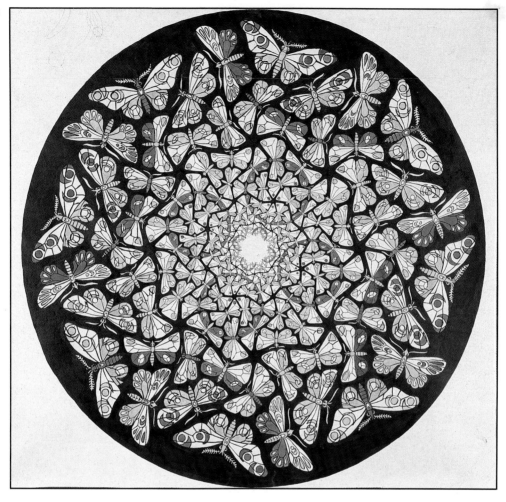

Butterflies by M. C. Escher. (© 1950 M. C. Escher/Cordon Art, Baarn, Holland.) This remarkable watercolor is rich in mathematical structure. In addition to black and white, Escher uses four other colors: red, yellow, blue and green. Each butterfly contains three of these colors (the body, the lower wings and the markings on the upper wings). There are 24 possible permutations of three colors and they are all present in equal proportions. Not only the butterflies but also the colored rings created by their arrangement exhibit *symmetry of scale* as they repeat themselves at smaller and smaller scales towards the center of the circle.

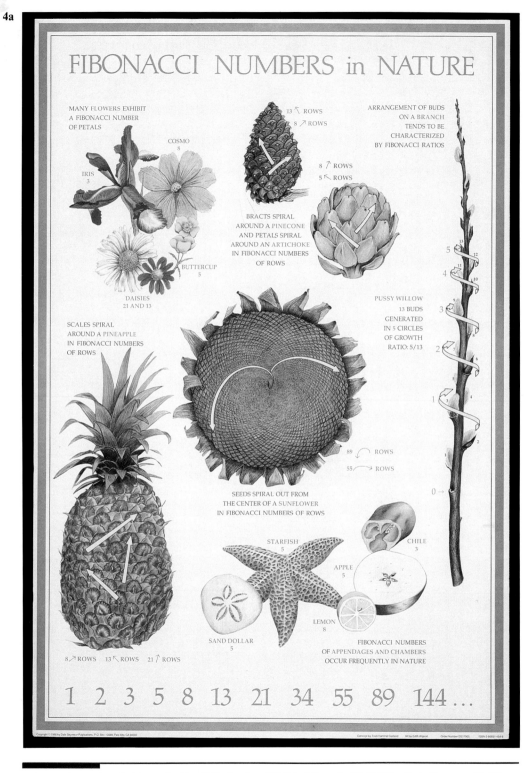

Fibonacci Numbers in Nature by Trudi Garland and Edith Allgood. (© 1988 by Dale Seymour Publications. Reprinted by permission.)

5a

5b

5c

A *logarithmic spiral* is exhibited by a cross section of the chambered nautilus shell (top) and some of its less glamorous cousins (bottom). Spirals of this type occur frequently in natural organisms whose growth is *gnomonic* (see chapter 9).

6a

The Queen of Spades: an example of 180 degree rotational symmetry. (Turn the book upside down—the card still looks the same.)

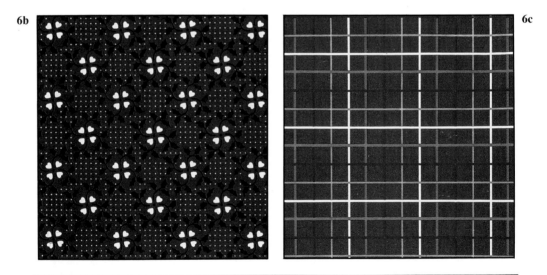

6b

6c

Although there are infinitely many different wallpaper designs, when classified according to their symmetries, wallpaper patterns can fall under one of only 17 different *pattern types* (see chapter 11). Two of these pattern types are illustrated by the "wallpaper patterns" shown above. To convince yourself that they are different turn the book upside down. One of them looks different now. Which one?

Although each queen by herself has 180 degree rotational symmetry, the pair of queens does not. What happens to the ladies' faces when you turn the book upside down?

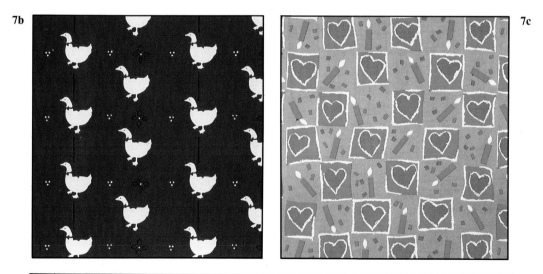

These are two very different looking wallpaper patterns which are nonetheless of the same pattern type: their only symmetries are translations (see chapter 11).

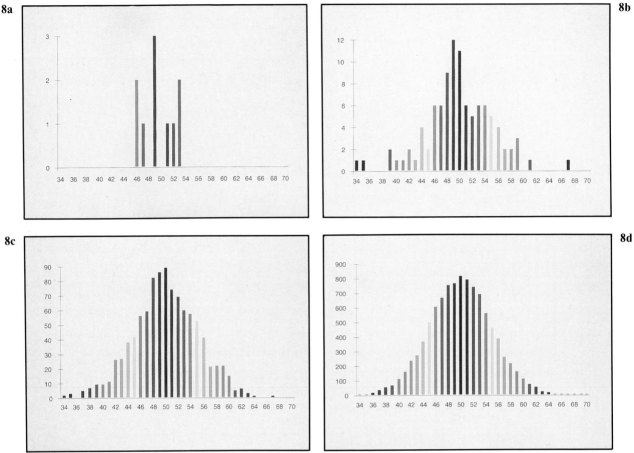

8a **8b** **8c** **8d**

Imagine playing the following game: An honest coin is tossed 100 times in a row. If the number of heads falls between 45 and 55 you win $1, otherwise you lose $1. Can you predict whether you will win or lose at this game? While the results of playing this game once, twice or even ten times are unpredictable, a predictable pattern emerges if we play the game a very large number of times. The bar graphs show a computer simulation for the results of playing the game (from left to right) 10, 100, 1000, and 10,000 times respectively. The graph on the lower right shows the inescapable pattern that the outcomes of this game must follow: a bell shape called a *normal distribution* (for details see chapter 16).

One question that immediately comes to mind is, What about a possible case 3: There are *no* black edges left at *Y*? If we have followed the instructions carefully, the algorithm guarantees that this can never happen, or to put in in a slightly different way: If it happens, we know that we are finished; there are no black edges left, and we are back at *X*.

One last point involves getting started. This algorithm works (i.e., it produces an Euler circuit) regardless of which vertex we choose as a starting point. In practice, of course, we have a specific place *X* that we must start from (home, the parking lot, etc.)

Example 5. Let's apply Fleury's algorithm to the graph in Fig. 5-16, using *A* as our starting point. (We know there is an Euler circuit because every vertex has even degree.)

Figure 5-16

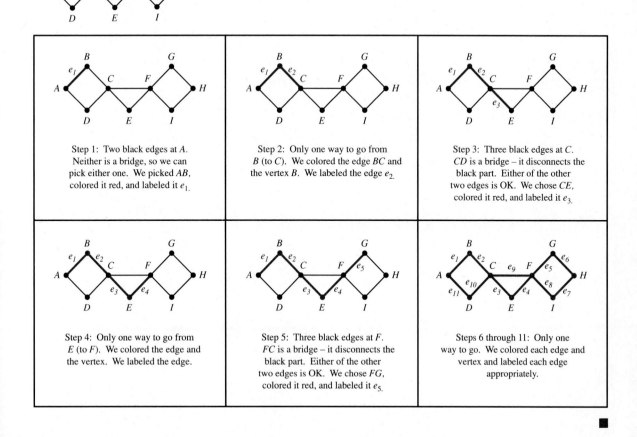

Step 1: Two black edges at *A*. Neither is a bridge, so we can pick either one. We picked *AB*, colored it red, and labeled it e_1.

Step 2: Only one way to go from *B* (to *C*). We colored the edge *BC* and the vertex *B*. We labeled the edge e_2.

Step 3: Three black edges at *C*. *CD* is a bridge – it disconnects the black part. Either of the other two edges is OK. We chose *CE*, colored it red, and labeled it e_3.

Step 4: Only one way to go from *E* (to *F*). We colored the edge and the vertex. We labeled the edge.

Step 5: Three black edges at *F*. *FC* is a bridge – it disconnects the black part. Either of the other two edges is OK. We chose *FG*, colored it red, and labeled it e_5.

Steps 6 through 11: Only one way to go. We colored each edge and vertex and labeled each edge appropriately.

Example 6.   Let's apply Fleury's algorithm to the graph in Fig. 5-17 (with starting point X). Being old pros, we will do this one without pictures of each stage of the algorithm. The reader is encouraged to trace the Euler circuit on top of Fig. 5-17 as we go along.

- ■ **Start.**   Our designated starting vertex is X.

- ■ **Step 1.**   There are two black edges coming out of X. Since neither one of them is a bridge, we can pick either one. We pick XJ, color it red, and label it e_1.

- ■ **Step 2.**   There are three black edges coming out of J, and none of them are bridges. We randomly choose JI, color it red, and label it e_2. (Be careful to leave the vertex J black.)

- ■ **Step 3.**   There is only one way to go from I (to H). We color the edge IH and the vertex I red and label IH with e_3.

- ■ **Step 4.**   There are several choices at H. One of them, HX, is a bridge—it disconnects the vertex X from the rest of the black part of the graph. Any choice other than HX is allowed. We choose HG, color the edge red, and label it e_4.

- ■ **Step 5.**   Color GF red, color vertex G red, and label GF as e_5.

- ■ **Step 6.**   There are several choices at F. None are bridges. We choose FE, color it red and label it e_6. (To speed things up we will indicate this by $e_6 \rightarrow FE$.)

- ■ **Step 7.**   There are no bridges at E. $e_7 \rightarrow EO$.

- ■ **Step 8.**   There are no bridges at O. $e_8 \rightarrow OK$.

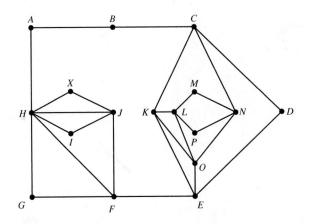

Figure 5-17

■ **Step 9.** There are no bridges at K. $e_9 \rightarrow KC$.

■ **Step 10.** One of the choices at C (CB) is a bridge (better not go that way!). $e_{10} \rightarrow CN$.

■ **Step 11.** There are no bridges at N. $e_{11} \rightarrow NP$.

■ **Step 12.** $e_{12} \rightarrow PL$ (no choice here!) Don't forget to color P red.

■ **Step 13.** LK is a bridge. We choose LO. $e_{13} \rightarrow LO$.

■ **Step 14.** $e_{14} \rightarrow ON$. We also color O red.

■ **Step 15.** $e_{15} \rightarrow NM$. We also color N red.

■ **Step 16.** $e_{16} \rightarrow ML$. We also color M red.

■ **Step 17.** $e_{17} \rightarrow LK$. We also color L red.

■ **Step 18.** $e_{18} \rightarrow KE$. We also color K red.

■ **Step 19.** $e_{19} \rightarrow ED$. We also color E red.

■ **Step 20.** $e_{20} \rightarrow DC$. We also color D red.

■ **Step 21.** $e_{21} \rightarrow CB$. We also color C red.

■ **Step 22.** $e_{22} \rightarrow BA$. We also color B red.

■ **Step 23.** $e_{23} \rightarrow AH$. We also color A red.

■ **Step 24.** We have some choices at H, but HX is a bridge. $e_{24} \rightarrow HJ$.

■ **Step 25.** $e_{25} \rightarrow JF$. We also color J red.

■ **Step 26.** $e_{26} \rightarrow FH$. We also color F red.

■ **Step 27.** $e_{27} \rightarrow HX$. We are finished!

Here is the Euler circuit we found:

X, J, I, H, G, F, E, O, K, C, N, P, L, O, N, M, L, K, E, D, C, B, A, H, J, F, H, X.   ■

Practice is the only way to really get the hang of how an algorithm works. Fleury's algorithm is no exception. The reader is strongly encouraged to try Exercises 23, 24, and 25 at this juncture.

For the record, here is a formal description of Fleury's algorithm.

Fleury's Algorithm.

■ Pick any vertex as the starting point.

■ Travel through an edge if
1. There is no other alternative, or
2. It is not a bridge for the untraveled part.

■ Label the edges in the order in which you travel them.

■ When you can't travel anymore, stop. (You are done!)

GRAPH MODELS

It should not be a great surprise at this point that the problem with which we started the chapter (finding the best route for collecting garbage along the streets of Cleansburg) is directly related to the problem of finding an Euler circuit in some graph. If we could only turn the map of Cleansburg into a graph, we would be halfway there. In fact, we can do this by making an edge out of each street block and a vertex out of each corner. When we do this, we get the graph in Fig. 5-18.

Let's take a moment to analyze exactly why turning the map of Cleansburg into a graph is a step in the right direction. We are after a good route for a garbage truck. What are the things that make a difference in trying

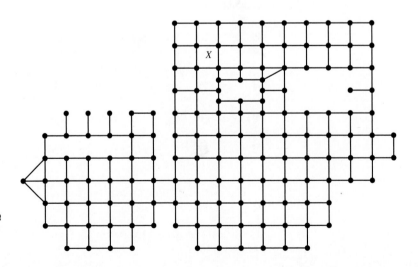

Figure 5-18 A graph describing the layouts of the streets in the city of Cleansburg.

to find such a route? Well, here is a list of things that don't make any difference: the names of the streets, the widths of the streets, and whether a block is straight or curvy. In fact when push comes to shove, in this problem there is only one thing that actually matters—which streets run into each other, and at which corners.[4] This is why the graph in Fig. 5-18 represents the ideal setting in which to analyze the original problem: It is an abstraction containing the essence of the problem itself.

When a mathematical structure such as a graph is used as the setting by means of which a real world problem can be described and studied, we call such a structure a *mathematical model* for the original problem. One of the most important things we will learn in this, and the next few chapters, is how to use a graph model as the basic analytical tool with which to study important types of applied problems from the world of management science.

■ The Routing of a Garbage Truck (Part II)

Recall that, ideally, Tony and Cleo would like a route that passes through each block of Cleansburg exactly once and returns to the starting point—in essence, an Euler circuit for the graph in Fig. 5-18. Euler's Theorem 1 tells us in a very unequivocal way that such a circuit cannot exist because there are vertices of odd degree. In Fig. 5-19 we have marked the vertices of odd degree for identification. For the sake of quick reference we need to give them a name, so we will informally call them **oddball vertices** (not the standard terminology, but who cares?).

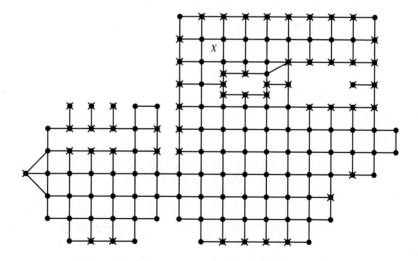

Figure 5-19

[4] Sometimes in practice there is one geometric consideration that can be relevant: the lengths of the blocks. When some blocks are much longer than others, counting blocks may not be the best criterion for optimization. In our example we will make the assumption that all the blocks are about the same length (so we can paraphrase Shakespeare and say ". . . a block is a block is a block").

It is exactly at each of these oddball vertices that our attempt to find an Euler circuit runs into trouble. We already know the reasons. Each visit to a vertex uses up two edges (one to get in, another to get out). When we come into an oddball vertex for the last time, we have painted ourselves into a corner, and the only way to get out is by retracing some edge, in other words, deadheading the garbage truck.

EULERIZING GRAPHS

In order to travel along each edge of the graph in Fig. 5-19 at least once and return to the starting point X, we will have to duplicate (i.e., deadhead) some blocks. We already know that much. The critical question is Which blocks should we deadhead? Well, we know that for each oddball vertex there must be one deadhead block either coming into it or going out of it. Sometimes, if we are lucky, we might be able to use a single deadhead block to simultaneously get out of one oddball vertex and into another. At other times we may not. Sounds complicated? Let's look first at a couple of simple examples.

Example 7.

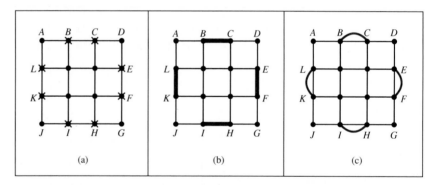

(a)   (b)   (c)

Figure 5-20

Figure 5-20(a) shows a graph representing a small subdivision (three blocks by three blocks) along which we want to route a garbage truck. The oddball vertices are marked, and there are eight of them (B, C, E, F, H, I, K, L). There must be a deadhead block at each one of these vertices, but that doesn't mean that we need eight deadhead blocks. Figure 5-20(b) shows how we can cleverly accomplish our purpose using a total of just four deadhead blocks. The deadhead blocks are BC, EF, HI, and KL and are shown in red in Fig. 5-20(b). Figure 5-20(c) shows what the actual graph representing the situation would look like: The deadhead blocks are now double edges of the graph, indicating that the garbage truck will make two passes along that block. Notice now that the graph in Fig. 20(c) has every vertex of even degree and therefore has an Euler circuit. If we apply Fleury's algorithm to find it, we will get an optimal routing for our garbage truck.   ■

Example 8.

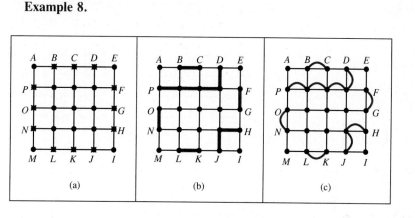

(a)   (b)   (c)

Figure 5-21

Figure 5-21(a) shows a graph representing a grid of streets that is four blocks by four blocks. Again, the oddball vertices are marked, and there are 12 of them. We need deadhead blocks at each of those vertices. How should we do it? An auspicious start is to make block *BC* a deadhead block (it worked in Example 7). Vertices *B* and *C* are now taken care of. We now need a deadhead block at *D*. Since there is no oddball vertex adjacent to *D* (other than *C* which is no longer an oddball), we need to connect *D* with another oddball vertex by means of a sequence of consecutive deadhead blocks which we will call a **deadhead path**. Let's say we decide to join *D* and *P* by means of the deadhead path shown in Fig. 5-21(b). We have now taken care of *D* and *P*. We can next take care of *F* and *G* by making *FG* a deadhead block. From *H* there are no immediate neighbors that are oddball vertices. As an alternative we take the deadhead path from *H* to *J* shown in Fig. 5-21(b). We can then take care of *K* and *L* by means of the deadhead block *KL*, and of *N* and *O* by means of the deadhead block *NO*. The modified graph showing the deadhead blocks as double edges is shown in Fig. 5-21(c). As before, every vertex in this graph has even degree, and we can therefore find an Euler circuit that would produce a routing of the garbage truck. The total number of deadhead blocks along this route is 10 [we can just count them from Fig. 5-21(c)].   ■

Everything we did in Example 8 is correct—the problem is that we weren't particularly clever in pairing up the oddball vertices. Figure 5-22 (on the following page) shows a different way of adding deadhead blocks and deadhead paths to the four-block-by-four-block street grid that results in fewer (eight) deadhead blocks. We leave it as an exercise for the student to verify that while there are several other ways to accomplish the same thing, eight deadhead blocks is the best one can do. (Exercise 28)

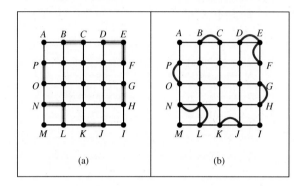

Figure 5-22

The process that we have illustrated in Examples 7 and 8 is called **eulerizing** the graph. It consists of adding deadhead edges and deadhead paths until every vertex of the graph has even degree. It is critical to mention that when eulerizing a graph, we can only duplicate edges that are already in the graph and that we can't put an edge where there was none before. Figure 5-23 illustrates what we mean.

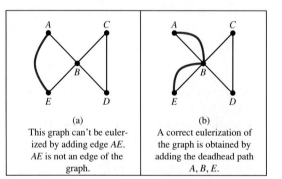

(a)
This graph can't be eulerized by adding edge *AE*.
AE is not an edge of the
graph.

(b)
A correct eulerization of
the graph is obtained by
adding the deadhead path
A, B, E.

Figure 5-23

Any eulerization that adds the fewest possible number of edges to the original graph is called an **optimal eulerization** of the graph.

■ **The Routing of a Garbage Truck (Part III)**

Figure 5-24 shows an optimal eulerization of the graph representing the city of Cleansburg. Fortunately for Tony and Cleo, almost all the oddball vertices can be paired up in such a way that a single deadhead block will simultaneously take care of both vertices. There are only two short dead-head paths: One is on the lower right hand side (*S, T, P*), and the other is on the left side (*U, V, W*). At any rate, it is not difficult to see that it's

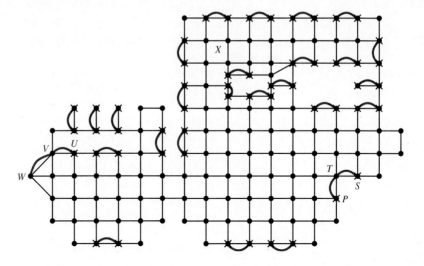

Figure 5-24

impossible to do any better. Tony and Cleo will have to deadhead a total
of 29 blocks on their daily trips through the city. The 29 blocks are exactly
the ones shown in red in Fig. 5-24. There is one last thing to do: Find the
actual Euler circuit for the graph in Fig. 5-24 that starts and ends at *X*
(that, of course, will be Tony and Cleo's route). We leave it to the reader
and Fleury's algorithm (Exercise 32) to take care of this last piece of
business.

CONCLUSION

In this chapter we learned several things, not the least of which is that
there is probably a mathematical reason behind the sometimes curious
manner in which the local garbage truck winds its way around the neigh-
borhood. In fact, many cities nowadays systematically route their garbage
trucks seeking maximum efficiency and using methods just like the ones
we developed in this chapter. For a large city, the efficient routing of
services (not only garbage collection but also street sweeping, snow removal, utility meter reading, mail delivery, etc.) can represent savings of
millions of dollars.

The efficient routing of garbage trucks is just one example of a broad
category of problems called *routing problems*. The name of the game in
routing problems is to find, among all the routes that accomplish a specified objective (pick up all the garbage, sweep every curb, inspect every
utility meter, call on every customer, etc.), one that is as efficient as
possible. In our garbage truck problem we measured efficiency by the
number of wasted (*deadhead*) blocks on the route, but in more complicated situations we might have to use other considerations such as total
distance traveled or total money spent. Routing problems are tricky in
this regard. While they all sound a lot alike and the broad objective is
always to do things in the most efficient possible way, what this means

in practice and how it is accomplished can differ greatly fiom problem to problem. This fact will become clear in the next chapter where we will explore a type of routing problem which marches to the tune of an altogether different drummer.

In this chapter we were introduced to the world of graphs. Graphs are to management science what squares, triangles, and circles are to traditional geometry: They provide an appropriate setting for analyzing (and hopefully solving) certain types of real-world problems. The study of graphs is in its own right an important branch of modern mathematics. Just like algebra and geometry, it has its own language and, more importantly, its own way of doing things. In this chapter we also had our first formal exposure to *algorithms* and learned about a specific algorithm called *Fleury's algorithm*. In the next three chapters we will get further glimpses of various parts of the world of graphs and discuss several new and surprising algorithms.

One final word about algorithms is in order: Carrying out an algorithm has often been compared to following a recipe or learning how to drive. In all these activities, intelligence is not a substitute for practice. Regardless of one's mathematical abilities, the only way to learn and understand an algorithm is to do it oneself as many times as is necessary.

KEY CONCEPTS

adjacent vertices	Euler path
algorithm	Euler's theorems
bridge	eulerizing graphs
circuit	Fleury's algorithm
connected graph	graph
deadhead block	loop
deadhead path	optimal eulerization
degree of a vertex	path
edge	routing problems
Euler circuit problem	vertices
Euler circuit	

EXERCISES

■ Walking

1. For each of the following graphs, list the vertices and edges.

(a)

(b)

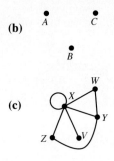

(c)

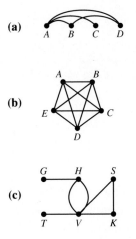

2. For each of the following graphs, list the vertices and edges.

(a)

(b)

(c)

3. For each of the following, draw two different pictures of the graph.
(a) Vertices: A, B, C, D
Edges: AB, BC, BD, CD
(b) Vertices: K, R, S, T, W
Edges: $RS, RT, TT, TS, SW, WW, WS.$

4. For each of the following, draw two different pictures of the graph.
(a) Vertices: L, M, N, P
Edges: LP, MM, PN, MN, PM
(b) Vertices: A, B, C, D, E
Edges: A is adjacent to C and E; B is adjacent to D and E; C is adjacent to $A, D,$ and E; D is adjacent to $B, C,$ and E; E is adjacent to $A, B, C,$ and $D.$

5. For each of the following graphs, find the degree of each vertex.

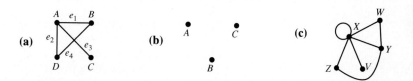

6. For each of the following graphs, find the degree of each vertex.

(a)   (b)   (c)

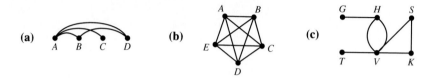

7. (a) Explain why the following figures represent the same graph.

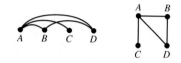

(b) Draw a third figure that represents the same graph.

8. (a) Explain why the following figures represent the same graph.

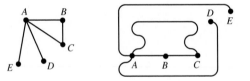

(b) Draw a third figure that represents the same graph.

9. (a) Draw a graph with four vertices and such that each vertex has degree 2.
(b) Draw a graph with six vertices and such that each vertex has degree 3.

10. (a) Draw a graph with four vertices and such that each vertex has degree 1.
(b) Draw a graph with eight vertices and such that each vertex has degree 3.

Exercises 11 through 14 refer to the following graph:

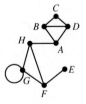

11. (a) Find a path from C to F passing through vertex B but not through vertex D.
(b) Find a path from C to F passing through both vertex B and vertex D.
(c) How many paths are there from C to A?
(d) How many paths are there from A to F?
(e) How many paths are there from C to F?

12. (a) Find a path from D to E passing through vertex G only once.
   (b) Find a path from D to E passing through vertex G twice.
   (c) How many paths are there from D to A?
   (d) How many paths are there from A to E?
   (e) How many paths are there from D to E?

13. (a) Find a circuit passing through the vertex D.
   (b) How many circuits start and end at vertex D?
   (c) Which edges in the graph are bridges?

14. (a) Find a circuit passing through the vertex H.
   (b) How many circuits start and end at vertex H?
   (c) Which edge can be added to this graph so that the resulting graph has no bridges?

15. (a) In the Königsberg bridge problem, which of the city bridges are bridges in the graph theory sense?
   (b) Give an example of a graph with four vertices in which every edge is a bridge.

16. (a) Give an example of a graph with five vertices and no bridges.
   (b) Give an example of a graph with five vertices in which every edge is a bridge.

17. For each of the following, determine if the graph has an Euler circuit, an Euler path, or neither of these. Explain your answer but do not find the actual path or circuit.

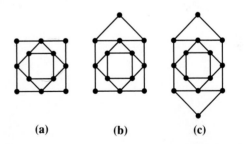

(a)   (b)   (c)

18. For each of the following, determine if the graph has an Euler circuit, an Euler path, or neither of these. Explain your answer but do not find the actual path or circuit.

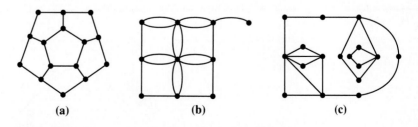

(a)   (b)   (c)

19. Trace each of the following graphs without lifting your pencil or retracing any lines. (Indicate your answer by labeling the edges 1, 2, 3, etc., in the order in which they are traced.) If it can't be done, explain why not.

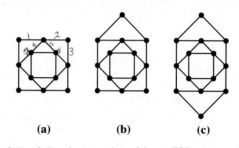

(a)   (b)   (c)

20. Trace each of the following graphs without lifting your pencil or retracing any lines. (Indicate your answer by labeling the edges 1, 2, 3, etc., in the order in which they are traced.) If it can't be done, explain why not.

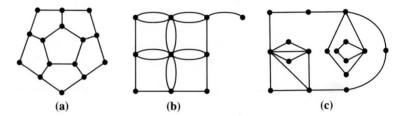

(a)   (b)   (c)

21. Eulerize the following graph.

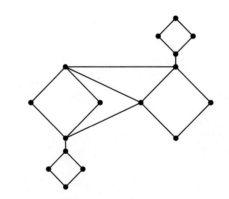

22. Eulerize the following graph.

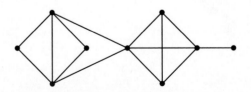

23. Use Fleury's algorithm to find an Euler circuit in the following graph.

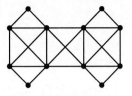

24. Use Fleury's algorithm to find an Euler circuit in the following graph.

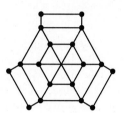

25. Use Fleury's algorithm to find an Euler circuit in the following graph.

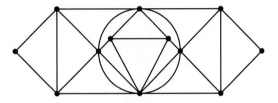

■ Jogging

26. The Handshaking Lemma.

   (a) For each of the graphs in Exercises 1 and 2, complete the following table.

Graph	Number of edges	Sum of the degrees of all vertices
Exercise 1(a)	4	3 + 2 + 1 + 2 = 8
Exercise 1(b)		
Exercise 1(c)		
Exercise 2(a)		
Exercise 2(b)		
Exercise 2(c)		

   (b) Explain why in any graph the sum of the degrees of all the vertices is double the number of edges.

27. A garbage truck must pick up garbage along all the streets of the grid shown (starting and ending at the garbage dump labeled *G*). All the streets are two-way streets, and garbage is picked up on both sides of the street simultaneously.

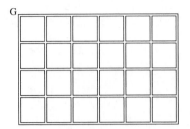

(a) Draw an appropriate graph representing this problem.
(b) Find an optimal eulerization of your graph.
(c) Describe an optimal route for a garbage truck using the above grid of streets. Describe the route by labeling the edges 1, 2, 3, . . ., in the order in which they are traveled.

28. (a) Find two different optimal eulerizations of the following graph:

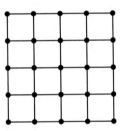

(b) Explain why any optimal eulerization of the graph above requires exactly eight deadhead edges.

29. Consider the following game: You must take a walk along the bridges of the city of Königsberg that starts and ends at the left bank (*L*) and crosses each bridge at least once. It costs $1 each time you cross a bridge. Describe the cheapest possible walk you can make.

Exercises 30 and 31 refer to a small town described by the following street map.

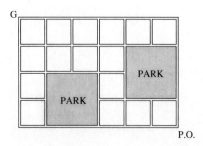

30. A garbage truck must pick up garbage along all the streets of the town (starting and ending at the garbage dump labeled *G*). All the streets are two way streets, and garbage is picked up on both sides of the street simultaneously.
   (a) Draw an appropriate graph representing this problem.
   (b) Find an optimal eulerization of your graph.
   (c) Describe an optimal route for collecting garbage in the town. Describe the route by labeling the edges 1, 2, 3, . . ., in the order in which they are traveled.

31. Suppose that we want to find an optimal route for a mail carrier to deliver mail along the streets of the town. In this case the mail carrier must walk along each block twice (once for each side of the street) except for blocks facing one of the parks (here only one pass is needed).
   (a) Draw the appropriate graph for this situation.
   (b) Find an optimal eulerization of your graph.
   (c) Describe an optimal route for the mail carrier (the starting and ending point is the post office labeled P.O.).

32. Use Fleury's algorithm to find an optimal route for the garbage collection problem in the city of Cleansburg. (Use the optimal eulerization of the city streets shown in Fig. 5-24.)

33. The following diagram is of a hypothetical city with a river running through the middle of the city. There are three islands and seven bridges as shown in the following figure. Is it possible to take a walk and cross each bridge exactly once? If so, show how; if not, explain why not.

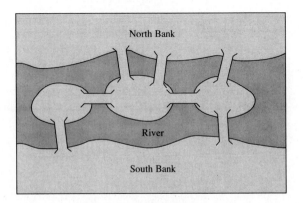

34. The following figure is the floorplan of an office complex.

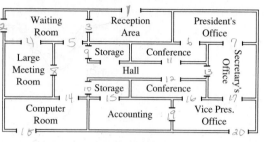

OFFICE COMPLEX

   (a) Show that it is impossible to start outside the complex and walk through each door of the complex exactly once and end up outside.
   (b) Show that it is possible to walk through every door of the complex exactly once (if you start and end at the right place).
   (c) Show that by removing exactly one door it would be possible to start outside the complex, walk through each door of the complex exactly once, and end up outside.

35. (a) Give an example of a graph with 15 vertices and no multiple edges (i.e., no more than one edge between any two vertices) that has an Euler circuit.
   (b) Give an example of a graph with 15 vertices and no multiple edges that has an Euler path but no Euler circuit.
   (c) Give an example of a graph with 15 vertices and no multiple edges that has neither an Euler circuit nor an Euler path.

■ Running

36. (a) Can a graph that has an Euler circuit have any bridges? If so, demonstrate it by showing an example. If not, explain why not.
   (b) Can a graph that has an Euler path have any bridges? If so, how many? Explain your answer.

37. Explain why there is always an even number of vertices of odd degree (what we called oddball vertices) in any graph.

38. Explain why in any graph in which the degree of each vertex is at least 2, there must be a circuit.

39. Suppose that we have a connected graph with exactly two vertices of odd degree. Describe (clearly and carefully) an algorithm for finding an Euler path in such a graph. (Use Fleury's algorithm as a blueprint.)

40. Suppose G and H are two graphs that have no common vertices and such that each graph has an Euler circuit. Let J be a (single) graph consisting of the graphs G, H, and one additional edge joining one of the vertices of G to one of the vertices of H. Explain why the graph J has no Euler circuit but does have an Euler path.

**REFERENCES
AND FURTHER
READINGS**

1. Beltrami, E., *Models for Public Systems Analysis*. New York: Academic Press, Inc., 1977.

2. Beltrami, E., and L. Bodin, "Networks and Vehicle Routing for Municipal Waste Collection," *Networks*, 4 (1973), 65–94.

3. Chartrand, Gary, *Graphs as Mathematical Models*. Belmont, Calif.: Wadsworth Publishing Co., Inc., 1977.

4. Euler, Leonhard, "The Koenigsberg Bridges," trans. James Newman, *Scientific American*, 189 (1953), 66–70.

5. Minieka, E., *Optimization Algorithms for Networks and Graphs*. New York: Marcel Dekker, Inc., 1978.

6. Newman, J. ed., *Mathematics—An Introduction to Its Sprit and Its Use*. New York: W. H. Freeman & Co., 1978.

7. Roberts, Fred S., "Graph Theory and Its Applications to Problems of Society," CBMS-NSF Monograph No. 29. Philadelphia: Society for Industrial and Applied Mathematics, 1978, chap. 8.

8. Tucker, A. C., "Perfect Graphs and an Application to Optimizing Municipal Services," *SIAM Review*, 15 (1973), 585–590.

9. Tucker, A. C., and L. Bodin, "A Model for Municipal Street-Sweeping Operations," in *Modules in Applied Mathematics*, vol. 3, ed. W. Lucas, F. Roberts, and R. M. Thrall. New York: Springer-Verlag, 1983, 76–111.

6 The Traveling Salesman Problem

Two roads diverged in a wood, and I— / I took the one less traveled by / And that has made all the difference...

ROBERT FROST

Hamilton Joins the Circuit

Sophie's Choices. Sophie, a salesperson, must call on customers in five different cities (A, B, C, D, and E). Each edge of the graph shown in Fig. 6-1 shows the cost of travel (gas, insurance, wear and tear on the car, etc.)[1] between any two cities. The trip must start and end at Sophie's home town (A). What is the *cheapest* possible route Sophie can take (starting and ending at A and visiting each of the other cities exactly once)?

We will take up Sophie's problem (and variations thereof) in this chapter. Before we do so we will introduce a little more graph terminology. Let's start with the observation that the graph in Fig. 6-1 is more than just a typical graph: In addition to vertices and edges, we have a number associated with each edge. Here the number represents the cost of traveling that particular edge—in another situation it could represent distance, time, the number of trees along the side of the road, or some other exotic variable. In general, when there are numbers associated with the edges of a graph, we call such numbers the **weights** of the edges[2] and we call

[1] In automobile travel, costs are generally taken to be directly proportional to distance (28 cents per mile, 21 cents per mile, etc.), but there are other variables (traffic, weather, road conditions, etc.) that can also affect costs.
[2] In scientific usage, the word "weight" is commonly used to generically represent variables such as cost, time, and distance—the kinds of things we usually want to minimize.

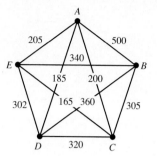

Figure 6-1

such a graph a **weighted graph**. So Fig. 6-1 is a weighted graph, the weight of edge AD is 185, the weight of edge BC is 305, etc. Notice that there is no requirement for the length of the edges to be proportional to the weights. As with ordinary graphs, the lengths and shapes of the edges in a weighted graph are irrelevant.

When the weights of the edges represent variables other than cost (say, distance or time), then the wording of the problem can be suitably modified (find the shortest route or the fastest route). We will use the term **optimal route** generically to describe the cheapest, shortest, fastest, etc., whatever the case may be.

HAMILTON CIRCUITS

We are now in a position to describe a generic version of our original problem: Given a weighted graph, we want to find an optimal route that starts and ends at some specified vertex and passes through *every other vertex exactly once*.

A circuit that starts at a vertex of the graph, passes through every other vertex of the graph exactly once, and returns to the starting vertex is called a **Hamilton circuit**.[3]

The distinction between an Euler circuit and a Hamilton circuit may appear minor on the surface (just a word of difference—substitute ''vertex'' for ''edge''). Mathematically speaking, however, there is a world of difference. Consider, for example, the two graphs in Fig. 6-2 shown on the next page. The graph in Fig. 6-2(a) has no Euler circuits (lots of rotten vertices of odd degree to spoil our fun), but it has lots of Hamilton circuits (such as A, B, D, C, E, F, G, A and A, D, C, E, B, G, F, A and others that the reader can readily find—see Exercise 1). Contrast this situation with the one in Fig. 6-2(b) where we know there are Euler circuits (every vertex has even degree) and yet the graph has no Hamilton circuit. (See Exercise 23.)

[3] Named after the great Irish mathematician and astronomer Sir William Rowan Hamilton (1805–1865). It is said that at the age of four Hamilton could read Latin, Greek, and Hebrew, as well as English. At the age of twenty-one he became a Professor of Astronomy at Trinity College in Dublin. Besides being a great scientist, Hamilton was an accomplished man of letters and a poet who counted Wordsworth and Coleridge among his closest friends.

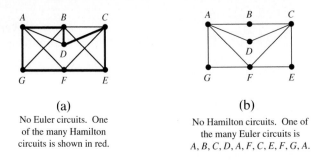

Figure 6-2

(a)

No Euler circuits. One
of the many Hamilton
circuits is shown in red.

(b)

No Hamilton circuits. One of
the many Euler circuits is
$A, B, C, D, A, F, C, E, F, G, A.$

While it would be nice to have a Hamilton's theorem (something like Euler's theorem for Hamilton circuits), there is unfortunately no such theorem. If we are given a graph and asked whether it has a Hamilton circuit or not, we have no surefire way to tell—unless of course we gut it out, try all possibilities, and either find one or give up. As a rule of thumb, the more "edge-rich" the graph, the more likely it is to have a Hamilton circuit. But that's just a rule of thumb. The graph in Fig. 6-3(a) has six vertices and six edges, and it has a Hamilton circuit (*A, B, C, D, E, F, A*). The graph in Fig. 6-3(b), on the other hand, has six vertices and eight edges, but it doesn't have a Hamilton circuit.

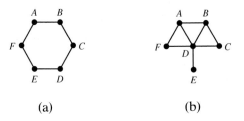

Figure 6-3

(a)   (b)

There are situations in which it is possible to know for a fact that the graph has a Hamilton circuit. The following theorem exemplifies one such situation. We give it as a helpful fact, and for an explanation or proof the reader is encouraged to consult any standard book on graph theory.[4]

> **Dirac's Theorem (It's a fact!).** Suppose that we have a connected graph with three or more vertices. If each vertex of the graph is adjacent to at least half of the vertices, then the graph has a Hamilton circuit.

[4] For example, references 2 and 8.

Example 1. The graph in Fig. 6-4 has six vertices each with degree 3. By Dirac's theorem it must have a Hamilton circuit. We leave it to the reader to find one.

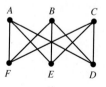

Figure 6-4

Example 2. *G* is a large graph (too large to draw). It is connected and has no loops or repeated edges. It has 20 vertices (V_1, V_2, V_3, . . ., V_{20}). The degrees of the vertices are $\deg(V_1) = 12$, $\deg(V_2) = 16$, $\deg(V_3) = 11$, $\deg(V_4) = 14$, $\deg(V_5) = 13$, $\deg(V_6) = 15$, $\deg(V_7) = 14$, $\deg(V_8) = 12$, $\deg(V_9) = 17$, $\deg(V_{10}) = 14$; all other vertices have degree 10. This graph has a Hamilton circuit (take it from Dirac—the man knew whereof he spoke!).

Example 3 **(Complete Graphs).**

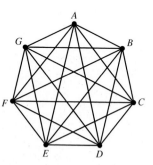

Figure 6-5

The graph shown in Fig. 6-5 has seven vertices, and every vertex is adjacent to every other vertex. A graph like this is called a **complete graph** (every pair of vertices is joined by an edge). Figure 6-6 shows complete graphs with four, five, and six vertices, respectively.

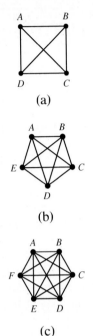

(a)

(b)

(c)

Figure 6-6

It is quite obvious that a complete graph has lots of Hamilton circuits. Let's consider, for example, the complete graph with four vertices shown in Fig. 6-6(a). It has the following Hamilton circuits:

1.   *A,B,C,D,A*

2.   *A,B,D,C,A*

3.   *A,C,B,D,A*

4.   *A,C,D,B,A*

5.   *A,D,B,C,A*

6.   *A,D,C,B,A.*

That's it! Notice that of the six Hamilton circuits listed above, the last three are repeats of the first three but read backward (circuit 4 is the same as circuit 2 but traveled backward, circuit 5 is circuit 3 traveled backward, and circuit 6 is circuit 1 traveled backward).   ∎

The complete graph with five vertices [shown in Fig. 6-6(b)] has 24 Hamilton circuits. As before, of these 24, half are repeats of the other half but traveled backward. We leave it to the enterprising reader to write them down.

There is a convenient formula that gives the number of Hamilton circuits in any complete graph. It uses the factorial, a concept we first came across in Chapter 2. Recall that if N is any positive integer the number $N! = N \times (N - 1) \times \cdots \times 3 \times 2 \times 1$ is called the **factorial** of N. The following table shows a few values of $N!$.

	Value
2!	$2 \times 1 = 2$
3!	$3 \times 2 \times 1 = 6$
4!	$4 \times 3 \times 2 \times 1 = 24$
5!	$5 \times 4 \times 3 \times 2 \times 1 = 120$
⋮	⋮
10!	$10 \times 9 \times 8 \times 7 \times 6 \times 5 \times 4 \times 3 \times 2 \times 1 = 3{,}628{,}800$

Here now is the formula for the number of Hamilton circuits in a complete graph:

> The complete graph with N vertices has $(N - 1)!$ Hamilton circuits. Of these, half are repeats of the other half but traveled backward.

THE TRAVELING SALESMAN PROBLEM

Our original problem for this chapter involved Sophie the traveling salesperson who wanted to find the cheapest trip that would allow her to visit each of the five cities shown in Fig. 6-1 once and return to the starting point (*A*). We can now rephrase the problem by saying we want to find an optimal Hamilton circuit in the complete weighted graph shown in Fig. 6-1.

The problem of finding an optimal Hamilton circuit in a complete weighted graph has many other important applications besides the example of a traveling salesperson. For reasons of tradition this type of problem has always been referred to as the **traveling salesman problem** regardless of the context in which it comes up. We will adhere to this tradition and refer to any problem in which we need to find an optimal Hamilton circuit in a complete weighted graph as a **TSP** (for traveling salesman problem). The following list gives a few examples of TSPs.

- **Package Deliveries.** Companies such as United Parcel Service (UPS) and Federal Express deal with this situation daily. Each truck has packages to deliver to a list of destinations. The travel time between any two delivery locations is known or can be estimated (experienced drivers always know such things). The object is to deliver the packages to each of the delivery locations and return to the starting point in the least amount of time—clearly an example of a TSP. On a typical day, a UPS truck delivers packages to somewhere between 100 and 200 locations, so we are dealing with a TSP involving a graph with that many vertices.

- **Fabricating Circuit Boards.** In the process of fabricating integrated circuit boards, tens of thousands of tiny holes must be drilled in each individual board. This is done by using a stationary laser beam and rotating the board around. Efficiency considerations require that the order in which the holes are drilled be such that the entire drilling sequence be completed in the least amount of time. This is an example of a TSP in which the vertices of the graph represent the holes on the circuit board and the weight of the edge connecting vertices *X* and *Y* represents the time needed to rotate the board from drilling position *X* to drilling position *Y*.

- **Scheduling Jobs on a Machine.** In many industries there are machines that perform many different jobs (they will be the vertices of the graph). After performing job *X* the machine needs to be set up to perform another job. The amount of time required to set up the machine to change from job *X* to job *Y* (or vice versa) is the weight of the edge connecting vertices *X* and *Y*. The problem is to schedule the machine to run through all the jobs in a cycle that requires the least total amount of time. This is another example of a TSP.

■ **Running Errands Around Town.**   If we have a lot of errands to do around town, organizing ourselves so that we can go from place to place following an optimal route and return home at the end of the day is an example of a TSP. For more details, the reader is encouraged to look at Exercise 19.

**SOLVING
THE TRAVELING
SALESMAN
PROBLEM**

OK, so let's say that we have a wonderful and exotic real-life problem which we cleverly reinterpret as one of finding an optimal Hamilton circuit in a complete weighted graph. How can we find such a circuit? This problem turns out to be a lot more involved than one would expect. Let's start with the obvious approach: trial and error or, as it is more commonly known among scholars, the **brute force algorithm**.

Brute Force Algorithm for Solving TSPs

■   List all possible Hamilton circuits for the weighted graph.

■   For each Hamilton circuit, add up the weights of the edges in the circuit (this total is called the **weight of the circuit**).

■   Of all the circuits, the one(s) with the least weight is optimal and therefore a solution to the problem.

Example 4.   Let's apply the brute force algorithm to solve Sophie's TSP which we introduced at the beginning of the chapter. For the reader's convenience we show the graph in Fig. 6-1 again (Fig. 6-7).

We already know that there are 24 possible Hamilton circuits and that 12 of them are duplicates of the other 12 but read backward (which allows us a shortcut in the computations). Since *A* is our starting point, we write down the Hamilton circuits as starting and ending at *A*. (If we had a different starting point, the circuits would still be exactly the same, but we would write them down differently. For example, if the starting point

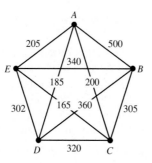

Figure 6-7

was *D*, then the circuit *A*, *B*, *C*, *D*, *E*, *A* would be written as *D*, *E*, *A*, *B*, *C*, *D*.) The computations (with the shortcut) are shown in Table 6-1.

	Hamilton Circuit	Circuit Weight	Same Circuit Reversed
1	*A*, *B*, *C*, *D*, *E*, *A*	500 + 305 + 320 + 302 + 205 = 1632	*A*, *E*, *D*, *C*, *B*, *A*
2	*A*, *B*, *C*, *E*, *D*, *A*	500 + 305 + 165 + 302 + 185 = 1457	*A*, *D*, *E*, *C*, *B*, *A*
3	*A*, *B*, *D*, *C*, *E*, *A*	500 + 360 + 320 + 165 + 205 = 1550	*A*, *E*, *C*, *D*, *B*, *A*
4	*A*, *B*, *D*, *E*, *C*, *A*	500 + 360 + 302 + 165 + 200 = 1527	*A*, *C*, *E*, *D*, *B*, *A*
5	*A*, *B*, *E*, *C*, *D*, *A*	500 + 340 + 165 + 320 + 185 = 1510	*A*, *D*, *C*, *E*, *B*, *A*
6	*A*, *B*, *E*, *D*, *C*, *A*	500 + 340 + 302 + 320 + 200 = 1662	*A*, *C*, *D*, *E*, *B*, *A*
7	*A*, *C*, *B*, *D*, *E*, *A*	200 + 305 + 360 + 302 + 205 = 1372	*A*, *E*, *D*, *B*, *C*, *A*
8	*A*, *C*, *B*, *E*, *D*, *A*	200 + 305 + 340 + 302 + 185 = 1332	*A*, *D*, *E*, *B*, *C*, *A*
9	*A*, *C*, *D*, *B*, *E*, *A*	200 + 320 + 360 + 340 + 205 = 1425	*A*, *E*, *B*, *D*, *C*, *A*
10	*A*, *C*, *E*, *B*, *D*, *A*	200 + 165 + 340 + 360 + 185 = 1250	*A*, *D*, *B*, *E*, *C*, *A*
11	*A*, *D*, *B*, *C*, *E*, *A*	185 + 360 + 305 + 165 + 205 = 1220	*A*, *E*, *C*, *B*, *D*, *A*
12	*A*, *D*, *C*, *B*, *E*, *A*	185 + 320 + 305 + 340 + 205 = 1355	*A*, *E*, *B*, *C*, *D*, *A*

Table 6-1

If we now look down the circuit weight column, we see that circuit 11 (*A*, *D*, *B*, *C*, *E*, *A*) and its partner (*A*, *E*, *C*, *B*, *D*, *A*) have the smallest circuit weight. We have now found what we were after: The optimal route for Sophie is given by the Hamilton circuit *A*, *D*, *B*, *C*, *E*, *A* (or alternatively the same route reversed, *A*, *E*, *C*, *B*, *D*, *A*). The cost of the optimal route is $1220. ∎

It is important to observe that when we talk about a solution to a TSP we mean a circuit and not a cost. In other words, the solution to Example 4 is not $1220. (It does poor Sophie little good to tell her that the cheapest possible route will cost her $1220 and not tell her what the route is.)

EFFICIENCY CONSIDERATIONS

So now we have a method of solution for any TSP—the brute force algorithm. It has only one minor flaw: It appears to be somewhat cumbersome. After all, it requires us to check the weight for each of the possible Hamilton circuits of the graph (or half of them if we use the shortcut). It is clear that this kind of checking can be done by a computer much better than by a human being because the brute force algorithm is essentially a mindless exercise in arithmetic with a little bookkeeping thrown in. Because computers are not only accurate but also extremely fast, we would expect that even if we throw a complicated graph at a computer, if the graph is within reason, the computer will quickly spit out an answer. (Let's arbitrarily decide that "within reason" means up to 1,000 vertices in the weighted graph. For many applications that's not a lot.)

Table 6-2 shows some time calculations for solving TSPs using the brute force algorithm. The critical variable that affects the amount of time needed to find a solution is the number of vertices (N) in the graph. We show the computation time for two different imaginary computers. Computer 1 is a personal computer, and we will assume that it can run through 10,000 Hamilton circuits per second. Computer 2 is a large computer that can run through 1 million Hamilton circuits per second.

Number of Vertices in Weighted Graph (N)	Number of Hamilton Circuits ($N - 1$)!	Number of Circuits to Check Using Shortcut $\frac{1}{2}(N - 1)$!	Time Needed by Computer 1 to Find a Solution	Time Needed by Computer 2 to Find a Solution
5	24	12	<1 second	<1 second
6	120	60	<1 second	<1 second
7	720	360	<1 second	<1 second
8	5,040	2,520	<1 second	<1 second
9	40,320	20,160	2 seconds	<1 second
10	362,880	181,440	18 seconds	<1 second
11	3,628,800	1,814,400	3 minutes	<2 seconds
12	39,916,800	19,958,400	33 minutes	<20 seconds
13	≈479 million	≈239.5 million	6 hr 39 min	≈4 minutes
14	≈6.2 billion	≈3.1 billion	≈86 hours	≈52 minutes
15	≈87 billion	≈43.6 billion	≈50 days	≈12 hr 6 min
16	≈1.3 trillion	≈654 billion	≈2 years	≈7 days 14 hr
17	≈20.9 trillion	≈10.5 trillion	≈33 years	≈121 days
18	≈356 trillion	≈178 trillion	≈564 years	≈5.64 years
19	≈6.4×10^{15}	≈3.2×10^{15}	≈10,100 years	≈101 years
20	≈1.2×10^{17}	≈6.1×10^{16}	≈193,000 years	≈1930 years
21	≈2.4×10^{18}	≈1.2×10^{18}	≈3.85 million years	≈38,500 years
⋮	⋮	⋮	⋮	⋮
100	≈10^{156}	≈5×10^{155}	≈1.5×10^{144} years	≈1.5×10^{142} years

Table 6-2

What Table 6-2 illustrates is an interesting phenomenon called the combinatorial explosion. Crudely put, this phenomenon says that any attempt to solve a problem such as a TSP by trial and error will sooner or later blow up in our faces. Even for a modest problem involving a weighted graph with 100 vertices, it would take computer 2 about 10^{142} years to run through all possible Hamilton circuits. This is considerably longer than the age of the universe, currently estimated to be about 20 billion (i.e., 2×10^{10}) years. This also means that we cannot count on improved technology to bail us out. Let's say we use an imaginary computer that is 1 trillion times faster (that's 10^{12} times faster) than computer 2. It would take this technological marvel 10^{130} years (give or take an eon or two) to check all the Hamilton circuits in a graph with 100 vertices.

The upshot of all of this is that the brute force algorithm has very limited practical value. For a typical graph with more than just a handful of vertices, this algorithm is useless. This situation, of course, is not generally true for all algorithms. Fleury's algorithm, discussed in Chapter 5, is a perfectly useful algorithm. Even when the number of vertices and edges in the graph is large, the algorithm can produce a solution in a reasonable amount of time.

We will not go into a detailed discussion of what we mean by all these vague expressions such as "a reasonable amount of time" and "the number of vertices in the graph is large." This would lead us into an exciting but difficult area of mathematics and computer science that deals with measuring the performance of algorithms in a precise way.[5] For our purposes, suffice it to say that we will draw a simplistic but helpful distinction between algorithms: There are *good* algorithms (which we will call **efficient** algorithms), and there are *bad* algorithms (which we will call **inefficient** algorithms). We have already seen one of each type: Fleury's algorithm for finding an Euler circuit is an efficient algorithm; the brute force algorithm for solving a TSP is an inefficient algorithm.

The theoretical basis for the distinction between efficient and inefficient graph algorithms is the relationship between the size of the graph (number of vertices and/or edges) and the number of steps needed to carry out the algorithm. When the number of steps needed to carry out the algorithm becomes disproportionately larger than the size of the graph, we have an inefficient algorithm. For example, in a complete weighted graph with 11 vertices, the brute force algorithm checks 10!/2 Hamilton circuits. If we double the number of vertices to 22, the brute force algorithm checks 21!/2 Hamilton circuits. The only problem is that 21!/2 is more than 100 billion times bigger than 10!/2 (Exercise 28). Thus, while the size of the problem doubled, the number of steps needed to solve it increased 100 billion times. This kind of explosion is characteristic of inefficient algorithms.

■ **Approximate Algorithms**

In a perfect world, our next step would be to present an algorithm that finds an optimal Hamilton circuit in any complete weighted graph and (unlike the brute force algorithm) is also an efficient algorithm. Unfortunately, no such algorithm is presently known. This can mean one of two things: (1) There is such an algorithm "out there," but no one has been clever enough to find it, or (2) such an algorithm is an impossibility (remember perfect voting methods?). For reasons that are too involved to explain here (but which obviously go beyond concern for the welfare and convenience of traveling salespeople), resolving this dilemma has become one of the most famous unsolved problems in modern mathematics. Most experts are inclined to believe that an efficient algorithm

[5] An excellent introduction to this subject is given in references 4 and 7.

that will solve any TSP is a mathematical impossibility. In fact, if such an algorithm were ever discovered, it would guarantee great fame and fortune to its discoverer. (For an excellent account of current progress on this famous unsolved mathematics problem see reference 5.)

In the meantime, we are faced with a quandary: In many industrial applications, it is important to solve a TSP for graphs that involve hundreds and even thousands of vertices, and to do so in real time (i.e., the boss wants an answer right away or, in the best of cases, by the end of the month). Since the brute force algorithm is out of the question and since no efficient algorithm that guarantees an optimal solution is known, the most reasonable strategy is one of partial retreat. We will give up on our requirement that the algorithm produce the very best solution and accept a solution that may not be optimal. In exchange, we ask for quick results. We will call any algorithm that produces solutions that are usually reasonably close to the optimal solution an **approximate algorithm**. The name of the game nowadays in dealing with a TSP, as well as with many other management science problems, is to find efficient approximate algorithms.

In the remaining sections of this chapter we will discuss two efficient approximate algorithms for a TSP: the nearest neighbor algorithm and the cheapest link algorithm.

THE NEAREST NEIGHBOR ALGORITHM

The **nearest neighbor algorithm** is almost completely described by its title. This is how the nearest neighbor algorithm would read (with embellishments) as a set of instructions to an imaginary traveling salesperson:

Nearest Neighbor Algorithm

■ Start at home (make sure you've packed your toothbrush and wallet).

■ Whenever you are in a city, pick the next city to visit from among the ones you haven't visited yet, and of all such cities pick the one that is cheapest to get to from where you are at (the nearest neighbor). In case of a tie, choose at random. Keep doing this until you've visited all the cities.

■ From the last city go home.

Example 5.   Let's apply the nearest neighbor algorithm to our original TSP. This is how it would go for Sophie: She starts at A (that's home). From A the cheapest trip is to D, so she goes there. From D the cheapest city to go to next (other than A) is E. From E the cheapest city to go to

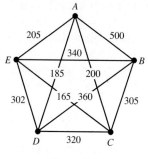

Figure 6-1

is *C*. From *C* the cheapest new city to go to is *B* (she doesn't want to go back to *E* or *A*!). Once she is at *B*, she has visited all the cities, so the last leg of the trip is back to *A*. Thus, the nearest neighbor algorithm produces the Hamilton circuit *A*, *D*, *E*, *C*, *B*, *A* with a total cost of $1457.

■

If we compare this algorithm (and the answer it produced) with the brute force algorithm (and the answer it produced), we observe two things:

1. The Hamilton circuit produced by the nearest neighbor algorithm is more expensive than the optimal solution ($1457 versus $1220). The relative percentage error is slightly under 20% ($\frac{237}{1220} \approx 0.194 = 19.4\%$).

2. As a reward for being willing to pay the penalty for a solution that is not optimal, we get an algorithm that is easily understood and quickly carried out.

To underscore this last point the reader is referred to Exercise 36 in which we need to route a traveling salesperson on a trip through 21 cities. We know from the last column in Table 6-2 that even with the fastest computers the brute force algorithm would require hundreds of years. With the nearest neighbor algorithm it takes the average student about 15 to 20 minutes to do it by hand and find an approximate solution.

■ **Changing the Starting Vertex**

There is a final twist to the nearest neighbor algorithm which we want to discuss next. It has to do with the choice of the starting vertex. Suppose that we choose *D* instead of *A* as the starting point in Example 5. We leave it as an exercise (Exercise 11) to verify that the solution produced by the nearest neighbor algorithm is *D*, *A*, *C*, *E*, *B*, *D* with a cost of $1250. This is a considerable improvement over the cost of our first solution based on the starting point *A*. If we choose *B* as the starting point for the nearest neighbor algorithm, we get a better solution yet: *B*, *C*, *E*, *A*, *D*, *B* with a cost of $1220 (Exercise 12).

The thoughtful reader might rightfully consider all of the above idle speculation—we can't choose *D* or *B* as a starting point because Sophie lives at *A* and must start her trip there. A different traveling salesperson living at *D* would (if using the nearest neighbor algorithm) be able to follow the better route *D*, *A*, *C*, *E*, *B*, *D*, and likewise a really lucky salesperson living at *B* could follow the route *B*, *C*, *E*, *A*, *D*, *B* which at a cost of $1220 happens to be the cheapest possible. This gets Sophie thinking— Why can't I start at *A* and then follow one of these better routes? In other words, the circuit *D*, *A*, *C*, *E*, *B*, *D* can also be traveled in the sequence *A*, *C*, *E*, *B*, *D*, *A* (we just start at a different spot). Likewise, the circuit *B*, *C*, *E*, *A*, *D*, *B* can be traveled in the sequence *A*, *D*, *B*, *C*, *E*, *A*. A circuit doesn't change just because we start at a different spot!

The point of all of the above is that the nearest neighbor algorithm can be started at any vertex of the graph (not just the home vertex) and will produce a Hamilton circuit. This same Hamilton circuit can then be re-written with the home vertex as the starting point. This leads to an improved strategy which we will call the **repetitive nearest neighbor algorithm**.

Repetitive Nearest Neighbor Algorithm

- **Step 1.** Pick an arbitrary vertex and apply the nearest neighbor algorithm with that vertex as the starting point.

- **Step 2.** Repeat the process with each of the vertices of the graph.

- **Step 3.** Of all the Hamilton circuits obtained in Step 2, keep the best as a solution.

- **Step 4.** Rewrite the solution obtained in step 3 using the home vertex as the starting point.

Let's illustrate the whole process with an example.

Example 6. It is the year 2020. A space probe is going to be launched from planet Earth scheduled to go to each of the planetary moons shown in Fig. 6-8 to collect mineral samples and return to Earth with all the loot. The weighted graph shows the time (in years) needed to travel between any two of these planetary bodies.

We want to use the repetitive nearest neighbor algorithm to find a Hamilton circuit in the weighted graph shown in Fig. 6-8. For each possible starting vertex Table 6-3 shows the circuit obtained by using the

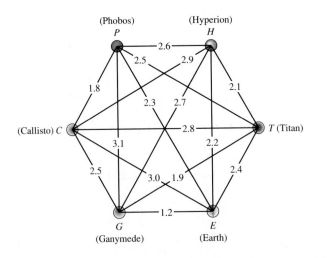

Figure 6-8

Circuit No.	Starting Point	Circuit Obtained by Nearest Neighbor Algorithm	Length of Trip (years)
1	Earth (*E*)	*E, G, T, H, P, C, E*	12.6
2	Phobos (*P*)	*P, C, G, E, H, T, P*	12.3
3	Hyperion (*H*)	*H, T, G, E, P, C, H*	12.2
4	Titan (*T*)	*T, G, E, H, P, C, T*	12.5
5	Callisto (*C*)	*C, P, E, G, T, H, C*	12.2
6	Ganymede (*G*)	*G, E, H, T, P, C, G*	12.3

Table 6-3

nearest neighbor algorithm and the total weight of the circuit. We leave it to the reader to verify the details.

Of the circuits in Table 6-3, circuits 3 and 5 tie for the best with a total length of 12.2 years. Of course, we know that we can't start the trip at Hyperion or Callisto—the actual trip described by circuit 3 is *E, P, C, H, T, G, E* while the trip described by circuit 5 is *E, G, T, H, C, P, E*. ∎

THE CHEAPEST LINK ALGORITHM

Our experience with the repetitive nearest neighbor algorithm has taught us that the order in which we build a Hamilton circuit and the order in which we actually travel it do not have to be the same. As a matter of fact, we can build a Hamilton circuit one edge at a time without even asking that the edges be consecutive. If we want to, we can first grab an edge here, then another edge over there, then another edge way yonder. This may look pretty disorganized at first, but there is no cause for concern as long as all the pieces come together at the end to form a Hamilton circuit. We often use this type of strategy in putting together a large jigsaw puzzle.

Our next algorithm, called the **cheapest link algorithm**, is based on this strategy. We look over the graph and grab the cheapest edge we can find, wherever it may be. We look again and grab the next cheapest. We keep doing this, each time grabbing the cheapest edge available but being careful that we don't close a circuit until the very end and that we don't have three edges coming out of the same vertex (either one of those two moves would make it impossible to put together a Hamilton circuit at the end).

Example 7. Before we give a more formal description of the cheapest link algorithm, let's apply it to Sophie's TSP.

The cheapest possible edge we can find anywhere in the graph in Fig. 6-9 is *EC* at $165. We'll grab it! (For convenient bookkeeping let's indicate

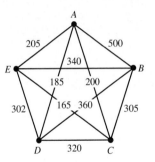

Figure 6-9

this by marking it in red.) The next cheapest edge is *DA* at $185. We'll take that one too! The next cheapest edge is *AC* with weight $200. This one is also OK, so we mark it in red. So far, so good! Figure 6-10 shows how things stand at this point.

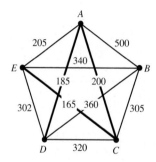

Figure 6-10

The next cheapest edge is *EA* at $205, but we can't use it because it would close the little circuit *E, A, C, E* and we can't have that! To remind ourselves that we have checked *EA* and can't use it, we can either erase it or cross it out. After doing that, the next cheapest edge available is *ED*, but we can't use it either because it creates the circuit *E, D, A, C, E*. We erase *ED* and try *BC*—the next cheapest available edge at $305. If we were to choose *BC* and mark it in red, we would have three different edges coming out of vertex *C* that would be part of our Hamilton circuit, so *BC* is also out. We next try edge *DC* at $320 but must also rule it out because it closes the circuit *D, C, A, D*. The next possible choice is edge *EB*, and this one is OK so we mark it in red. Our last choice is edge *BD* which closes the Hamilton circuit. We can now read off our solution using any starting vertex we want. In this case our Hamilton circuit is *A, D, B, E, C, A* (or *A, C, E, B, D, A*) at a cost of $1250.   ■

A formal description of the cheapest link algorithm can be given as follows:

Cheapest Link Algorithm

- **Step 1.** Pick the edge with the smallest weight first (in case of a tie pick one at random). Mark the edge.

- **Step 2.** Pick the next cheapest unmarked edge and mark it unless
  (a) it closes a smaller circuit or
  (b) it results in three marked edges coming out of a single vertex.
  If there are ties, break them arbitrarily.

- **Steps 3, 4, etc.** Repeat step 2 until the Hamilton circuit is complete.

Example 8. If we use the cheapest link algorithm to find a Hamilton circuit for the space probe in Example 6 (as shown in Fig. 6-11), then the sequence of steps goes as follows:

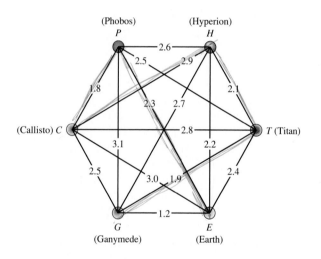

Figure 6-11

- **Step 1.** We pick (and mark) the edge *GE* (1.2 years). It is the cheapest edge in the graph.

- **Step 2.** Choose edge *CP* (1.8 years).

- **Step 3.** Choose edge *GT* (1.9 years).

- **Step 4.** Choose edge *HT* (2.1 years).

- **Step 5.** The next cheapest edge is *EH* (2.2 years), but it can't be used (it closes a circuit). We next choose *EP* (2.3 years).

- **Step 6.** There is only one way to close up the Hamilton circuit, and that's with edge *CH* (2.9 years).

The Hamilton circuit we get is *E, P, C, H, T, G, E* with total weight 12.2 years. By coincidence this is the same circuit we got in Example 6 using the repetitive nearest neighbor algorithm. ■

CONCLUSION

The nearest neighbor and cheapest link algorithms are two fairly simplistic approaches to the notoriously difficult problem of finding an optimal Hamilton circuit in a complete weighted graph (commonly known by the generic name of the *traveling salesman problem*). Both of them are based on what is aptly described as a *strategy of greed*: Go for whatever is cheapest at the moment without worrying about the long-term consequences. (Algorithms based on this kind of strategy are commonly referred to as *greedy algorithms*.) We are often reminded by well-meaning people that greed doesn't pay in the long run, and indeed it is possible to concoct examples in which the nearest neighbor or the cheapest link algorithm leads to just about the worst possible choice for a Hamilton circuit (see Exercises 31 and 32). What the nearest neighbor and cheapest link algorithms do have going for them is that they are easy to understand, can be carried out very efficiently, and in most typical problems give an approximate solution that is within a reasonable margin of error.

Many other, more sophisticated algorithms for finding approximate solutions to the traveling salesman problem are known, and in most cases (unlike the nearest neighbor and cheapest link algorithms) they come with a performance guarantee (i.e., the approximate solution is guaranteed never to be off by more than a certain percent). One such algorithm is given in Exercise 35. At present, the best algorithms produce approximate solutions that are off by no more than 1% of the optimal solution for TSPs involving about 100,000 vertices and off by not more than 3.5% of the optimal solution for TSPs involving about 1 million vertices. Constant refinements of these algorithms and improvements in technology guarantee that such levels of performance are going to get even better (see reference 5).

In the meantime, the fundamental question remains unsolved: Is there an efficient algorithm for the traveling salesman problem guaranteed to always produce the optimal route or is such an algorithm a mathematical impossibility? This problem is still waiting for the next Euler to come along.

KEY CONCEPTS

approximate algorithms	inefficient algorithms
brute force algorithm	nearest neighbor algorithm
cheapest link algorithm	optimal route
complete graph	repetitive nearest neighbor
Dirac's Theorem	algorithm
efficient algorithms	traveling salesman problem (TSP)
factorial	weighted graph
Hamilton circuit	weights

EXERCISES

■ Walking

1. Two of the Hamilton circuits in the following graph are *A, B, D, C, E, F, G, A* and *A, D, C, E, B, G, F, A*. Find two more (not these circuits reversed). Use *A* as the starting point.

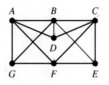

2. Find two Hamilton circuits in the following graph.

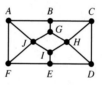

3. List all possible Hamilton circuits in the following graph.

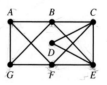

4. List all possible Hamilton circuits in the following graph.

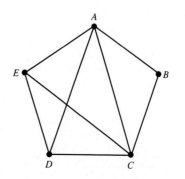

5.

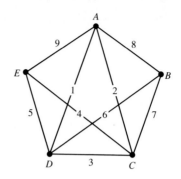

(a) In the weighted graph above, find the weight of edge *BD*.
(b) In the weighted graph above, find the weight of edge *EC*.
(c) Find a Hamilton circuit in the graph and give its weight.
(d) Find a different Hamilton circuit in the graph and give its weight.

6.

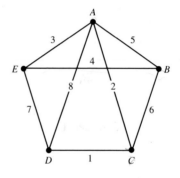

(a) In the weighted graph above, find the weight of edge *AD*.
(b) In the weighted graph above, find the weight of edge *AC*.
(c) Find a Hamilton circuit in the graph and give its weight.
(d) Find a different Hamilton circuit in the graph and give its weight.

7. (a) 5! = 120. Use this fact to painlessly find 6! (No calculators please.)
  (b) Given that 10! = 3,628,800, find 9! (No calculators please.)
  (c) How many different Hamilton circuits are there in a complete graph with 10 vertices?

8. (a) 7! = 5040. Use this fact to painlessly find 8! (No calculators please.)
  (b) How many different Hamilton circuits are there in a complete graph with 9 vertices?

9. Consider the following weighted graph.

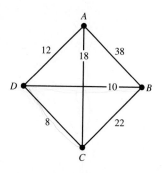

   (a) Use the brute force algorithm to find an optimal Hamilton circuit.
   (b) Use the nearest neighbor algorithm with starting vertex A to find a Hamilton circuit.
   (c) Use the cheapest link algorithm to find a Hamilton circuit.
   (d) Compare the optimal solution obtained in part (a) with the Hamilton circuits obtained in parts (b) and (c). Give the relative error in each of these solutions.

$$\text{Relative error} = \frac{\text{(cost of approximate solution)} - \text{(cost of optimal solution)}}{\text{cost of optimal solution}}$$

10. Consider the following weighted graph.

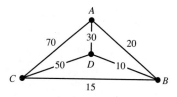

   (a) Use the brute force algorithm to find an optimal Hamilton circuit.
   (b) Use the nearest neighbor algorithm with starting vertex A to find a Hamilton circuit.
   (c) Use the cheapest link algorithm to find a Hamilton circuit.
   (d) Compare the optimal solution obtained in part (a) with the Hamilton circuits obtained in parts (b) and (c). Give the relative error in each of these solutions.

$$\text{Relative error} = \frac{\text{(cost of approximate solution)} - \text{(cost of optimal solution)}}{\text{cost of optimal solution}}$$

Exercises 11 and 12 refer to the following weighted graph.

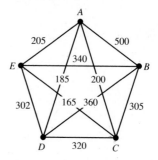

11. Apply the nearest neighbor algorithm with starting vertex D to find a Hamilton circuit in the graph.

12. Apply the nearest neighbor algorithm with starting vertex B to find a Hamilton circuit in the graph.

13. **(a)** Use the cheapest link algorithm to find a Hamilton circuit in the following weighted graph.

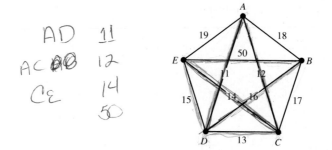

(b) Is there a cheaper Hamilton circuit? If so, find one. If not, explain why.

14. **(a)** Use the nearest neighbor algorithm to find a Hamilton circuit in the following weighted graph.

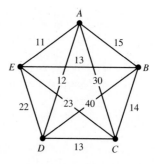

(b) Is there a cheaper Hamilton circuit? If so, find one. If not, explain why.

15. Use the repetitive nearest neighbor algorithm to find a Hamilton circuit in the following weighted graph.

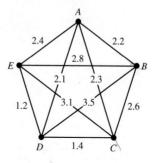

Exercises 16 through 18 refer to the following weighted graph.

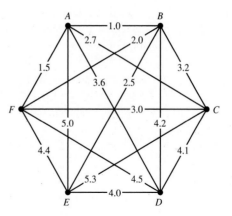

16. Use the cheapest link algorithm to find a Hamilton circuit.

17. Use the nearest neighbor algorithm starting at vertex *A* to find a Hamilton circuit.

18. Use the repetitive nearest neighbor algorithm to find a Hamilton circuit.

19. You have a busy day ahead of you. You must run the following errands (in no particular order): go to the post office, deposit a check at the bank, pick up some French bread at the deli, visit a friend at the hospital, and get a haircut at Karl's Beauty Salon. You must start and end at home. Each block on the map is exactly 1 mile.

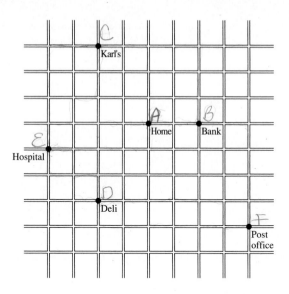

(a) Draw a weighted graph corresponding to this problem.
(b) Find the optimal (shortest) way to run all the errands. (Use any algorithm you think is appropriate.)

20. Rosa's Floral must deliver flowers to each of the five locations *A*, *B*, *C*, *D*, and *E* shown on the map. The trip must start and end at the flower shop, which is located at *X*. Each block on the map is exactly 1 mile.

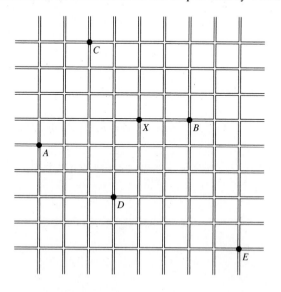

(a) Draw a weighted graph corresponding to this problem.
(b) Find the optimal (shortest) way to run all the errands. (Use any algorithm you think is appropriate.)

■ Jogging

21. Hamilton's puzzle. Find a Hamilton circuit in the following graph. Indicate your solution by marking the circuit right on the graph.

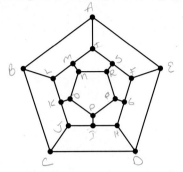

Historical footnote. Hamilton made up and marketed a game which was a three-dimensional version of this exercise using a regular dodecahedron (see figure) in which the vertices were different cities. The purpose of the game

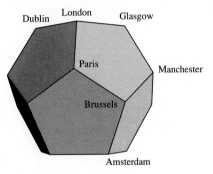

was to find a "trip around the world" going from city to city along the edges of the dodecahedron without going back to any city (except for the return to the starting point). When a regular dodecahedron is flattened, we get the graph shown in this exercise.

22. Find a Hamilton circuit in the following graph.

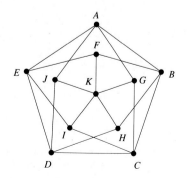

23. Explain why the following graph has no Hamilton circuit.

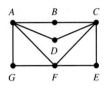

24. The Petersen graph. The following graph is called the Petersen graph. Explain why the Petersen graph does not have a Hamilton circuit.

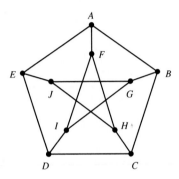

25. A Hamilton path. This is a path that passes through every vertex exactly once without necessarily returning to the starting vertex.

   (a) Find a Hamilton path in the following graph.

   (b) Find a Hamilton path in the Petersen graph (see Exercise 24).

26. (a) Give an example of a graph with four vertices in which the same circuit can be both an Euler circuit and a Hamilton circuit.

   (b) Give an example of a graph with N vertices in which the same circuit can be both a Euler circuit and a Hamilton circuit.

27. (a) Compare 2^3 with 3! Which is bigger?

   (b) Compare 2^4 with 4! Which is bigger?

   (c) Suppose N is more than 5. If you had a choice between 2^N or $N!$ dollars, which would you choose? Explain why the one you picked is the bigger number.

28. Explain why 21! is more than 100 billion times bigger than 10! (i.e., show that $21! > 10^{11} \times 10!$).

Exercises 29 and 30 refer to the following situation: A traveling salesperson's territory consists of the 11 cities shown on the following mileage chart. The salesperson must organize a round trip that starts and ends in Dallas (that's home) and will pass through each of the other 10 cities exactly once.

Mileage Chart

	Atlanta	Boston	Buffalo	Chicago	Columbus	Dallas	Denver	Houston	Kansas City	Louisville	Memphis
Atlanta	*	1037	859	674	533	795	1398	789	798	382	371
Boston		*	446	963	735	1748	1949	1804	1391	941	1293
Buffalo			*	522	326	1346	1508	1460	966	532	899
Chicago				*	308	917	996	1067	499	292	530
Columbus					*	1028	1229	1137	656	209	576
Dallas						*	781	243	489	819	452
Denver							*	1019	600	1120	1040
Houston								*	710	928	561
Kansas City									*	520	451
Louisville										*	367
Memphis											*

29. Use the nearest neighbor algorithm with starting city Dallas to find a Hamilton circuit for the traveling salesperson. You should apply the algorithm using the data directly from the chart. *Do not draw a graph!*

30. Use the cheapest link algorithm to find a Hamilton circuit for the traveling salesperson. You should apply the algorithm using the data directly from the chart. (You may wish to draw a partial graph as you apply the algorithm so that you do not close any circuits before the end.)

■ Running

31. Make up an example of a complete weighted graph such that the Hamilton circuit produced by the nearest neighbor algorithm (you may pick the starting vertex) gives the worst possible choice of a circuit (in other words one whose weight is bigger than any other).

32. Make up an example of a complete weighted graph such that the Hamilton circuit produced by the cheapest link algorithm gives the worst possible choice of a circuit.

33. Make up an example of a complete weighted graph such that the Hamilton circuit produced by the nearest neighbor algorithm (you can pick the starting vertex) has a relative percentage error of at least 100% (in other words, the weight of the Hamilton circuit produced by the nearest neighbor algorithm is at least twice as much as the weight of the optimal Hamilton circuit).

34. Make up an example of a complete weighted graph such that the Hamilton circuit produced by the cheapest link algorithm has a relative percentage error of at least 100% (in other words, the weight of the Hamilton circuit produced by the cheapest link algorithm is at least twice as great as the weight of the optimal Hamilton circuit).

35. **The nearest insertion algorithm.** Here is a description of a different approximate algorithm for the traveling salesman problem. The basic idea is to start with a small subcircuit of the graph and enlarge it one vertex at a time until all the vertices are included and it is a full-fledged Hamilton circuit.

   ■ Step 1.   Pick any vertex as a starting circuit (consisting of one vertex and zero edges). Mark it red.

   ■ Next step.   Suppose that at step k we have already built a red subcircuit with k vertices (call it C_k). We look for a black vertex in the graph that is as close as possible to some vertex of C_k. Let's call this black vertex B, and the vertex of C_k it is nearest to, R. We now create a new red circuit C_{k+1} which is the same as C_k except that B is inserted immediately after R in the sequence. Repeat until you have a Hamilton circuit.

   (a) Verify that when the nearest insertion algorithm is applied to the original traveling salesperson problem in this chapter the following sequence of circuits is produced (we use A as our starting vertex):

   ■ C_1: A

   ■ C_2: A, D, A (D is the nearest vertex to C_1)

   ■ C_3: A, C, D, A (C is nearest to A in C_2)

   ■ C_4: A, C, E, D, A (E is nearest to C in C_3)

   ■ C_5: A, C, B, E, D, A (B is nearest to C in C_4).

   (b) Use the nearest insertion algorithm to find a Hamilton circuit for the graph in Exercise 15. Use A as the starting vertex.
   (c) Use the nearest insertion algorithm to find a Hamilton circuit for the graph in Exercise 15. Use B as the starting vertex.
   (d) Use the nearest insertion algorithm to find a Hamilton circuit for the graph in Exercise 15. Use C as the starting vertex.

36. A traveling salesperson's territory consists of the 21 cities shown on the accompanying mileage chart. The salesperson must organize a round trip that starts and ends in Miami (that's home) and passes through each of the other 20 cities exactly once.

Mileage Chart

	Atlanta	Boston	Buffalo	Chicago	Columbus	Dallas	Denver	Houston	Kansas City	Louisville	Memphis	Miami	Minneapolis	Nashville	New York	Omaha	Pierre	Pittsburgh	Raleigh	St. Louis	Tulsa
Atlanta	*																				
Boston	1037	*																			
Buffalo	859	446	*																		
Chicago	674	963	522	*																	
Columbus	533	735	326	308	*																
Dallas	795	1748	1346	917	1028	*															
Denver	1398	1949	1508	996	1229	781	*														
Houston	789	1804	1460	1067	1137	243	1019	*													
Kansas City	798	1391	966	499	656	489	600	710	*												
Louisville	382	941	532	292	209	819	1120	928	520	*											
Memphis	371	1293	899	530	576	452	1040	561	451	367	*										
Miami	655	1504	1409	1329	1160	1300	2037	1190	1448	1037	997	*									
Minneapolis	1068	1368	927	405	713	936	841	1157	447	697	826	1723	*								
Nashville	242	1088	700	446	377	660	1156	769	556	168	208	897	826	*							
New York	841	206	372	802	542	1552	1771	1608	1198	748	1100	1308	1207	892	*						
Omaha	986	1412	971	459	750	644	537	865	201	687	652	1641	357	744	1251	*					
Pierre	1361	1726	1285	763	1071	943	518	1186	592	1055	1043	2016	394	1119	1565	391	*				
Pittsburgh	687	561	216	452	182	1204	1411	1313	838	388	752	1200	857	553	368	895	1215	*			
Raleigh	372	685	605	784	491	1166	1661	1160	1061	541	728	819	1189	521	489	1214	1547	445	*		
St. Louis	541	1141	716	289	406	630	857	779	257	263	285	1196	552	299	948	449	824	588	804	*	
Tulsa	772	1537	1112	683	802	257	681	478	248	659	401	1398	695	609	1344	387	760	984	1129	396	*

(a) Use the nearest neighbor algorithm to find a Hamilton circuit for the 21 cities.

(b) Use the cheapest link algorithm to find a Hamilton circuit for the 21 cities.

REFERENCES AND FURTHER READINGS

1. Bellman, R., K. L. Cooke, and J. A. Lockett, *Algorithms, Graphs and Computers*. New York: Academic Press, Inc., 1970, chap. 8.

2. Chartrand, Gary, *Graphs as Mathematical Models*. Belmont, Calif.: Wadsworth, Publishing Co., Inc., 1977, chap. 3.

3. Knuth, Donald, "Mathematics and Computer Science: Coping with Finiteness," *Science*, 194 (December 1976), 1235–1242.

4. Kolata, Gina, "Analysis of Algorithms: Coping With Hard Problems," *Science*, 186 (November 1974), 520–521.

5. Kolata, Gina, "Math Problem, Long Baffling, Slowly Yields," *The New York Times*, March 12, 1991, B8–B10.

6. Lawler, E. L., J. K. Lenstra, A. H. G. Rinooy Kan, and D. B. Shmoys, *The Traveling Salesman Problem*. New York: John Wiley & Sons, Inc., 1985.

7. Lewis, H. R., and C. H. Papadimitriou, "The Efficiency of Algorithms," *Scientific American*, 238 (January 1978), 96–109.

8. Wilson, Robin, and John J. Watkins, *Graphs: An Introductory Approach*. New York: John Wiley & Sons, Inc., 1990.

7

Minimum Network Problems

Spanning Trees and Steiner Trees

The Amazon telephone network problem. The Amazon Telephone and Telegraph Company (AT & T) wants to connect six cities deep in the Amazon jungle by means of a network of underground telephone lines. Roads directly connecting any two of these cities already exist, and for this reason the cheapest way to connect a pair of cities with telephone lines is to bury the telephone cable right along the roads. The weighted graph in Fig. 7-1 shows the cost (in millions of dollars) of laying down the possible telephone links between the various cities. AT & T wants to link all the cities so that a call can be made from any city to any other city either directly or by routing the call indirectly through one or more intermediate cities. At the same time, the company wants to save as much money as possible. What is the *cheapest* way to link up the six cities?

Problems just like the Amazon telephone network problem occur in many types of applications besides the construction of telephone networks: designing an irrigation system, laying down tracks for a high-speed rail system, designing the layout of a computer chip, linking up a network of computers, etc. The common thread in all these problems is the need to connect several locations (the vertices of a weighted graph) in such a way that it is always possible to get from any location to any other location and so that the total weight of the network is *minimal*. In connecting the

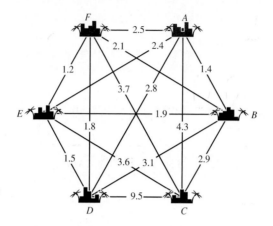

Figure 7-1 The cost (in millions) of laying down telephone lines between any two cities.

various locations, no consideration is given to how roundabout or inconvenient the connections may be. The single-minded objective in determining the solution is to make the total weight of the network as small as possible. Problems of this type are known as **minimum network problems**, and in this chapter we will discuss two basic variations on this theme.

TREES

Let's start by asking (without getting into specifics), What should we expect the solution of the Amazon telephone network problem to be like?

First, we note that the solution to the problem is itself a graph that *lives inside* the original graph in Fig. 7-1. After all, Fig. 7-1 shows each of the potential linkages between cities (the edges of the graph) and in finding the solution our job will be to select the right set of linkages from among these. In the language of graph theory we say that the solution is a **subgraph** of the original graph. Moreover, since we are trying to build a telephone network that includes each of the cities on the original graph, it is clear that the vertices of the solution subgraph must include each vertex of the original graph. In addition, the solution subgraph should have the following two characteristics:

1. It should be *connected*. This is an obvious consequence of the fact that our network must link every city with every other city.

2. It should *not contain any circuits* (i.e., nowhere in the solution subgraph should we be able to start at some vertex, travel along the edges of the solution subgraph, and return to the starting vertex). The reason for this is that in looking for an optimal (cheapest) subgraph we should never have more edges than are absolutely necessary to connect the various cities. Suppose, for example, that the solution subgraph contains a circuit as shown in Fig. 7-2. It is clear that in this case we

Figure 7-2 Laying telephone lines along the segment *XY* is a waste of money.

could delete any one of the edges of the circuit and still keep the telephone calls going through the network.

Any graph that is connected and has no circuits is called a **tree**. It is clear from our preceding discussion that the solution to the Amazon telephone network problem must be a tree.

Example 1. Figure 7-3 shows several examples of trees.

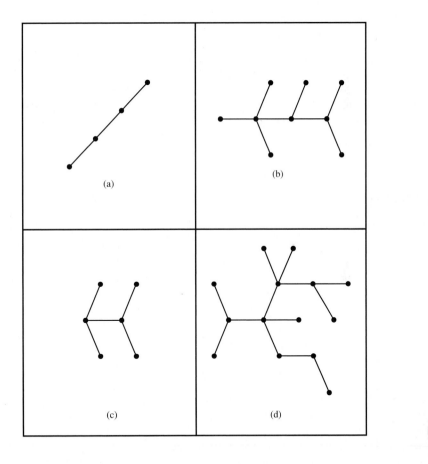

Figure 7-3

Example 2. Figure 7-4 shows examples of graphs that are *not* trees. Figures 7-4(a) and (b) have circuits; Fig. 7-4(c) has no circuits but is not connected; and Fig. 7-4(d) fails to be a tree on both accounts—it has circuits and it is not connected.

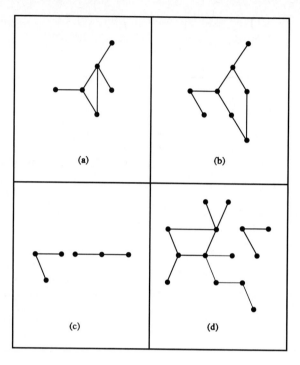

(a)

(b)

(c)

(d)

Figure 7-4

■ Properties of Trees

Trees are extremely important and useful structures. Not only do they help reduce carbon dioxide levels in the atmosphere, but they also show up in many mathematical applications, not the least of which are the minimum network problems we are discussing in this chapter. In what follows, we will briefly examine some of the main features of (mathematical) trees.

Let's start with the fact that in any connected graph there is always a path joining any vertex to any other vertex. Sometimes, as in Fig. 7-5, there is more than one path joining a pair of vertices. If that is the case, we can be sure the graph is not a tree. This follows from the simple observation that if there is more than one path joining two vertices, the vertices must be in a circuit (Fig. 7-6).

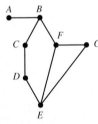

Figure 7-5 Vertices *B* and *E* can be joined by several different paths such as *B*, *C*, *D*, *E* and *B*, *F*, *E* and *B*, *F*, *G*, *E*.

Property 1. If a graph is a tree, there is one and only one path joining any two vertices. Conversely, if there is one and only one path joining any two vertices of a graph, the graph must be a tree.

One practical consequence of property 1 is that a tree is connected in a very precarious way: The removal of *any* edge of a tree will disconnect

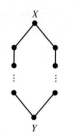

Figure 7-6 Two different paths joining vertices *X* and *Y* make a circuit.

it (Exercise 23). We can restate this by saying that in a tree, every edge is a *bridge*.

Property 2. In a tree, every edge is a bridge. Conversely, if every edge of a connected graph is a bridge, then the graph must be a tree.

From our preceding discussion, it seems intuitively obvious that a tree is "edge-poor." Having lots of edges is in some sense contrary to the nature of being a tree. At the same time, a tree must be connected, so a certain minimum number of edges is going to be necessary. A very important property of trees is that in a tree there is a very precise numerical relation between the number of edges and the number of vertices: The total number of edges is always 1 less than the number of vertices.

Property 3. A tree with N vertices must have $N-1$ edges.

Property 3 also has a converse, but we must be a little careful. Can we say outright that if a graph has one less edge than it does vertices, then it must be a tree? The graph in Fig. 7-7 shows that this need not be the case—it has 14 vertices and 13 edges, and yet it is not a tree. The rub is that it is not connected. If we require the graph to be connected, then the converse of property 3 is indeed true.

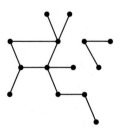

Figure 7-7 A graph with 14 vertices and 13 edges that is not a tree.

Property 4. A connected graph with N vertices and $N-1$ edges must be a tree.

Example 3.   This example illustrates some of the ideas discussed so far. Let's say that we have five vertices, and let's start putting edges on these vertices. At first, with one, two, or three edges [Figs. 7-8(a) through (c)] we just don't have enough edges to make the graph connected. When

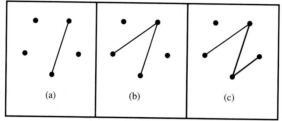

Figure 7-8 The graphs are disconnected. There are not enough edges.

we get to four edges, we can, for the first time, make the graph connected. If we do so, we have a tree [Figs. 7-9(a) through (c)]. As we add more

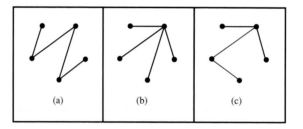

Figure 7-9 Just enough edges (four) to make a tree. Three different trees based on the same vertices.

edges (five, six, etc.), the connected graph starts picking up circuits and stops being a tree [Figs. 7-10(a), through (c)].

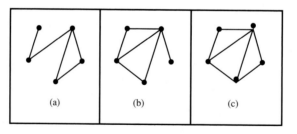

Figure 7-10 These graphs have circuits. They have too many edges to be trees.

MINIMUM SPANNING TREES

In Fig. 7-10 we have three examples of graphs that are connected but are not trees; they have more than the requisite number of edges. Within such a graph we can always find a tree reaching out to each of the vertices—somewhat like a skeleton holding up the rest of the body. We call such a tree a spanning tree of the original graph. Before we formally define a spanning tree, let's look at some examples.

Example 4.   Consider the graph shown in Fig. 7-11(a). It is connected, it has seven vertices and nine edges, and we know it is not a tree. Inside

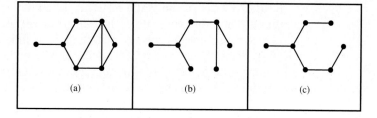

Figure 7-11

such a graph we can find subgraphs like the one shown in Fig. 7-11(b) with the same seven vertices but with only six of the edges. Such a subgraph is a tree that spans (reaches out to) all the vertices of the original graph and, as such, is called a **spanning tree** for the graph in Fig. 7-11(a). A graph may have more than one spanning tree: Fig. 7-11(c) shows a different spanning tree for the graph in Fig. 7-11(a). We leave it to the reader [Exercise 6(a)] to find other spanning trees for the graph in Fig. 7-11(a).   ■

Any connected graph G has at least one spanning tree. If G is a tree, then it is its own spanning tree; otherwise, any tree contained within G and with the same vertices as G is a spanning tree. In general, the number of different spanning trees in a connected graph can be quite large. Consider the following example.

Example 5.
The weighted graph in Fig. 7-12 has eight different spanning trees, T_1, T_2, . . ., T_8, as shown in Fig. 7-13.

Figure 7-12

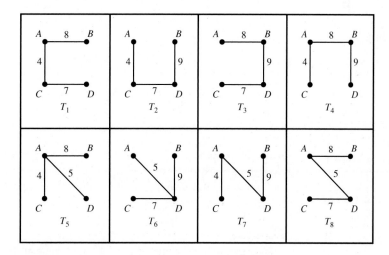

Figure 7-13

If we think of the graph in Fig. 7-12 as having edges which represent the possible linkages between four cities (*A*, *B*, *C*, and *D*) and weights which represent the costs of installing these linkages, then the eight spanning trees shown in Fig. 7-13 will represent the eight possible ways of connecting the cities in a treelike network. Of these, we can readily see (by just checking them one by one) that T_5 is the cheapest. Such a tree is called the minimum spanning tree of the original weighted graph. ■

With all this background, we should be ready now for a formal statement.

1. Suppose that *G* is an arbitrary connected graph. Then *G* has at least one (usually more) spanning tree. A **spanning tree** of *G* is a subgraph of *G* such that
    (a) Its vertices are exactly the vertices of *G*.
    (b) Its edges are some of the edges of *G*.
    (c) It is a tree.

2. Suppose that *G* is a connected weighted graph. Among all the spanning trees of *G*, there is one (maybe more) with the least total weight. Such a tree is called a **minimum spanning tree** of the weighted graph *G*.

Figure 7-14

Example 6. Note that it is possible for a weighted graph to have more than one minimum spanning tree.
The weighted graph in Fig. 7-14 has three different minimum spanning trees, each of which has a total weight of 4. They are shown in Fig. 7-15.

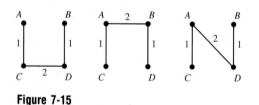

Figure 7-15

      ■

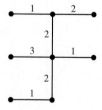

Figure 7-16

Example 7.
The weighted graph in Fig. 7-16 is a tree, and therefore it has only one spanning tree—itself. Of course since it is the only one, it is a minimum spanning tree as well. ■

Using our new terminology, we can restate the Amazon telephone network problem at the beginning of the chapter (and any other problem like it) as one of finding a minimum spanning tree for the given graph. Not surprisingly, problems of this type are known as **minimum spanning tree problems**.

KRUSKAL'S ALGORITHM

We will now discuss a simple algorithm that will *always* find a minimum spanning tree for a connected weighted graph. This algorithm, known as **Kruskal's algorithm**,[1] is almost identical to the cheapest link algorithm in Chapter 6. It is nothing more than the result of combining greed with common sense.

Kruskal's Algorithm.

- ■ **Step 1.** Find the edge with the smallest weight in the graph (if there is more than one, pick one at random). Mark it using your favorite color. (Ours is red.)

- ■ **Step 2.** Find the next smallest unmarked (i.e., not red) edge in the graph. If it forms a circuit with already marked (red) edges, discard it. (Remember: We don't want any circuits!) If it doesn't, then mark it.

- ■ **Steps 3, 4, etc.** Repeat step 2 until there are no more unmarked edges to choose from. The marked edges form the desired minimum cost spanning tree.

Example 8. Let's apply Kruskal's algorithm to the original Amazon telephone network problem (Fig. 7-17).

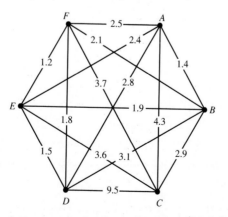

Figure 7-17 The costs of linking the various cities.

[1] The algorithm is named after Joseph Kruskal, a mathematician working at the Bell Laboratories, who discovered it in 1956, although there is evidence that unbeknown to Kruskal the algorithm had already been discovered by several other mathematicians in Czechoslovakia, Poland, and France.

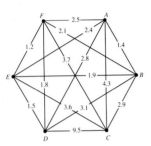

■ **Step 1.** The edge joining E and F at \$1.2 million is the cheapest edge in the graph. We mark it in red. (Note that this does not mean that the actual telephone lines between E and F have to be laid first. We are putting the network together on paper and will do so in a certain sequence. Once we have the network layout we can build it in any order we want.)

■ **Step 2.** The edge joining A and B is chosen next (\$1.4 million). We mark it in red.

■ **Step 3.** The edge DE (\$1.5 million) comes next. By all means, let's mark it in red.

■ **Step 4.** The next cheapest edge is DF (\$1.8 million), but DF closes a circuit with previously marked edges. We delete it, as shown in Fig. 7-18.

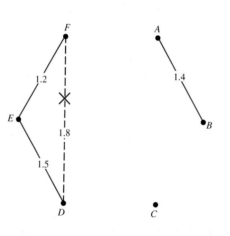

Figure 7-18 The network after step 4. Cities D, E, and F can communicate with each other without the link DF.

■ **Step 5.** Having deleted DF, we consider the next cheapest edge—BE (\$1.9 million). It closes no red circuits so we mark it in red.

■ **Step 6.** The next cheapest edge is BF (\$2.1 million), but BF closes a circuit with BE and EF and so we delete it.

■ **Step 7.** The next cheapest edge is AE (\$2.4 million), but it closes a circuit with AB and BE and so we delete it.

■ **Step 8.** The next cheapest edge is AF (\$2.5 million). It closes the circuit A, F, E, B, A and so we delete it.

■ **Step 9.** The next cheapest edge is AD (\$2.8 million). It also closes a circuit—in this case A, D, E, B, A. No dice!

■ **Step 10.** The next cheapest edge is BC (\$2.9 million). This edge does not close any circuits, so we mark it in red.

At this point all six cities are connected, and we are done. The actual solution is shown in Fig. 7-19 and has a total cost of $8.9 million. We call

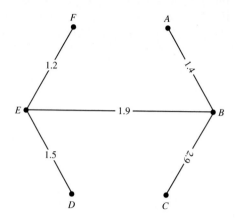

Figure 7-19 A minimum-cost spanning tree for the telephone network problem. Junction points for the network are *E* and *B*.

the reader's attention to a fact that will become relevant later in the chapter. Our solution network has two **junction points**: They are the cities *E* and *B*. As the reader has no doubt guessed, by a junction point we mean a place where several (i.e., more than one) lines in the network meet. ∎

Before concluding this section, let's apply Kruskal's algorithm one more time, in this case using a slightly more elaborate example. To speed things up we will use the word "redundant" to describe any edge that closes a circuit in our network-to-be.

Example 9. The graph in Fig. 7-20 shows all the possible connections (with the corresponding costs) for laying down a drip irrigation system.

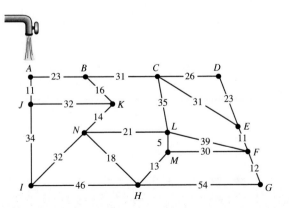

Figure 7-20

Vertex *A* is the main water source—all other vertices represent sprinkler-head locations. We want to lay down a network of pipes so that water from *A* can get to each of the sprinkler-heads. We also want to do the job as cheaply as possible.

Clearly, we are looking for a minimum spanning tree for the weighted graph in Fig. 7-20. We will use Kruskal's algorithm to find it.

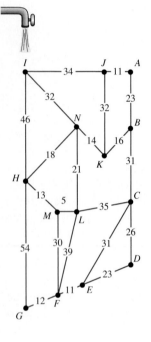

- **Step 1.** Mark *LM*. No redundant edges to delete.
- **Step 2.** Mark *AJ*. No redundant edges to delete.
- **Step 3.** Mark *EF*. No redundant edges to delete.
- **Step 4.** Mark *FG*. No redundant edges to delete.
- **Step 5.** Mark *HM*. No redundant edges to delete.
- **Step 6.** Mark *NK*. No redundant edges to delete.
- **Step 7.** Mark *BK*. No redundant edges to delete.
- **Step 8.** Mark *HN*. Delete *NL*.
- **Step 9.** Mark *AB*. Delete *JK*.
- **Step 10.** Mark *DE*. No redundant edges to delete.
- **Step 11.** Mark *CD*. Delete *CE*.
- **Step 12.** Mark *FM*. Delete *CL*, *FL*, *GH*, and *BC*.
- **Step 13.** Mark *IN*. Delete *IJ* and *HI*.

We are finished! The minimum spanning tree describing the cheapest irrigation system[2] (at a total cost of $234) is shown in Fig. 7-21. ■

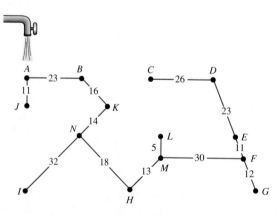

Figure 7-21

[2] The reader is warned that there are important practical aspects of laying down irrigation pipe (such as water pressure) which we have neglected to take into account.

The reader has no doubt noticed that there is a great deal of similarity between Kruskal's algorithm and the cheapest link algorithm we discussed in Chapter 6. In fact, they are both based on an identical strategy: Be cheap (i.e., always make the most inexpensive choice possible), but follow the rules for putting together the particular structure you need (a spanning tree or a Hamilton circuit, respectively). Our previous experience with this strategy is that there usually is a price to pay—greed in the short term can produce bad results in the long term. Kruskal's algorithm is, however a pleasant surprise. In this case the solutions we get with our greedy strategy are guaranteed to be optimal: There is no cheaper spanning tree than the one we get from Kruskal's algorithm, which means therefore that Kruskal's algorithm is an optimal algorithm.

With regard to its practicality, Kruskal's algorithm is an efficient algorithm. Just as with the cheapest link algorithm, it is not unreasonable to attempt a problem with hundreds of vertices by hand and one with hundreds of thousands of vertices by computer.

There are several other algorithms for finding a minimum spanning tree that are both efficient and optimal. (One of them, known as **Prim's algorithm**, is a modified version of the nearest neighbor algorithm we studied in Chapter 6—see Exercise 34 for details.)

SHORTEST NETWORKS AND STEINER POINTS

The general problem of finding minimum spanning trees is one of those rare examples of a management science problem in which things work out just right—we can find efficient algorithms that always give optimal solutions (who could ask for better karma?). We will now discuss a subtle variation of minimum spanning trees and discover how susceptible these kinds of problems are to minor changes.

Let's say that once again we want to construct a network that connects a set of cities (by means of highways, telephone lines, whatever) in the most efficient way possible. Up until now, in constructing such a network we have made the assumption that the only way to connect two vertices (i.e., cities) is either by a direct link (represented by one of the edges of the original graph) or through a sequence of links passing through other vertices. We can rephrase this by means of two equivalent statements: (1) Our solution network is a subgraph of the original graph, and (2) the only places in our solution where we can have a junction point are the original cities. Now suppose that we change the scenario a little bit and, in building the optimal network, we are allowed to introduce junction points at places other than the original vertices. Does this change the problem a lot? Let's look at a few examples.

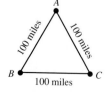

Figure 7-22

Example 10.   Three cities (*A*, *B*, and *C*) must be connected by a network of highways in an optimal way. Figure 7-22 shows the three cities

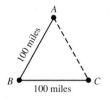

Figure 7-23 A minimum spanning tree network for the three cities. The total length of the network is 200 miles. The junction point (*B*) is one of the original cities.

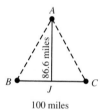

Figure 7-24 A highway network with junction point *J*. The total length of the network is approximately 186.6 miles.

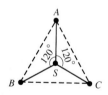

Figure 7-25 The shortest network connecting the cities *A*, *B*, and *C*. The junction point *S* is a Steiner point (three roads come in at 120° angles to each other). The total length of the network is approximately 173.2 miles.

as the vertices of a triangle, with the distances between them given along the sides of the triangle. To keep things simple we will start with an example in which the three cities are all equidistant from each other.

If we were to use our original approach to the problem (*A*, *B*, and *C* are the only possible junction points), then we would find a minimum spanning tree of the graph. In this example the solution is quite trivial— any two of the three edges can be chosen to form a minimum spanning tree. Figure 7-23 shows one possible solution with a total length of 200 miles. In this solution the only junction point is *B*.

Suppose now that we are allowed to choose junction points other than the original cities *A*, *B*, and *C*. Is there a better (shorter) highway network connecting these cities? It is not hard to see that the highway network shown in Fig. 7-24 is actually shorter than the one in Fig. 7-23. In fact, we know from elementary geometry that the length of this network is approximately 186.6 miles (Exercise 19(a)). In this network there is one junction point (*J*).

With a small investment of time and effort we can actually do better than Fig. 7-24. The network shown in Fig. 7-25 has a junction point *S* located in the center of triangle *ABC*. Each of the segments *SA*, *SB*, and *SC* has equal length (approximately 57.7 miles), so that the total length of the network is approximately 173.2 miles (Exercise 20). More significantly, the three branches of the network form 120° angles at the junction point *S*, and because of this the network cannot be made shorter. The network in Fig. 7-25 is the shortest possible highway network connecting the cities *A*, *B*, and *C*. (While this fact may seem intuitively clear, the proof is far from obvious. For details the reader is encouraged to pursue Exercise 36.)

A network such as the one shown in Fig. 7-25 is called a **shortest network** (there is no shorter one). A junction point such as *S* in which there are three edges forming 120° angles is called a **Steiner point**.[3] ■

To summarize what we learned in Example 10: If we do not require that the junction points of the network be located at one of the cities, then we may be able to do better than the minimum spanning tree solution. In the situation illustrated by this example the shortest network is one in which there is an interior junction point at which three roads come in at 120° angles (i.e., a Steiner point). This solution represents a savings of 26.8 miles or 13.4% over the 200 miles given by the minimum spanning tree solution shown in Fig. 7-23. (Notice that in finding our solution, geometry plays an important role, and therefore our figures must be drawn to exact scale.)

[3] Named after Jacob Steiner (1796–1863), a mathematician at the University of Berlin in the early nineteenth century.

Example 11. Suppose that once again we have three cities (*A*, *B*, and *C*) that must be connected by an optimal network of highways, but this time they are not equidistant from each other. Let's say for the sake of argument that the distances are *AB* = 100 miles, *AC* = 150 miles, and *BC* = 165 miles, as shown in Fig. 7-26(a). Can we improve on the minimum spanning tree solution [Fig. 7-26(b)] by introducing new junction points? Just as in Example 10, the secret is to find a Steiner point *S* from which we can fork three straight-line segments *SA*, *SB*, and *SC* at 120° angles, as shown in Fig. 7-26(c). Some straightforward geometric considerations guarantee that in this case the network will be as short as possible (Exercise 36). A somewhat more sophisticated argument is needed to compute the actual length of the network, which turns out to be approx-

Figure 7-26 (a) A map showing the original cities and the distances between them. (b) The minimum spanning tree solution with a total length of 250 miles. (c) The shortest network solution with a total distance of approximately 235.5 miles. The junction point *S* is a Steiner point. The savings over the solution shown in (b) is about 5.8%.

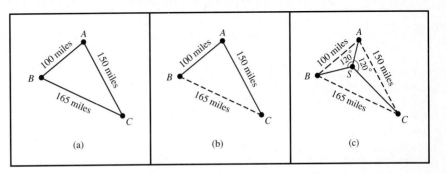

imately 235.5 miles. In this example, the shortest network solution represents a savings of 14.5 miles or about 5.8% over the minimum spanning tree solution (250 miles). ∎

Example 12. For the third (and last) time let's connect three cities *A*, *B*, and *C* by means of an optimal highway network. This time the distances are *AB* = 100, *AC* = 150, and *BC* = 220, as shown in Fig. 7-27(a). In

Figure 7-27 (a) A map showing the original cities and the distances between them. Angle *A* is approximately 122°. (b) The minimum spanning tree solution with a total length of 250 miles. (c) The shortest network *is the same as* the minimum spanning tree solution. No Steiner point is possible because angle *A* is more than 120°.

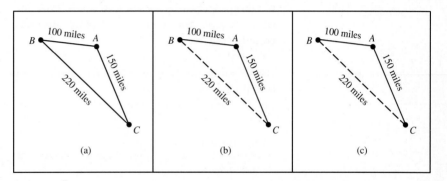

this case, because of the fact that the triangle *ABC* has one angle (angle *BAC*) that is more than 120°, it is impossible to find a Steiner junction point for the three cities. Here the minimum spanning tree solution [Fig. 7-27(b)] turns out to be the shortest network solution as well! ■

Let's summarize what we have learned from Examples 10 through 12 (not necessarily in order) about the shortest network connecting three vertices *A*, *B*, and *C*.

1. When one of the angles of the triangle *ABC* is 120° or more, then the shortest network solution is the same as the minimum spanning tree solution [Fig. 7-28(a)] and finding such a network is a piece of cake (pick the two shortest sides of the triangle).

Figure 7-28 The shortest network connecting *A, B,* and *C.* (a) Case 1: an angle of 120° or more. The junction point is one of the vertices. (b) Case 2: all angles less than 120°. The junction point *S* is a Steiner point.

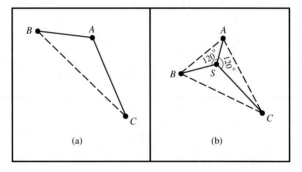

2. When all the angles of the triangle are less than 120°, then the shortest network solution and the minimum spanning tree solution part company. The shortest network solution is obtained by finding a Steiner junction point *S* and joining *S* to each of the vertices *A*, *B*, and *C* [Fig. 7-28(b)]. ■

There is one more detail that we need to deal with involving case 2: how to pinpoint the location of the Steiner junction point *S*. We will now describe a simple, elegant geometric procedure that will allow us to do exactly that.[4]

Procedure for finding a Steiner point for a triangle *ABC* with all angles less than 120°:

---

[4] This procedure can be traced back to the Italian physicist-mathematician Evangelista Torricelli (1608–1647). Torricelli was a disciple of Galileo and is best remembered for discovering the barometer.

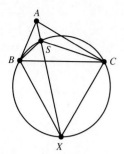

Figure 7-29 Because the triangle *BCX* is equilateral, the angle *BXC* is 60°. Because *S* is on the same circle as *B, C,* and *X* but "opposite" *X,* the angle *BSC* is 120°.

- ■ **Step 1.** On the largest side of the triangle (in Fig. 7-29 it is *BC*), build an equilateral triangle *BCX* as shown in Fig. 7-29.

- ■ **Step 2.** Circumscribe a circle around the equilateral triangle *BCX*.

- ■ **Step 3.** Join *X* to *A*. The point where the line segment *XA* intersects the circle is the desired Steiner point *S*.

The critical geometric facts that justify this construction are that the angle *BSC* equals 180° minus angle *BXC* and that angle *XBC* equals angle *XSC*. Since angles *BXC* and *XBC* are both 60°, it follows that not only angle *BSC* but also angles *BSA* and *CSA* are 120° and *S* is the desired Steiner point. As an additional point of interest, it turns out that the length of the entire network (*SA* + *SB* + *SC*) is exactly the same as the length of the segment *AX*.

STEINER TREES

Before we move on to a few more sophisticated examples, it is time to formalize our terminology. Suppose we have a set of points A_1, $A_2, . . ., A_N$ located in a plane (let's say on a map), and suppose we know the exact distances between any two points.

1. If we want to connect the points without introducing any new junction points other than the points $A_1, A_2, . . ., A_N$ themselves, the optimal way to do it is obtained by finding a **minimum spanning tree** for the points $A_1, A_2, . . ., A_N$. We learned how to do this in the first part of this chapter.

2. If we want an optimal network that connects these points in the shortest possible way while allowing (if necessary) the introduction of junction points other than the original points $A_1, A_2, . . ., A_N$, then we are talking about finding a **shortest network** solution.

3. A junction point at which three segments of a network join at 120° angles is called a **Steiner point** of the network.

4. A **Steiner tree** is a tree network connecting the original points A_1, $A_2, . . ., A_N$, such that all of the additional junction points are Steiner points.

In our last three examples we discovered that if we start with three points, the shortest network solution is either a Steiner tree with one Steiner point (Examples 10 and 11) or a minimum spanning tree with no Steiner points (Example 12). We will now look at some examples with four points.

Example 13. Four cities (A, B, C, and D) are to be connected by an optimal highway network. For starters, let's assume that the four cities form the vertices of a square as shown in Fig. 7-30(a). The minimum

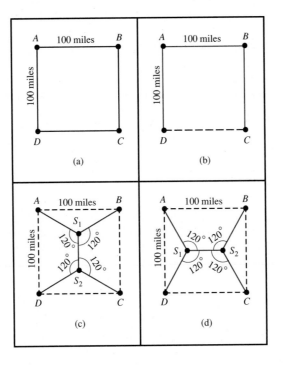

Figure 7-30 (a) Four cities located at the vertices of a square. (b) A minimum spanning tree solution with a total length of 300 miles. (c) A shortest network solution with a total length of approximately 273.2 miles. The junction points S_1 and S_2 are both Steiner points. (d) A different but obviously equivalent shortest network solution.

spanning tree for A, B, C, D is trivial—choose any three of the four sides of the square as shown, for example, in Fig. 7-30(b). The length of the minimum spanning tree is 300 miles. The shortest network solution is somewhat more difficult to find and is shown in Fig. 7-30(c). The total length of this network is approximately 273.2 miles (Exercise 29), which represents a savings of 26.8 miles or about 8.9% over the minimum spanning tree solution. We should not be surprised to see that the solution is once again a Steiner tree, this time with two Steiner points. It is also worth noting that this is the first time we have come across a situation in which the shortest network solution is not unique. Figure 7-30(d) shows another, obviously equivalent, shortest network solution for the vertices of the square. ∎

Example 14. Suppose once again that we have four cities A, B, C, and D, this time forming the vertices of a rectangle of sides $AB = CD = 160$ miles and $AD = BC = 120$ miles as shown in Fig. 7-31(a). The

Figure 7-31 (a) Four cities located at the vertices of a rectangle. (b) A minimum spanning tree solution with a total length of 400 miles. (c) A Steiner tree with two Steiner junction points but not a shortest network. The total length of this network is approximately 397.1 miles. (d) A different Steiner tree with two Steiner points. This is the shortest network solution. The total length of this network is approximately 367.8 miles.

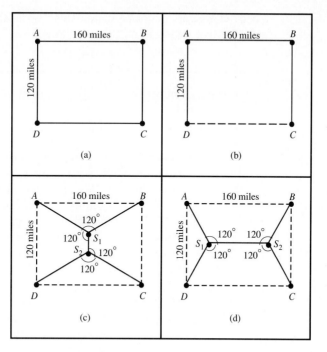

minimum spanning tree solution is (as usual) quite trivial and is given in Fig. 7-31(b). As for the shortest network solution, it is only natural that we should take a cue from what happened in Example 13 and try to find a Steiner tree with two Steiner points. It turns out that there are two different ways that we can do this, and they are shown in Figs. 7-31(c) and (d), respectively. What's interesting in this case is that these two networks are not equivalent: The total length of Fig. 7-31(c) is approximately 397.1 miles [Exercise 30(a)], while the total length of Fig. 7-31(d) is approximately 367.8 miles [Exercise 30(b)]. The network shown in Fig. 7-31(d) is the shortest network solution. It represents a savings of 32.2 miles or about 8% over the minimum spanning tree solution. ∎

Example 14 illustrates a new wrinkle in the search for a shortest network: Just because we have found a Steiner tree, there is no guarantee that we have also found a shortest network solution. In general, there are many different Steiner trees connecting the same set of points.

Armed with all these examples, a little more wisdom, and new terminology, we are ready (and presumably eager) to face the big question: Given an arbitrary set of points $A_1, A_2, A_3, \ldots, A_N$ located to scale in their exact geographical positions in a plane, how do we find the *shortest network* connecting all these points. The preceding examples seem to suggest that we should be looking to either a minimum spanning tree or a Steiner tree as the prime candidates for the shortest network. There is indeed a true general fact that guarantees this:

> The shortest network connecting a set of points is either a Steiner tree or a minimum spanning tree.

So far, so good—things are looking promising. In theory, all we have to do now is find all possible Steiner trees, find the minimum spanning tree, and determine which of these gives the shortest network. The problem, however, is that while finding the minimum spanning tree is easy (just use Kruskal's algorithm with the distances between points as the weights of the edges), finding all the Steiner trees is not. Nature provides a particularly slick way to find Steiner trees by using a soap bubble solution (see the Appendix), but there are no guarantees that every possible Steiner tree can be found this way. Even when the number of points is as small as 10, the number of possible Steiner trees can be in the thousands. Finding the one that gives the shortest network is like finding a needle in a haystack, and no efficient algorithm for doing this is currently known. At present, for example, finding the shortest network connecting 1000 points can be beyond the reach of even the most powerful computers.

The similarity between the shortest network problem and the traveling salesman problem (Chapter 6) is not entirely coincidental. Finding the shortest network and finding the optimal Hamilton circuit connecting a set of points are problems of a similar ilk, and they stand shoulder to shoulder as two of the great unmet challenges of modern mathematics.

CONCLUSION

In this chapter we discussed the problem of interconnecting a set of points in a network in an optimal way, where "optimal" usually means least expensive or shortest. In practice, the points represent geographical locations (cities, pumping stations, sprinkler heads, etc.), and the connections can be highways, pipelines, telephone lines, etc. Depending on the circumstances, we considered two different ways of doing this.

Version 1. In the first half of the chapter we were required to build the network in such a way that no junction points other than the original locations were allowed, in which case the optimal network is a *minimum spanning tree.* Finding a minimum spanning tree was the great success story of the chapter—Kruskal's algorithm provides the answer in an efficient and optimal way.

Version 2. In the second half of the chapter we considered what on the surface appeared to be a minor modification by removing the prohibition against new junction points. In this case the problem becomes one of finding the shortest possible network connecting the points, and we found that the solution can once again be the minimum spanning tree but most often is a new type of network called a Steiner tree. No efficient optimal algorithm is presently known for finding the shortest network connecting an arbitrary set of points, but for practical purposes there are several ways of getting good approximate solutions.

As a matter of fact, in 1990, two Chinese mathematicians, Frank Hwang of Bell Laboratories and Ding-Zhu Du of Princeton University were able to prove that the shortest network connecting any set of points is never more than 13.4% shorter than the minimum spanning tree connecting the same set of points (and in most cases the savings is much less than that).[5] This means that Kruskal's algorithm not only provides an optimal solution in version 1 of the problem but also does a very good job of providing an approximate solution to the notoriously more difficult problem in version 2.

KEY CONCEPTS

junction points	**spanning tree**
Kruskal's algorithm	**Steiner points**
minimum network problems	**Steiner tree**
minimum spanning tree	**subgraph**
shortest network	**tree**

EXERCISES

■ **Walking**

1. For each of the following graphs determine whether the graph is a tree. If it is not a tree, give a reason why.

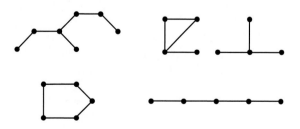

[5] See reference 6.

2. For each of the following graphs determine whether the graph is a tree. If it is not a tree, give a reason why.

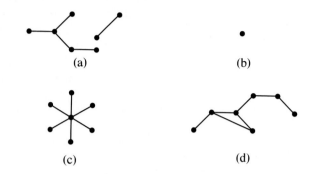

(a)   (b)

(c)   (d)

3. Find all the *different* (unlabeled) trees with
   (a) two vertices
   (b) four vertices
   (c) five vertices.

4. Find all the *different* (unlabeled) trees with
   (a) three vertices
   (b) six vertices.

5. Find a spanning tree for each of the following graphs.

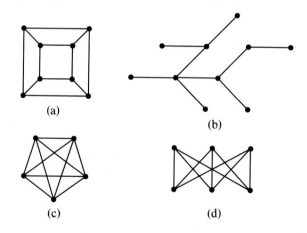

(a)   (b)

(c)   (d)

6. Find two different spanning trees for each of the following graphs.

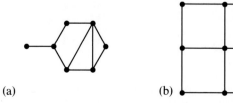

(a)   (b)

7. Find all the possible spanning trees of the following graph.

8. Find all the possible spanning trees of the following graph.

9. Use Kruskal's algorithm to find a minimum spanning tree of the following weighted graph.

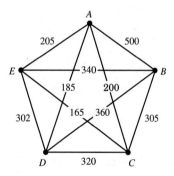

10. Use Kruskal's algorithm to find a minimum spanning tree of the following weighted graph.

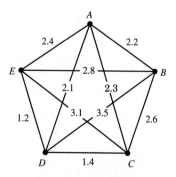

11. Use Kruskal's algorithm to find a minimum spanning tree of the following weighted graph.

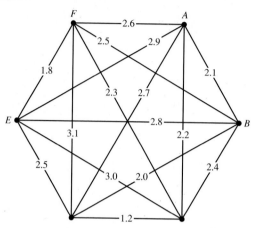

12. Use Kruskal's algorithm to find a minimum spanning tree of the following weighted graph.

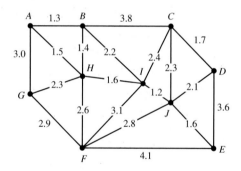

Exercises 13 and 14 refer to the following map. In the figure, AC = AB, DC = DB, and angle CDB = 120°.

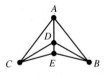

13. (a) Which is larger, *CD* + *DB* or *CE* + *ED* + *EB*? Explain.
    (b) What is the shortest network connecting the points *C, D,* and *B*? Explain.
    (c) What is the shortest network connecting the points *C, E,* and *B*? Explain.

14. (a) Which is larger, *CA* + *AB* or *DC* + *DA* + *DB*? Explain.
    (b) Which is larger, *EC* + *EA* + *EB* or *DC* + *DA* + *DB*? Explain.
    (c) What is the shortest network connecting the points *A, B,* and *C*? Explain.

15. Find the length of the shortest network connecting the points *A, B,* and *C* shown in the map at the top of the next page.

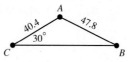

16. Find the length of the shortest network connecting the points A, B, and C shown in the following map.

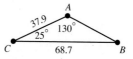

Exercises 17 through 20 refer to an equilateral triangle ABC with sides of length 100 and Steiner point S as shown in the following figure.

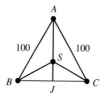

17. Show that triangles ASB, BSC and CSA are congruent.

18. Show that $SA = SB = SC$. (*Hint*: Use Exercise 17.)

19. Use your knowledge of $30°$–$60°$–$90°$ triangles to find the lengths of
(a) JA
(b) SA

20. Show that the length of the shortest network connecting A, B and C is approximately 173.2. (*Hint*: Use Exercise 19(b).)

■ **Jogging**

21. Five cities (A, B, C, D, E) are located as shown on the following map. The five cities need to be connected by a railroad, and the cost of building the railroad system connecting any two cities is proportional to the distance between the cities. Find the length of the railroad network of cheapest cost (assuming that no additional junction points can be added).

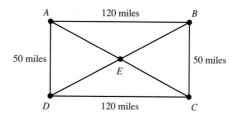

22. Explain why if G is a connected graph with N vertices, Kruskal's algorithm will require exactly $N-1$ steps.

23. Explain why in a tree, every edge is a bridge. (Recall that a bridge is an edge whose removal disconnects the graph.)

24. (a) How many different spanning trees does the following graph have?

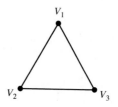

(b) How many different spanning trees does the following graph have?

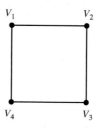

(c) How many different spanning trees does the following graph have?

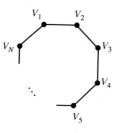

25. (a) Can you give an example of a tree with four vertices such that the degrees of the vertices are 2, 2, 3, and 3? If yes, do so. If not, explain why not.
   (b) If you have a tree with four vertices and you add up the degrees of all the vertices, what do you get?
   (c) If you have a tree with five vertices and you add up the degrees of all the vertices, what do you get?
   (d) If you have a tree with N vertices and you add up the degrees of all the vertices, what do you get?

26. (a) Give an example of a tree with four vertices such that the degrees of the vertices are 1, 1, 2, and 2.
   (b) Give an example of a tree with six vertices such that the degrees of the vertices are 1, 1, 2, 2, 2, and 2.
   (c) Give an example of a tree with N vertices such that the degrees of the vertices are 1, 1, 2, 2, 2, . . ., 2.

27. **(a)** Give an example of a tree with four vertices such that the degrees of the vertices are 1, 1, 1, and 3.

   (b) Give an example of a tree with five vertices such that the degrees of the vertices are 1, 1, 1, 1, 4.

   (c) Give an example of a tree with N vertices such that the degrees of the vertices are 1, 1, 1, . . ., 1, $N-1$.

28. A highway system connecting nine cities C_1, C_2, C_3, . . ., C_9 is to be built. Use Kruskal's algorithm to find a minimum spanning tree for this problem. The accompanying table shows the cost (in millions of dollars) of putting a highway between any two cities. (Draw the weighted graph only if you have to!)

	C_1	C_2	C_3	C_4	C_5	C_6	C_7	C_8	C_9
C_1	*	1.3	3.4	6.6	2.6	3.5	5.7	1.1	3.8
C_2	1.3	*	2.4	7.9	1.7	2.3	7.0	2.4	3.9
C_3	3.4	2.4	*	9.9	3.4	1.0	9.1	4.4	6.5
C_4	6.6	7.9	9.9	*	8.2	9.7	0.9	5.5	4.9
C_5	2.6	1.7	3.4	8.2	*	4.8	7.4	3.7	3.5
C_6	3.5	2.3	1.0	9.7	4.8	*	8.9	4.4	5.8
C_7	5.7	7.0	9.1	0.9	7.4	8.9	*	4.7	3.9
C_8	1.1	2.4	4.4	5.5	3.7	4.4	4.7	*	2.8
C_9	3.8	3.9	6.5	4.9	3.5	5.8	3.9	2.8	*

29. Use 30°-60°-90° triangles to show that the length of the following network is $100\sqrt{3} + 100 \approx 273.2$.

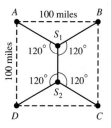

30. **(a)** Use 30°-60°-90° triangles to show that the length of the following network is $160\sqrt{3} + 120 \approx 397.1$.

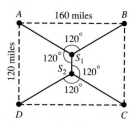

(b) Show that the length of the following network is $120\sqrt{3} + 160 \approx 367.8$.

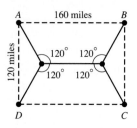

31. Consider triangle *ABC* with equilateral triangle *EFG* inside as shown in the following figure.
   (a) Find angles *BFA*, *AEC*, and *CGB*.
   (b) Explain why all the angles of triangle *ABC* are less than 120°.
   (c) Explain why the Steiner point for triangle *ABC* lies inside triangle *EFG*.

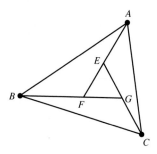

■ Running

32. (a) Show that in any tree with two or more vertices, there are at least two (maybe more) vertices of degree 1. [*Hint:* Use Exercise 25(d).]
   (b) Explain why in any tree with three or more vertices, it is impossible for all the vertices to have the same degree.

33. (a) Suppose that you are asked to find a minimum spanning tree for a weighted graph that must contain a given edge. Describe a modification of Kruskal's algorithm that accomplishes this.
   (b) Consider Exercise 28 again. Suppose that C_3 and C_4 are the two largest cities in the area and the chamber of commerce insists that the section of highway directly connecting them must be built (or heads will roll). Find the minimum spanning tree that includes the section of highway between C_3 and C_4.

34. Prim's algorithm for finding a minimum spanning tree:

   ■ Step 0. Pick any vertex as a starting vertex. (Call it *S*). Mark it red.

   ■ Step 1. Find the nearest neighbor of *S* (call it P_1). Mark both P_1 and the edge SP_1 in red.

■ Step 2. Find the nearest black neighbor to the red subgraph (i.e., the closest vertex to *any* red vertex). Mark it and the edge connecting the vertex to the red subgraph in red. Delete all black edges in the graph that connect two red vertices.

Repeat Step 2 until all the vertices are marked red. The red subgraph is a minimum spanning tree.

Let's quickly show how Prim's algorithm works with the graph in Fig. 7-1. Let's arbitrarily choose *B* as the starting point and mark it red.

■ Step 1. The nearest neighbor to *B* is *A*. Mark vertex *A* and edge *AB* in red.

■ Step 2. The nearest black neighbor to a red vertex is *E* because of all the edges coming out of either *A* or *B*, *EB* is the nearest (at $1.9 million). Mark vertex *E* and edge *EB* red. Delete *EA*.

■ Step 3. The nearest black neighbor to a red vertex is *F* at $1.2 million (*FE*). Mark *F* and *FE* in red. Delete *FA* and *FB*.

■ Step 4. The nearest black neighbor to a red vertex is *D* at $1.5 million (*DE*). Mark *D* and *DE* in red. Delete *DA*, *DB*, and *DF*.

■ Step 5. The nearest black neighbor to a red vertex is *C* at $2.9 million (*CB*). Mark *C* and *CB* in red. All the vertices are now marked in red, so we are finished.

Use Prim's algorithm to find a minimum spanning tree for the drip irrigation system problem given in Example 9.

35. Consider four points (*A*, *B*, *C*, and *D*) forming the vertices of a rectangle with length *b* and width *a* as shown in the following figure ($a < b$). Determine the conditions on *a* and *b* so that the length of the minimum spanning tree is less than the length of *one* of the (two) Steiner trees.

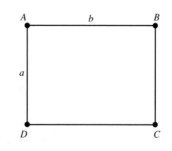

36. (a) Let *P* be an arbitrary point inside the equilateral triangle *RQT* as shown below. Suppose also that perpendiculars are drawn from *P* to the three sides as shown in the figure. Show that the sum of the lengths of *PA*, *PB*, and *PC* is equal to the length of the altitude of the triangle. (*Hint:* Compute the areas of triangle *RPQ*, triangle *QPT*, and triangle *TPR*. How does the sum of these three areas compare with the area of triangle *RQT*?)

(b) Let triangle ABC be an arbitrary triangle with all angles less than 120°, and let S be a Steiner point inside the triangle. Draw lines perpendicular to SA, SB, and SC. Explain why these lines intersect to form an equilateral triangle RQT.

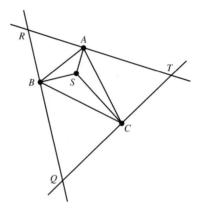

(c) Let P be any point other than S in triangle ABC and draw perpendiculars from P to the three sides of triangle RQT as shown. Use part (a) to conclude that $PA' + PB' + PC' = SA + SB + SC$ and yet $PA \geq PA'$, $PB \geq PB'$, and $PC \geq PC'$ (why?), and so $PA + PB + PC \geq SA + SB + SC$.

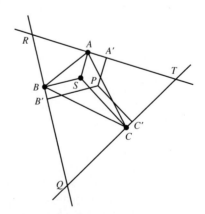

(d) Use part (c) to conclude that no junction point P can give a shorter network than the Steiner point S.

APPENDIX:
The Soap Bubble
Solution

Every child knows about the magic of soap bubbles: Take a wire or plastic ring, dip it into a soap-and-water solution, blow, and presto—beautiful iridescent geometric shapes magically materialize to delight and inspire our fantasy. Adults are not adverse to a puff or two themselves.

What's special about soapy water that makes this happen? A very simplistic understanding of the forces of nature that create soap bubbles will help us understand how these same forces can be used to find (imagine of all things) Steiner trees connecting a given set of points.

Take a liquid (any liquid) and put it into a container. When the liquid is at rest, there are two categories of molecules: those that are on the surface and those that are below the surface. The molecules below the surface are surrounded on all sides by other molecules like themselves and are therefore in perfect balance—the forces of attraction between molecules all cancel each other out. The molecules on the surface, however, are only partly surrounded by other molecules and are therefore unbalanced. Because of this, an additional force called **surface tension** comes into play for these molecules. As a result of this surface tension, the surface layer of any liquid behaves exactly as if it were made of a very thin, elastic material. The amount of elasticity of this surface layer depends on the structure of the molecules in the liquid. Soap or detergent molecules are particularly well suited to create an extremely elastic surface layer. An ideal soap film solution can be obtained by mixing approximately equal parts of dishwashing detergent and water and adding a small amount of glycerin (not nitro please!) to the solution.

The connection between the preceding brief lesson in soapy solutions and the material in this chapter is made through one of the fundamental principles of physics: A physical system will remain in a certain configuration only if it cannot *easily* change to another configuration that uses less energy. Because of its extreme elasticity, the surface layer of a soapy solution has no trouble changing its shape until it feels perfectly comfortable—i.e., at a position of relatively minimal energy. Ergo Steiner trees.

Suppose that we have a set of points $(A_1, A_2, \ldots, A_N)$ for which we want to find a shortest network. We can find a Steiner tree that connects these points by means of an ingenious device which we will call a *soap bubble computer*. To begin with, we draw the points $A_1, A_2, \ldots, A_N$ to exact scale on a piece of paper. The scale should be such that points are not too close to each other (a minimum of 1 or 2 inches apart will do just fine). We now take two sheets of Plexiglas or Lucite and, using the paper map as a template, drill small holes on both sheets of Plexiglass at the exact locations of the points. Then we put thin metal or plastic pegs through the holes in such a way that the two sheets are held about an inch apart. The whole device should look something like the following figure.

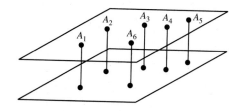

When we dip our device into a soap-and-water solution and pull it out, the soap bubble computer goes to work. The film layer that is formed between the plates connects the various pegs. For a while it will move seeking a configuration of minimal energy. Very shortly, it settles into a Steiner tree.

It is a bit of a disappointment that the Steiner tree we get is not necessarily the shortest network. The reasons for this are beyond the scope of our discussion. (The interested reader is referred to the excellent technical discussion of soap film computers in reference 1) At the same time, we should be thankful for what nature has provided: a simple device that can compute in seconds what might take hours to do with pencil and paper.

REFERENCES AND FURTHER READINGS

1. Almgren, Fred J., Jr., and Jean E. Taylor, "The Geometry of Soap Films and Soap Bubbles," *Scientific American*, 235 (July 1976), 82—93.

2. Bern, M., and R. L. Graham, "The Shortest Network Problem," *Scientific American*, 260 (January 1989), 84—89.

3. Cheriton, D., and R. E. Tarjan, "Finding Minimum Spanning Trees," *SIAM Journal on Computing*, 5 (1976), 724—742.

4. Cockayne, E. J., and D. E. Hewgill, "Exact Computation of Steiner Minimal Trees in the Plane," *Information Processing Letters*, 22 (1986), 151—156.

5. Courant, R., and H. Robbins, *What is Mathematics?* New York: Oxford University Press, 1941.

6. Du, D.-Z. and F. K. Hwang, "The Steiner ratio conjecture of Gilbert and Pollak is true," Proceedings of the National Academy of Sciences, USA, 87 (December 1990), 9464—9466.

7. Gilbert, E. N., and H. O. Pollak, "Steiner Minimal Trees," *SIAM Journal of Applied Mathematics*, 16 (1968), 1—29.

8. Graham, R. L., and P. Hell, "On the History of the Minimum Spanning Tree Problem," *Annals of the History of Computing*, 7 (January 1985), 43—57.

9. Kruskal, J. B., Jr., "On the Shortest Spanning Subtree of a Graph and the Traveling Salesman Problem," *Proceedings of the American Mathematical Society*, 7 (1956), 48—50.

10. Pierce, A. R., "Bibliography on Algorithms for Shortest Path, Shortest Spanning Tree and Related Circuit Routing Problems (1956—1974)," *Networks*, 5 (1975), 129—149.

8 Scheduling Problems

What is the hardest task in the world? / To think.

RALPH WALDO EMERSON

Directed Graphs and Critical Paths

Modular Homes Corporation has just announced their new line of modular homes called WIGWAMs. According to the company these luxurious new models represent a major breakthrough in modular home building—only 15 basic individual steps are needed for the construction of one of these revolutionary new homes. The company expects to announce its marketing plans and prices soon.

BUSINESS BRIEFS (September **2025**)

General Robotics Corporation has just announced the development of their latest special-purpose robot—the Home Builder 100 (HB100). HB100 robots are specially designed to perform any of the jobs required to build a house and are particularly effective when used for the building of modular homes. The company expects to announce its marketing plans and prices soon.

BUSINESS BRIEFS (October **2025**)

General Robotics Corporation and Modular Homes Corporation have just announced their merger. The new company, Robotic Homes Corporation, will build both the components that go into assembling WIG-

WAM homes as well as HB100 robots that can be used to do the assembling. The new company has announced that its marketing strategy will be to sell the WIGWAM components and rent the HB100 robots to the general public. The company has also announced that it will launch a major ad campaign based on the theme "Build your own dream home, but let our robots do the work." Prices for the rental of HB100 robots have been set as follows: one robot will rent for $1000 an hour; two robots will rent for $1800 an hour, and three robots will rent for $2400 an hour.

BUSINESS BRIEFS (November **2025**)

It is now the year 2026 and we are in the market for a new home. The prospect of no more plumbers, carpenters, electricians, and sundry other contractors to deal with sounds enticing, so we decide to build ourselves a WIGWAM home. Actually this is just a figure of speech because we are really going to let the HB100 robots do the building. All *we* have to do is decide how many robots we are going to rent and then organize the little critters to carry out the various steps required in putting the home together. Needless to say, we want to do the job as efficiently (cheaply) as possible. Sounds simple enough, but once again, appearances are deceptive.

SCHEDULING PROBLEMS

The problem of scheduling a certain number of machines (in our case HB100 robots) to carry out a series of interrelated jobs that make up some sort of complex project (in our case building a WIGWAM home) is typical of a type of problem in management science called a **scheduling problem**. The elements of a typical scheduling problem are

1. A set of **processors**. These are the objects that do the work. Processors can be machines, humans, and even teams consisting of a combination of machines and/or people.

2. A set of **tasks**. These are the individual jobs that make up the project. (By an individual job we mean a job that cannot be further broken down into smaller jobs in any reasonable or convenient way.) Associated with each task is a number called the **processing time**. It represents the amount of time needed by one processor to carry out that task.

3. A set of **precedence relations**. These are restrictions on the order in which the tasks can be carried out. A typical precedence relation is of the form "task *X* precedes task *Y*," and it means that task *Y* cannot be started until task X is completed. This precedence relation can be conveniently abbreviated by writing $X \rightarrow Y$.

Based on the above three elements, there are different kinds of scheduling problems that can come up. In our example (once we make the decision on how many robots we will rent), we have a fixed number of processors at our disposal, and the name of the game is to schedule the tasks so that the entire project is finished as early as possible. Problems of this type are called **minimal-length scheduling problems**. A different type of scheduling problem arises when we are given a certain time deadline by which we must complete a project and we want to find the fewest number of processors that will be needed to ensure that the project is completed by that deadline. A scheduling problem of this type is called a **bin-packing problem**. For the sake of simplicity, in this chapter we will focus on minimal-length scheduling only. Bin-packing problems, while very similar on the surface, require altogether different methods of solution. To keep things brief and simple from now on, we will use the term "scheduling problem" to describe minimal-length scheduling. Since this is the only kind of scheduling we will discuss in the chapter, this should cause no confusion.

To summarize, for the rest of this chapter we will discuss scheduling problems of the following type: We are given a certain number of processors, a certain number of tasks (and with each task a processing time), and a set of precedence relations governing the tasks. We are asked to find a **schedule** (i.e., a complete description of which processors will perform which tasks and in what order) such that the **project length** (i.e., the amount of time that elapses from the start of the project to the completion of the last task) is as short as possible.

Problems of this type can be (and often are) very difficult to solve, and even with the most powerful computers, finding an optimal schedule can be hopelessly complicated. To make things a bit easier, it is customary to make some simplifying assumptions. They are as follows (the titles are for fun only):

1. All processors are created equal. Assumption 1 says that there is no difference between the processors. Any given task can be carried out by any of the processors in exactly the same amount of processing time. (Generally, this is true when dealing with robots, machines, and even highly automated assembly-line types of human activities.)

2. Too many cooks spoil the broth. Assumption 2 says that each task is to be carried out by one and only one processor.

3. No preempting. Assumption 3 says that once a processor starts a task, it will continue with that task until it is completed.

4. Processors don't procrastinate. Assumption 4 says that a processor should not be idle unless absolutely necessary. In other words, if a processor finishes a task and there are one or more tasks that can be started at that time, the processor should go ahead and start one of them.

Real-life scheduling situations that (more or less) satisfy these assumptions can range from the deadly serious (such as the scheduling of computer programs in an antiballistic missile defense system) to the somewhat mundane (such as the problem of scheduling mechanics in a garage) to the frivolous (such as the problem of scheduling ovens in a bakery).

■ The Making of a WIGWAM

From page 1 of the WIGWAM manual:

Building a WIGWAM involves 15 basic steps as described below. The number next to each individual task indicates the number of hours needed by an HB100 robot to carry out that task.

F:	SET FOUNDATION	7 HOURS
AF:	ASSEMBLE FLOORING	5 HOURS
AW:	ASSEMBLE WALLS	6 HOURS
AR:	ASSEMBLE ROOF	8 HOURS
IF:	INSTALL FLOORS	5 HOURS
IW:	INSTALL WALLS	7 HOURS
IR:	INSTALL ROOF	5 HOURS
PL:	PLUMBING	4 HOURS
E:	ELECTRICAL	4 HOURS
HC:	INSTALL HEATING AND COOLING SYSTEM	3 HOURS
PT:	PAINTING	4 HOURS
C:	INSTALL CABINETS	1 HOUR
FC:	FINISH WORK, CARPENTRY	6 HOURS
FM:	FINISH WORK, MASONRY	3 HOURS
EC:	INSTALL ENVIRONMENT CONTROL SYSTEM (lighting, stereo, TV, computers, alarm)	2 HOURS
	TOTAL	70 HOURS

From page 2 of the WIGWAM manual:

The following table shows the way in which the 15 steps are inter-related.

Steps That Must Be Finished	Before	This Step Can Be Started
F, AF		IF
IF, AW		IW
AR, IW		IR
IF		PL
IW		E
E, IR		HC
E		PT
PL, PT		C
HC, PT		EC
C		FC
PT		FM

Good luck!

DIRECTED GRAPHS

There is a convenient way to describe all the information presented on pages 1 and 2 of the WIGWAM manual and it is shown in Fig. 8-1. In this figure, the tasks are represented by the vertices and the precedence relations are represented by the arrows, so that an arrow pointing from vertex *X* to vertex *Y* indicates that task *X* must be completed before task *Y* can be started.

Figure 8-1

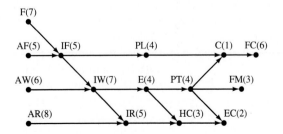

Notice that Fig. 8-1 looks a lot like an ordinary graph except that the edges point in a certain direction. A graph in which the edges have a direction associated with them is called a **directed graph**, and more commonly a **digraph**.

To distinguish digraphs from ordinary graphs, we use a slightly different terminology. Instead of "edge," we use the word **arc** to indicate the edge has a certain direction; instead of saying that vertices X and Y are adjacent, we say that X is **incident to** Y if the arc's arrowhead points toward Y, and that X is **incident from** Y if the arc's arrowhead points toward X; in addition to the degree of a vertex we speak about the indegree and the outdegree of a vertex (the **indegree** is the number of arrowheads pointing toward the vertex; the **outdegree** is the number of arrowheads coming out of the vertex.)

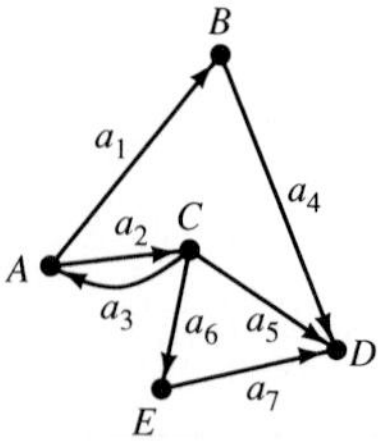

Figure 8-2

Example 1. Consider the digraph in Fig. 8-2. This is a digraph with five vertices (A, B, C, D, and E) and seven arcs (a_1, a_2, a_3, a_4, a_5, a_6, and a_7). An arc such as a_4, for example, indicates that a vertex (B) is incident to another vertex (D).

Notice that it is perfectly OK to have a situation like the one involving vertices A and C where A is incident to C (a_2) and C is incident to A (a_3). The indegrees and outdegrees of each vertex in Fig. 8-2 are shown in the following table.

Vertex	Indegree	Outdegree	Total Degree
A	1	2	3
B	1	1	2
C	1	3	4
D	3	0	3
E	1	1	2

In a digraph, a **path** from vertex X to vertex Y is a directed sequence of arcs starting at X and ending at Y. In Fig. 8-2, there are several paths from vertex A to vertex D (a_1, a_4, or a_2, a_6, a_7, or even a_2, a_3, a_1, a_4), but there is no path from vertex B to vertex E (there is only one way to *get out* of B—going to D—and there is no way to get out of D).

Just like graphs, digraphs are used to describe relationships between objects, but in this case the relationships do not necessarily flow both ways. When object X is related to object Y, we show this by making vertex X incident to vertex Y (i.e., there is an arc going from X to Y). The digraph in Fig. 8-2 shows, for example, that C is related to D but D

is not related to C. On the other hand, the relationship between A and C is mutual.

There are many real-life situations that can be represented by digraphs. Here are a few examples:

1. **Transportation.** Here the vertices might represent locations within a city, and the arcs might represent one-way streets.

2. **Communication.** Here the vertices might represent message centers or message sources, and the arcs the possible flow of information.

3. **Pipelines.** Here the vertices might represent pumping stations, and the arcs the direction of flow.

4. **Chain of command.** In a corporation or in the military we can use a digraph to describe the chain of command. The vertices are individuals, and an arc from X to Y indicates that X can give orders (is a superior) to Y.

5. **Asymmetric relationships.** Asymmetric relationships are those that are not always reciprocated. Being in love is a good example. We can describe asymmetric relationships by a digraph. In the case of being in love, the vertices are individuals, and we put an arc from X to Y if X is "in love" with Y. Sometimes there might also be an arc from Y to X, and sometimes (sigh) there might not.

PROJECT DIGRAPHS

Let's return now to the problem of scheduling the construction of a WIG-WAM home. As indicated earlier, the relevant information is summarized in the digraph shown in Fig. 8-1. In this case the vertices of the digraph represent the various tasks, and the arcs the precedence relations among the tasks. Next to each vertex label we have added in parentheses the processing time for that particular task based on the use of HB100 robots as processors. Figure 8-3 shows a slight modification of Fig. 8-1. Here we have added two fictitious tasks, START and END, with processing time 0. This is just a convenience that allows us to visualize the entire project as a flow that begins at START and concludes at END. By giving

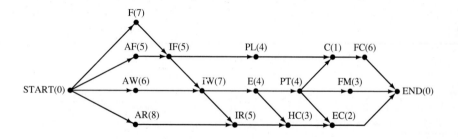

Figure 8-3

these fictitious tasks zero processing time we are not affecting the time calculations for the project. A digraph of the kind shown in Fig. 8-3 is called a **project digraph**.

The project digraph is the basic tool that will help us create a suitable schedule for the completion of our project. As we mentioned, we can visualize the project as a time flow, starting at the vertex START at time 0 and ending at the vertex END at some time X which is to be determined. The game has one basic rule: We cannot move to a new task until all the other tasks incident to the new task have been completed.

We will organize our analysis based on the number of processors available.

■ The Case of the Single Processor

If we are working with just one processor, there isn't much to making out a schedule. We make our single processor (in our example, it's an HB100 robot named Max) do all the tasks, one after the other, making sure that all the precedence relations are respected. Figures 8-4 and 8-5 show two different possible schedules for Max.

Figure 8-4 A single-processor schedule for building a WIGWAM.

Tasks	AR	AW	AF	F	IF	IW	IR	PL	E	PT	HC	EC	FM	C	FC
Time 0	8	14	19	26	31	38	43	47	51	55	58	60	63	64	70

Figure 8-5 A different schedule for a single processor.

Tasks	AW	AF	AR	F	IF	PL	IW	E	IR	HC	PT	EC	FM	C	FC
Time 0	6	11	19	26	31	35	42	46	51	54	58	60	63	64	70

In both schedules, the total time required for the completion of the project is 70 hours. (At $1000 per hour, the robot rental cost for building a WIGWAM would be $70,000.) This is the best one can do, since there is a total of 70 hours worth of work and only one processor to do it all. This situation holds in all scheduling problems involving only one processor: *The project length is always the sum of the processing times for all the tasks.*

Just because we have lots of freedom in making up a schedule for a single processor doesn't mean we can do anything we want. Figure 8-6

Figure 8-6 An *illegal* schedule for building a WIGWAM.

Tasks	AR	AW	AF	F	IF	IR	IW	PL	E	PT	HC	EC	FM	C	FC
Time 0	8	14	19	26	31	36	43	47	51	55	58	60	63	64	70

shows a schedule that is illegal. (Task IR is listed ahead of task IW, in violation of one of the precedence relations.)

■ Two Processors

Let's consider what happens if we try to schedule the building of a WIGWAM using two HB100 robots named Xavier and Zygmund (X and Z for short). Right off the bat we know that the best we can hope for is a project length of 35 hours. (A total of 70 hours of work and two robots to do it.) At the same time, because of the precedence relations, we might not be able to keep both robots working continuously.

Let's start by making up a schedule without using any strategy at all—what we might call the flying by the seat of the pants approach. Even in this situation we must be careful not to violate any of the precedence relations. Figure 8-7 shows a schedule one might come up with.

Figure 8-7 A possible schedule for two processors.

Time	0		8	15	20		32	3536	38		44
X		AR	F	IF	IW		IR	HC | c | EC		FC	
Z		AW	AF	idle	idle	PL	idle	E	PT	FM	idle
Time	0		6	11	15	20	24	27	31	35	38	44

The schedule shown in Fig. 8-7 has a project length of 44 hours. The labor cost for building the WIGWAM (at a rental price of $1800 per hour for a pair of robots) would be 44 × $1800 = $79,200. We can see that while the schedule is perfectly legal, it is not a very good one: Zygmund is sitting around doing nothing for a significant number of hours (18 to be exact). Moreover, it is clear that this is not the result of laziness on the part of the robot (we won't put up with that!) but rather of poor scheduling on our part. Look, for example, at what happens at the eleventh hour. Zygmund has just finished assembling the floor (AF) and is ready for a new task. Unfortunately there are no jobs available at the moment, and there won't be any until Xavier finishes with the foundation (F) and installing the floors (IF). To compound our bad luck, it so happens that F and IF are pretty long tasks, so Zygmund will have to sit idle for a relatively long time (9 hours *is* a long time to waste when you are paying $900 per robot per hour). A similar situation comes up near the end when there is just one job to be done (FC) and Zygmund has to sit idle for a full 6 hours.

THE DECREASING TIME ALGORITHM

The first lesson that we learn from Fig. 8-7 is that it might make sense to schedule the longer jobs as early as possible when there is a little more elbow room, so that if a processor needs to sit idle later, it will at least be idle for a shorter period of time. This simple strategy is the basis for an actual scheduling algorithm called the decreasing time algorithm.

Basically, the **decreasing time algorithm** says that whenever a processor is free, it should choose from among all the tasks that are available at that moment the one with the longest processing time. (In case of a tie we will make things simple and break the tie at random.)

Let's apply the decreasing time algorithm to schedule the building of a WIGWAM with our two robot friends, but before we do so let's discuss a little bookkeeping. When applying this algorithm, it will be helpful to keep track of the following things: (1) the time, (2) which processors are free to start a new task at that time, and (3) which tasks are eligible to be started at that time. To keep track of items 1 and 2 we will use a **time table** like the one shown in Fig. 8-7. As a task is assigned to a processor, we enter that task and its processing time on the time table. To keep track of item 3 we will use the project digraph. Whenever a task is completed, we will mark in red the corresponding vertex of the project digraph. Eligible tasks can then be read directly from the project digraph—they are those that are not currently being worked on and are incident from only red vertices. So here we go.

■ **Time:** **0 hours** (Color vertex START red.)

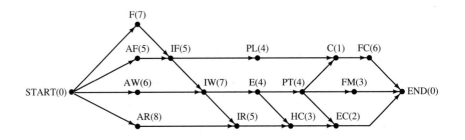

Free processors: X and Z.
Eligible tasks: F(7), AF(5), AW(6), AR(8).
Assignments: X does AR; Z does F.

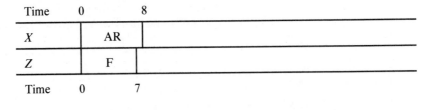

Time	0		8	
X		AR		
Z		F		
Time	0		7	

■ **Time:** **7 hours** (Color vertex F red.)

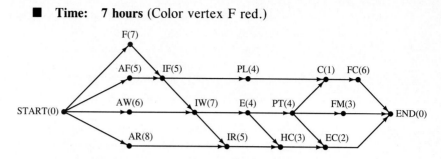

Free processors: Z (X is still working on AR).
Eligible tasks: AF(5), AW(6).
Assignment: Z does AW.

Time	0	8		
X	AR			
Z	F	AW		
Time	0	7	13	

■ **Time:** **8 hours** (Color vertex AR red.)

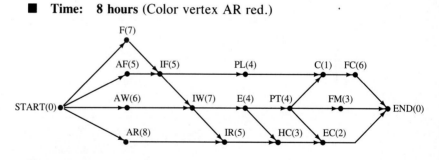

Free processor: X.
Eligible task: AF(5).
Assignment: X does AF.

Time	0	8	13	
X	AR	AF		
Z	F	AW		
Time	0	7	13	

■ **Time:** **13 hours** (Color vertices AW and AF red.)

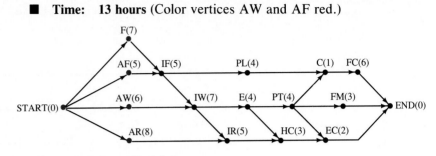

Free processors: X and Z.

Eligible task: IF(5).

Assignments: X does IF; Z is idle until IF is completed (there are no more eligible tasks at the moment).

Time	0		8		13	18	
X		AR		AF		IF	
Z		F		AW		idle	
Time	0		7		13	18	

■ **Time:** **18 hours** (Color vertex IF red.)

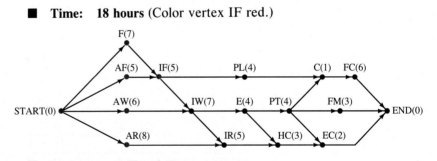

Free processors: X and Z.

Eligible tasks: IW(7), PL(4).

Assignments: X does IW; Z does PL.

Time	0		8		13		18		25	
X		AR		AF		IF		IW		
Z		F		AW		idle		PL		
Time	0		7		13		18	22		

■ **Time: 22 hours** (Color vertex PL red.)

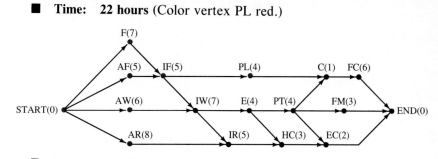

Free processor: Z.

Eligible tasks: None.

Assignment: Z is idle.

Time	0		8		13	18		25
X		AR		AF	IF		IW	
Z		F		AW	idle		PL	idle
Time	0		7		13	18	22	25

■ **Time: 25 hours** (Color vertex IW red.)

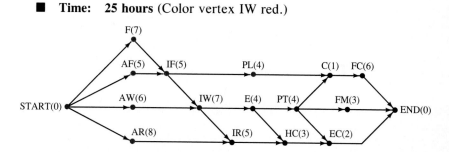

Free processors: X and Z.

Eligible tasks: E(4), IR(5).

Assignments: X does IR; Z does E.

Time	0		8		13	18		25	30
X		AR		AF	IF		IW	IR	
Z		F		AW	idle		PL	idle	E
Time	0		7		13	18	22	25	29

■ **Time: 29 hours** (Color vertex E red.)

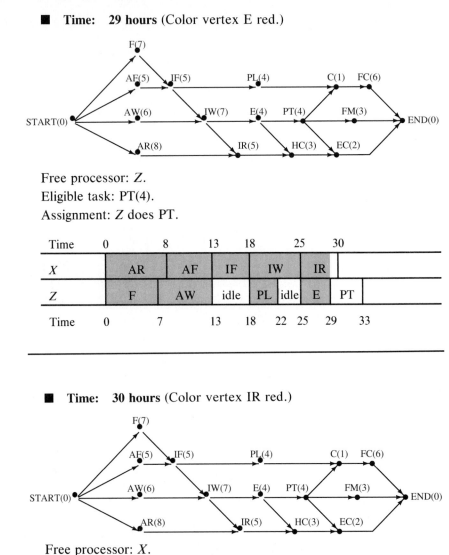

Free processor: Z.
Eligible task: PT(4).
Assignment: Z does PT.

Time	0		8	13	18		25	30	
X		AR		AF	IF	IW		IR		
Z		F		AW	idle	PL	idle	E	PT	
Time	0		7		13	18	22	25	29	33

■ **Time: 30 hours** (Color vertex IR red.)

Free processor: X.
Eligible task: HC(3).
Assignment: X does HC.

Time	0		8	13	18		25	30	33
X		AR		AF	IF	IW		IR	HC	
Z		F		AW	idle	PL	idle	E	PT	
Time	0		7		13	18	22	25	29	33

■ **Time:** **33 hours** (Color vertices HC and PT red.)

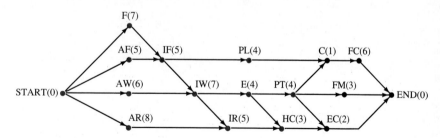

Free processors: X and Z.
Eligible tasks: C(1), FM(3), EC(2).
Assignments: X does FM; Z does EC.

Time	0		8	13	18		25	30	33	36
X		AR		AF	IF	IW		IR	HC	FM	
Z		F		AW	idle	PL	idle	E		PT	EC
Time	0		7		13	18	22	25	29	33 35

■ **Time:** **35 hours** (Color vertex EC red.)

F(7)

AF(5)   IF(5)   PL(4)   C(1) FC(6)

AW(6)   IW(7)   E(4) PT(4)   FM(3)

START(0)   END(0)

AR(8)   IR(5)   HC(3)   EC(2)

Free processor: Z.
Eligible task: C(1).
Assignment: Z does C.

Time	0		8	13	18		25	30	33	36
X		AR		AF	IF	IW		IR	HC	FM	
Z		F		AW	idle	PL	idle	E		PT	EC	c
Time	0		7		13	18	22	25	29	33 35 36

■ **Time: 36 hours** (Color vertices C and FM red.)

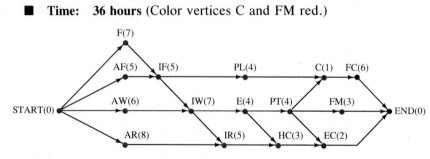

Free processors: *X* and *Z*.

Eligible task: FC(6).

Assignments: *X* does FC; *Z* is idle.

Time	0		8	13	18		25	30	33	36		42
X		AR		AF	IF	IW		IR	HC	FM	FC	
Z		F		AW	idle	PL	idle	E	PT	EC	c	idle
Time	0		7		13	18	22	25	29		33 35 36	42

■ **Time: 42 hours** (Color vertex FC red.)

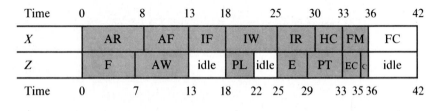

The project is complete. (As a symbolic gesture of triumph, we color vertex END red.) The final schedule is shown in Fig. 8-8. The project length is 42 hours.

Time	0		8	13	18		25	30	33	36		42
X		AR		AF	IF	IW		IR	HC	FM	FC	
Z		F		AW	idle	PL	idle	E	PT	EC	c	idle
Time	0		7		13	18	22	25	29		33 35 36	42

Figure 8-8

We can see that using the decreasing time algorithm is not difficult but requires a fair amount of bookkeeping. We can also see that in spite of all the effort we were able to make only a small improvement on our original flying by the seat of the pants schedule in Fig. 8-7 (which is admittedly a pretty bad schedule).

X and *Z* celebrate the completion of the project.

CRITICAL PATHS

The decreasing time algorithm, which in principle sounded like such a promising idea, seems to be somewhat of a bust! What went wrong? If we work our way backward from the end, we can first see that we made a bad choice at time 33 hours. At this point there were three eligible tasks [C(1), FM(3), and EC(2)] and two free processors. The decreasing time algorithm told us to choose the two longest tasks, FM and EC. This was a very shortsighted strategy: If we had looked at what was down the road, we would have seen that C is a much more critical task than the other two because we can't start FC—a task that requires 6 hours—until we finish C. We see once again that looking at the immediate rather than the long-term consequences of our actions can result in bad decisions.

An even more blatant example of how the decreasing time algorithm can lead to bad choices in scheduling occurs at the very start: We failed to notice that it is critical to start F(7) and AF(5) as early as possible. Until we finish F and AF, we cannot start IF(5), and until we finish IF we cannot start IW(7), and until we finish IW we cannot start E(4) and IR(5), and so on down the line. The lesson to be learned here is that we might be better off if the priority we give to a task is based on the length of all the tasks that lie ahead of it. Simply put, the more the total amount of work lying ahead of a task, the sooner that task should be started.

To formalize this idea we introduce the concept of a critical path from a vertex. Given a vertex X of a project digraph, of all the paths from vertex X to vertex END, the one that has the longest total sum of processing times is called the **critical path from** X.

Here are some examples using our original project digraph (Fig. 8-9):

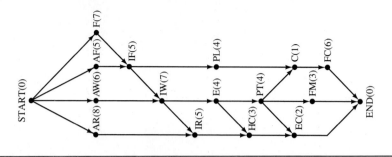

Figure 8-9

Example 2.   Let's suppose we want to find the critical path from the vertex PT. There are three paths from PT to END. They are (a) PT, C, FC, END; (b) PT, FM, END; and (c) PT, EC, END. The sum of the processing times for the first path (a) is $4 + 1 + 6 = 11$; the sum for the second path (b) is $4 + 3 = 7$; and the sum for the third path (c) is $4 + 2 = 6$. Of the three, the first path (a) has the largest sum, so it is the critical path from PT.   ■

Example 3.   If we use vertex AR as our starting point, there is only one path from AR to END—it is AR, IR, HC, EC, END. Since this is the only path, it is also the critical path from AR. Its total processing time is 18.   ■

Example 4.   There are quite a few paths from START to END, but after looking at the project digraph closely we can see that the one with the longest total processing time is START, F, IF, IW, E, PT, C, FC, END (with a total time of 34 hours). This, of course, is the critical path from START, and it is usually referred to as just the **critical path** of the project digraph. The critical path can be thought of as the *longest* path (in terms of total completion time) in the entire project digraph. We will come back to it soon.   ■

■ **The Backflow Algorithm**

When the project digraph is more complicated (lots of vertices, lots of arcs, and therefore lots of paths), finding critical paths by just "eyeballing" the digraph (as we did in Example 4) might not be so easy. We will now describe a simple procedure called the **backflow algorithm** which will allow us to find the length of the critical paths from each and every vertex of a project digraph. The basic idea is to start at END and move backward toward START. At each vertex we record (we will use square brackets for lengths of critical paths) the length of the critical path from that vertex. When we move backward to the "next" vertex, we simply add the processing time of the vertex to the length of the longest critical path "ahead" of it. Let's concentrate on the last part of the WIGWAM project digraph (Fig. 8-10) and use it to illustrate how the backflow algorithm works.

■ In step 1 we record the lengths of the critical paths from FC[6], FM[3], and EC[2].

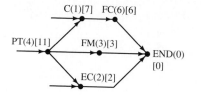

Figure 8-10 Steps 1 through 3 of the backflow algorithm.

■ In step 2 we record the length of the critical path from C [6 + 1 = 7].

■ In step 3 we record the length of the critical path from PT [7 + 4 = 11]. We note that the reason we got 11 is that of the three critical paths ahead of PT (at C, FM, and EC), the longest one has length 7, which added to the 4 for PT gives us 11. Continuing in this way, we get the critical paths from each vertex of the digraph as shown in Fig. 8-11. We leave it to the reader to verify the details.

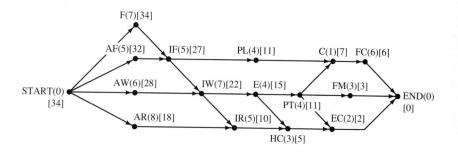

Figure 8-11 Lengths of critical paths from each vertex are shown inside the square brackets.

THE CRITICAL PATH ALGORITHM

With something like Fig. 8-11 at our disposal, we can introduce an improved algorithm for scheduling called the critical path algorithm. In a nutshell, the **critical path algorithm** says that whenever a processor is free, it should look at all the tasks that are eligible at that moment and choose the one for which the critical path is (yes, you guessed it!) the longest. (As before, in case of a tie we break it randomly.)

The process we will use for implementing the critical path algorithm is very similar to the one we used for the decreasing time algorithm: We keep track of the time, the free processors, and the eligible tasks in exactly the same way, but we now use critical path lengths [in square brackets] instead of processing times (in parentheses) to choose which tasks are given priority.

We will now illustrate the critical path algorithm using the same two robots (Xavier and Zygmund).

■   **Time: 0 hours** (Color vertex START red.)

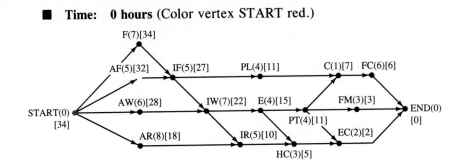

Free processors: X and Z.
Eligible tasks: F[34], AF[32], AW[28], AR[18].
Assignments: X does F; Z does AF.

Time 0		7	
X		F	
Z	AF		

Time   0   5

■   **Time: 5 hours** (Color vertex AF red.)

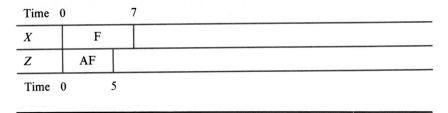

Free processor: Z.
Eligible tasks: AW[28], AR[18].
Assignment: Z does AW.

Time 0		7	
X		F	
Z	AF	AW	

Time   0   5   11

■ **Time:** **7 hours** (Color vertex F red.)

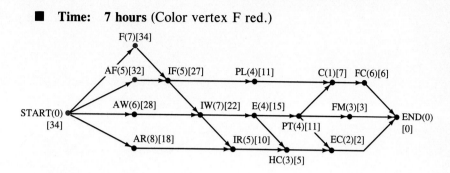

Free processor: *X*.

Eligible tasks: AR[18], IF[27].

Assignment: *X* does IF.

Time 0		7	12
X	F	IF	
Z	AF	AW	
Time 0	5	11	

■ **Time:** **11 hours** (Color vertex AW red.)

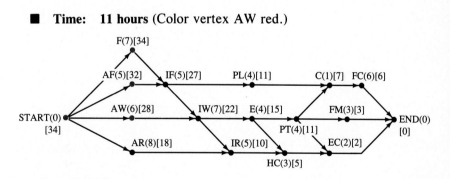

Free processor: *Z*.

Eligible task: AR[18].

Assignment: *Z* does AR.

Time 0		7	12	
X	F	IF		
Z	AF	AW	AR	
Time 0	5	11	19	

■ **Time: 12 hours** (Color vertex IF red.)

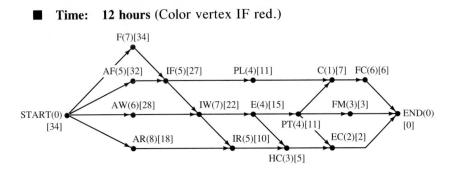

Free processor: X.

Eligible tasks: IW[22], PL[11].

Assignment: X does IW.

Time	0		7	12	19	
X		F		IF	IW	
Z		AF	AW		AR	
Time	0		5	11	19	

■ **Time: 19 hours** (Color vertices IW and AR red.)

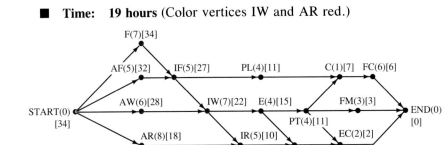

Free processors: X and Z.

Eligible tasks: PL[11], E[15], IR[10].

Assignments: X does PL; Z does E.

Time	0		7	12	19	23	
X		F		IF	IW	PL	
Z		AF	AW		AR	E	
Time	0		5	11	19	23	

■ **Time: 23 hours** (Color vertices PL and E red.)

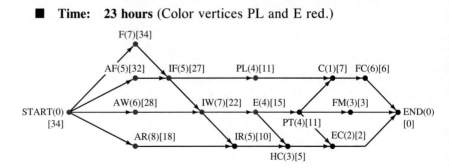

Free processors: X and Z.
Eligible tasks: IR[10], PT[11].
Assignments: X does IR; Z does PT.

Time 0		7		12		19	23		28
X	F		IF		IW		PL	IR	
Z	AF		AW		AR		E	PT	
Time 0		5		11			19	23	27

■ **Time: 27 hours** (Color vertex PT red.)

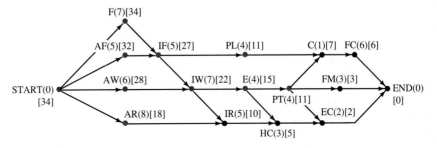

Free processor: Z.
Eligible tasks: C[7], FM[3].
Assignment: Z does C.

Time 0		7		12		19	23		28
X	F		IF		IW		PL	IR	
Z	AF		AW		AR		E	PT	C
Time 0		5		11			19	23	2728

■ **Time:** **28 hours** (Color vertices C and IR red.)

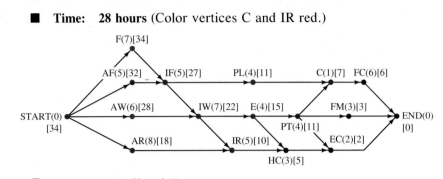

Free processors: X and Z.

Eligible tasks: FC[6], FM[3], HC[5].

Assignments: X does FC; Z does HC.

Time	0		7	12		19	23		28		34
X		F		IF	IW		PL		IR	FC	
Z		AF	AW		AR		E		PT	C	HC	
Time	0		5		11		19	23		27 28	31	

■ **Time:** **31 hours** (Color vertex HC red.)

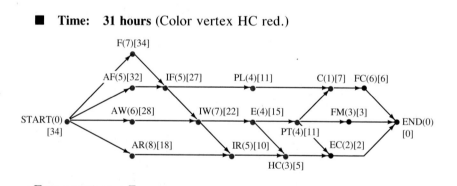

Free processor: Z.

Eligible tasks: FM[3], EC[2].

Assignment: Z does FM.

Time	0		7	12		19	23		28		34
X		F		IF	IW		PL		IR	FC	
Z		AF	AW		AR		E		PT	C	HC	FM
Time	0		5		11		19	23		27 28	31	34

■ **Time: 34 hours** (Color vertices FM and FC red.)

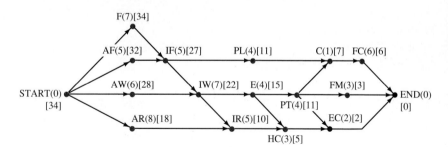

Free processors: X and Z.

Eligible task: EC[2].

Assignments: X does EC; Z is idle.

Time 0		7	12		19	23		28		34	36
X	F	IF	IW		PL	IR		FC		EC	
Z	AF	AW	AR		E	PT	c	HC	FM	idle	
Time 0	5		11		19	23		27 28	31	34	36

■ **Time: 36 hours** (Color vertex EC red.)

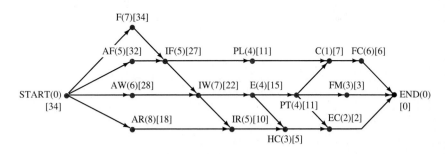

Project is complete. (Color vertex END red.)

The schedule produced by the critical path algorithm (Fig. 8-12) has a project length of 36 hours and is a considerable improvement over the one produced by the decreasing time algorithm (with a project length of 42 hours). We are pleased with the savings (6 hours at $1800 per hour is $10,800!) but we are curious—is this the optimal answer? We know that there are 2 hours at the end of the schedule during which Zygmund is

Figure 8-12 Schedule for WIGWAM project obtained using the critical path algorithm.

Time 0		7	12	19	23	28	34 36
X	F	IF	IW	PL	IR	FC	EC
Z	AF	AW	AR	E	PT	c HC	FM idle
Time 0	5	11		19	23	2728	31 34 36

idle, and if possible we would like to eliminate even that little bit of waste. It turns out that it is indeed possible to improve on the schedule produced by the critical path algorithm. Shown in Fig. 8-13 is a schedule with a project length of 35 hours.

Figure 8-13 An optimal schedule for the WIGWAM project using two processors.

Time 0		7	12	19	23	28 29	35 36
X	F	IF	IW	PL	IR	c FC	
Z	AF	AW	AR	E	PT	FM HC EC	
Time 0	5	11		19	23	27 30	33 3536

It is clear that this schedule is optimal and cannot be improved on. How did we come up with this optimal schedule? The answer, unfortunately, is plain luck. We were fortunate that the problem was small, and we were able to come up with an optimal solution using a trial-and-error approach. Once again, we are in familiar territory: We know of an algorithm (the critical path algorithm) that in most cases produces a good approximate solution to a scheduling problem. If we want an optimal solution we had better be prepared to spend a long time looking for it, because no efficient algorithm is presently known that will always give an optimal solution.

Let's summarize our conclusions about the cost of building a WIG-WAM when we use two processors. Of the four schedules we discussed, the first two (flying by the seat of the pants with a completion time of 44 hours, and the decreasing time algorithm with a completion time of 42 hours) were quite bad. The schedule produced by the critical path algorithm was good but not optimal (completion time of 36 hours). At the end we found an optimal schedule (completion time of 35 hours) by trial and error, something that we cannot count on happening in general. Under the last two schedules, robot rental costs would be $1800 × 36 = $64,800 and $1800 × 35 = $63,000, respectively. In either case this is a nice chunk of savings over the $70,000 it would cost for renting a single robot.

This leads to the next obvious question: Can we save even more if we rent three robots? (Three robots rent for $800 per robot per hour for a total cost of $2400 per hour.) Let's check it out.

■ **Scheduling Three Processors**

Let's play our best card and apply the critical path algorithm to schedule the WIGWAM project using three HB100 robots (Xavier, Yolanda, and Zygmund). Once again we must have in front of us the project digraph with both processing times ( ) and critical path lengths [ ] showing, as in Fig. 8-14. We will also need a blank three-processor timetable as shown in Fig. 8-15.

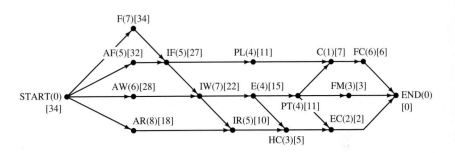

Figure 8-14

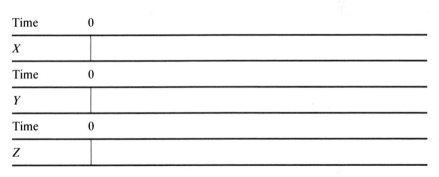

Figure 8-15

■ **Time: 0 hours** (Color vertex START red.)

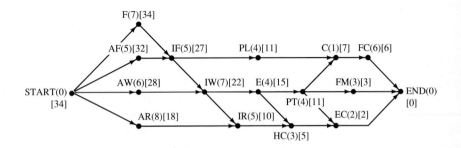

Free processors: *X*, *Y*, and *Z*.

Eligible tasks: F[34], AF[32], AW[28], AR[18].

Assignments: *X* does F; *Y* does AW; *Z* does AF.

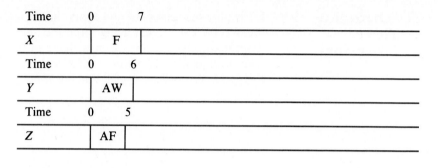

■ **Time: 5 hours** (Color vertex AF red.)

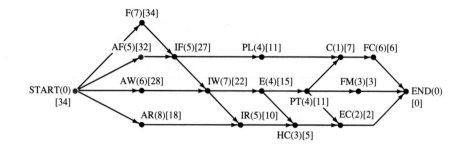

Free processor: Z.
Eligible task: AR[18].
Assignment: Z does AR.

Time	0	7						
X		F						

Time	0	6						
Y		AW						

Time	0	5	13					
Z		AF	AR					

■ **Time:** **6 hours** (Color vertex AW red.)

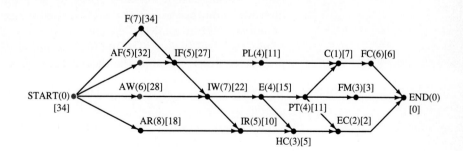

Free processor: Y.

Eligible tasks: none.

Assignment: Y is idle.

Time	0		7
X		F	

Time	0		6
Y		AW	idle

Time	0	5		13
Z		AF	AR	

By now we are pretty familiar with this song-and-dance routine, so we will leave it to the reader to fill in the rest of the details. The complete schedule produced by the critical path algorithm is shown in Fig. 8-16.

Time	0		7	12		19	23	27 28		34
X		F	IF	IW		E	PT	C	FC	

Time	0		6	12	16	19		24	27	30	34
Y		AW	idle	PL	idle	IR		HC	FM	idle	

Time	0	5		13		27	29		34
Z		AF	AR		idle		EC	idle	

Figure 8-16

The schedule shown in Fig. 8-16 turns out to be a bit of a disappointment. We increased our labor force by 50% but we were only able to improve upon the results of the critical path algorithm from 36 hours to 34 hours. With a total of 70 hours of work to be divided among three robots, we had higher expectations—something like a schedule with a project length of 24 hours or at the most 25 hours. What went wrong here? The answer is nothing. The fact of the matter is that we cannot finish the WIGWAM project in less than 34 hours regardless of how many robots we put on the job!

The reason for this has to do with the critical path of the project digraph (remember, this was the longest path from START to END). We know that the critical path in the WIGWAM project digraph is START, F, IF, IW, E, PT, C, FC, END, with a total length of 34 hours. This path (like any other path) represents a sequence of tasks that must be done in succession, and no task in the sequence can be started until the preceding task has been completed. It follows that the project length cannot be less than the length of a path (any path) and, in particular, it cannot be less than the length of the critical path. Moreover, this is a requirement that has nothing to do with the number of processors at our disposal—it is intrinsic to the project digraph itself.

Our reluctant conclusion is that the three-processor schedule shown in Fig. 8-16 is optimal—we simply cannot improve on the 34-hour completion time. This also means that renting three robots would cost at least $81,600 ($2400 × 34) and would obviously be a rather bad business decision.

CONCLUSION

In this chapter we discussed *minimal-length scheduling problems*, that is, problems in which we are given a fixed number of processors, a set of tasks with their respective processing times, and a set of precedence relations governing the tasks, and where the object is to find a schedule that completes all the tasks in the shortest length of time. Problems of this type occur in many applications: home construction, scheduling classes at a university, scheduling operations at a hospital, preparing a banquet, etc. The two algorithms we discussed for solving this type of problem are the *decreasing time algorithm* and the *critical path algorithm*, both of which in general only give approximate solutions. At present, no efficient algorithm that always produces an optimal schedule is known.

In the process of discussing methods for solving these problems, we learned about a new mathematical structure called a *directed graph* (digraph) and some related concepts such as *arcs, paths,* and *critical paths*. Directed graphs are an essential and powerful tool in many disciplines beyond mathematics and management science, and many of the concepts we discussed in this chapter have found their way into important applications in fields as diverse as linguistics, anthropology, and sociology. (A detailed discussion of the many applications of directed graphs in these and other fields can be found in reference 8.)

KEY CONCEPTS

arc
backflow algorithm
critical path
critical path algorithm
critical path from a vertex
decreasing time algorithm
digraph
incident (to and from)
indegree (outdegree)
minimal-length scheduling
    problem

path
precedence relation
processing time
processor
project digraph
project length
tasks
time table

EXERCISES

■ Walking

1. For each of the following digraphs, make and complete a table similar to the one shown here.

Vertex	Degree	Indegree	Outdegree	Vertex is incident to	Vertex is incident from
A					
B					
⋮					

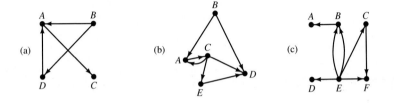

2. For each of the following digraphs list the vertices and arcs. (Use $\overrightarrow{XY}$ to represent an arc from X to Y.) Give the indegrees and outdegrees of each vertex and list the vertices incident to and incident from each vertex.

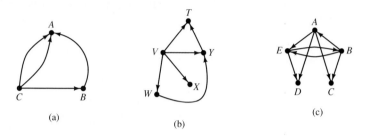

(a)   (b)   (c)

3. For each of the following, draw a picture of the digraph.
   (a) Vertices: *A, B, C, D*
       Arcs: *A* is incident to *B* and *C*; *D* is incident from *A* and *B*.
   (b) Vertices: *A, B, C, D, E*
       Arcs: *A* is incident to *C* and *E*; *B* is incident to *D* and *E*; *C* is incident from *D* and *E*; *D* is incident from *C* and *E*.

4. For each of the following, draw a picture of the digraph.
   (a) Vertices: *A, B, C, D*
       Arcs: *A* is incident to *B, C,* and *D*; *C* is incident from *B* and *D*.
   (b) Vertices: *V, W, X, Y, Z*
       Arcs: *X* is incident to *V, Z,* and *Y*; *W* is incident from *V, Y,* and *Z*; *Z* is incident to *Y* and incident from *W* and *V*.

5. Make a schedule for building a WIGWAM with one processor that has building the foundation as the first task and has the floors installed before either the roof or walls are assembled.

6. Make a schedule for building a WIGWAM with one processor that has assembling the floors as the first task and has the masonry finish work completed before the roof is assembled.

7. Explain what is illegal about the following schedule for building a WIGWAM with one processor.

Tasks	AF	F	AR	AW	IF	IW	PL	IR	HC	E	PT	EC	FM	C	FC

Time 0   5   12   20   26   31   38   42   47   50   54   58   60   63 64   70

8. Explain what is illegal about the following schedule for building a WIGWAM with one processor.

Tasks	AF	AW	AR	F	IF	PL	IW	IR	E	HC	EC	PT	FM	C	FC

Time 0   5   11   19   26   31   35   42   47   51 54   56   60   63 64   70

9. Explain what is illegal about the following schedule for building a WIGWAM with two processors.

Time	0		7		15		23		28	31		35	36		42
X		F		AR		idle			IR	HC	PL	c		FC	
Z		AF	AW		IF		IW		E		PT	EC	FM	idle	
Time	0	5		11		16			23	27		31	33	36		42

10. Explain what is illegal about the following schedule for building a WIGWAM with three processors.

Time	0		8		17		24		28	29		32	34	35		41
X		AR		idle		IW		E	idle	HC	idle	c		FC		

Time	0		6		12		17		21	24		29		34	37		41
Y		AW		idle		IF		PL	idle		IR		PT		FM	idle	

Time	0		5		12									34	36		41
Z		AF		F			idle							EC	idle		

Exercises 11, and 12 refer to the problem of scheduling seven tasks (A, B, C, D, E, F, G) in accordance with the following project digraph.

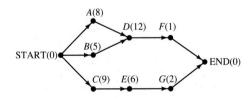

11. Use the decreasing time algorithm to schedule the project using two processors.

12. (a) Find the length of the critical path from each vertex.
    (b) What is the length of the critical path of the project digraph?
    (c) Use the critical path algorithm to schedule the project using two processors.
    (d) Explain why the schedule obtained in this case is optimal.

Exercises 13 through 16 refer to the problem of scheduling 11 tasks in accordance with the following project digraph.

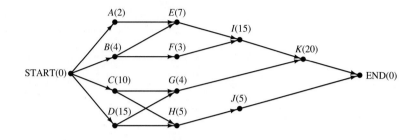

13. Use the decreasing time algorithm to schedule the project using two processors.

14. (a) Find the length of the critical path from each vertex.
    (b) What is the length of the critical path of the project digraph?
    (c) Use the critical path algorithm to schedule the project using two processors.
    (d) Explain why the schedule obtained in this case is optimal.

15. Use the critical path algorithm to schedule the project using three processors.

16. Use the decreasing time algorithm to schedule the project using three processors.

17. Make a schedule for building a WIGWAM with two processors that has the walls assembled and the roof assembled before the foundation is started.

18. Make a schedule for building a WIGWAM with three processors that has the floors assembled before the foundation is started.

19. Four identical processors are available to perform a set of 10 tasks. There are 3 tasks that require 4 minutes each to complete, 3 tasks that require 7 minutes each to complete, and 4 tasks that require 15 minutes each to complete. *None* of the 15-minute tasks can be started until *all* the 7-minute tasks have been completed.
    (a) Draw a project digraph for this scheduling problem.
    (b) Use the decreasing time algorithm to schedule the project using the four processors available.

20. Three identical processors are available to perform a set of eight tasks. There is one task that requires 10 minutes to complete, two tasks that require 7 minutes each to complete, two tasks that require 12 minutes each to complete, and three tasks that require 20 minutes each to complete. *None* of the 20-minute tasks can be started until *both* of the 7-minute tasks are completed.
    (a) Draw a project digraph for this scheduling problem.
    (b) Use the critical path algorithm to schedule the project using the three processors available.

■ **Jogging**

21. (a) Draw a project digraph for a project consisting of the eight tasks described by the following table

Task	Length of Task	Tasks that must be completed before the task can start
A	5	C
B	5	C, D
C	5	
D	2	G
E	15	A, B
F	6	D, H
G	2	
H	2	G

(b) Use the critical path algorithm to schedule this project using two processors.

22. (a) Draw a project digraph for a project consisting of the eight tasks described by the following table

Task	Length of Task	Tasks that must be completed before the task can start
A	3	
B	10	C, F, G
C	2	A
D	4	G
E	5	C
F	8	A, H
G	7	H
H	5	

(b) Use the critical path algorithm to schedule this project using two processors.

23. In 1961, T.C. Hu of the University of California showed that in any scheduling problem in which all the tasks have equal processing times and in which the original project digraph (without the START and END vertexes) is a tree, the critical path algorithm will give an optimal schedule. Using this result, find an optimal schedule for the scheduling problem with the following project digraph using three processors. Assume each task takes 3 days. (Notice that we have omitted the START and END vertexes.)

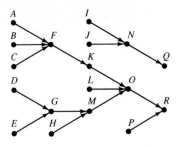

Exercises 24 through 27 involve the concept of **independent tasks**. Tasks are said to be independent when there are no precedence relations among them.

24. Explain why in a project where all the tasks are independent, the decreasing time algorithm and the critical path algorithm are the same.

25. The following is a schedule using four processors for a project where all the tasks are independent. Schedule this project using the decreasing time algorithm with four processors.

Processor 1	A(7)		B(5)	
Processor 2	C(7)		D(5)	
Processor 3	E(6)		F(6)	
Processor 4	G(4)	H(4)	I(4)	

26. The following is a schedule using three processors for a project where all the tasks are independent. Schedule this project using the decreasing time algorithm with three processors.

Processor 1	A(8)		B(2)
Processor 2	C(5)	D(5)	
Processor 3	E(4)	F(3)	G(3)

27. Use the decreasing time algorithm to schedule independent tasks of length 1, 1, 2, 2, 5, 7, 9, 13, 14, 16, 18, and 20 using three processors. Is this schedule optimal? Explain.

28. The Faster Processor Paradox.

   (a) Use the critical path algorithm to schedule a project with nine tasks using two processors according to the following project digraph.

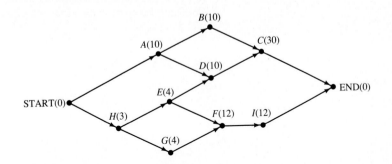

   (b) Explain why the schedule found in part (a) is optimal.

   (c) Now suppose that the processing times for each of the tasks are decreased by 1 (a faster model of processor is used), giving the following project digraph. Use the critical path algorithm to reschedule this project using two processors.

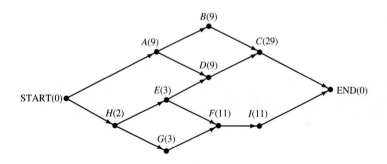

   (d) Explain why the schedule found in part (c) is optimal.

   (e) Compare the completion times obtained in parts (a) and (c). Explain how this can happen.

29. The More is Less Paradox.

   (a) Use the critical path algorithm to schedule a project with eight tasks using two processors according to the following project digraph.

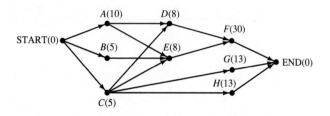

(b) Explain why the schedule found in part (a) is optimal.

(c) Now suppose that the number of processors is increased to three processors. Use the critical path algorithm again to schedule the project.

(d) Explain why the schedule found in part (c) is optimal.

(e) Compare the completion times obtained in parts (a) and (c). Explain how this can happen.

30. Consider the problem of scheduling nine tasks in accordance with the following project digraph.

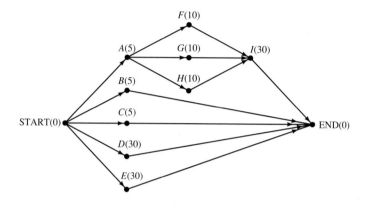

(a) Use the critical path algorithm to schedule the project with three processors.

(b) Find an optimal schedule for the project using three processors.

■ **Running**

Exercises 31 through 33 refer to the following fact: In 1966, Ronald L. Graham of Bell Laboratories showed that if T_{OPT} is the optimal completion time for a given scheduling problem, then the completion time T for any other schedule for the problem must satisfy the inequality $T \le [2 - (1/m)]T_{OPT}$, where m is the number of processors. For example, with two processors (m = 2) the completion time T for any schedule must satisfy the inequality $T \le \frac{3}{2} T_{OPT}$ so that no scheduling will result in a time longer than $1\frac{1}{2}$ times the optimal time.

31. Show that if T_1 = 21 hours and T_2 = 12 hours are completion times of two different schedules for the same scheduling problem with 4 processors, then T_2 is the optimal completion time for the scheduling problem and T_1 is the longest possible completion time for the problem.

32. Suppose we have a scheduling problem with two processors and we came up with a schedule with completion time T_1 = 9 hours. Explain why the optimal completion time for this scheduling problem cannot be less than 6 hours.

33. Suppose we have a scheduling problem with three processors and we came up with two different schedules with completion times T_1 = 12 hours and T_2 = 15 hours. Explain why the optimal completion time for this scheduling problem has to be somewhere between 9 and 12 hours.

Exercises 34 and 35 refer to the following: It has been shown that if T_{OPT} is the optimal completion time for a given scheduling problem in which all the tasks are independent, then the completion time T for any schedule for the problem obtained by using the critical-path algorithm must satisfy the inequality $T \leq (\frac{4}{3} - \frac{1}{3m})T_{OPT}$, where m is the number of processors. For example, with four processors ($m = 4$) the completion time T for any schedule obtained using the critical path algorithm must satisfy the inequality $T \leq \frac{5}{4} T_{OPT}$. (See Exercise 25 for an example where the equality holds.)

34. Give an example of a scheduling problem using three processors in which all the tasks are independent and such that the completion time using the critical path algorithm is $\frac{11}{9}$ of the optimal completion time.

35. Give an example of a scheduling problem using five processors in which all the tasks are independent and such that the completion time using the critical path algorithm is $\frac{19}{15}$ of the optimal completion time.

REFERENCES AND FURTHER READINGS

1. Baker, K. R., *Introduction to Sequencing and Scheduling*. New York: John Wiley & Sons, Inc., 1974.

2. Coffman, E. G., *Computer and Jobshop Scheduling Theory*. New York: John Wiley & Sons, Inc., 1976, chaps. 2 and 5.

3. Conway, R. W., W. L. Maxwell, and L. W. Miller, *Theory of Scheduling*. Reading, Mass.: Addison-Wesley Publishing Co., Inc., 1967.

4. Garey, M. R., R. L. Graham, and D. S. Johnson, "Performance Guarantees for Scheduling Algorithms," *Operations Research*, 26 (1978), 3–21.

5. Graham, R. L., "Combinatorial Scheduling Theory," in *Mathematics Today*, ed. L. Steen. New York: Springer-Verlag, Inc., 1978, 183–211.

6. Graham, R. L., E. L. Lawler, J. K. Lenstra, and A. H. G. Rinnooy Kan, "Optimization and Approximation in Deterministic Sequencing and Scheduling: A Survey," *Annals of Discrete Mathematics*, 5 (1979), 287–326.

7. Hillier, F. S., and G. J. Lieberman, *Introduction to Operations Research* (3rd ed.). San Francisco: Holden-Day, Inc., 1980, chap. 6.

8. Roberts, Fred S., *Graph Theory and Its Applications to Problems of Society*, CBMS-NSF Monograph No. 29. Philadelphia: Society for Industrial and Applied Mathematics, 1978.

9

Spiral Growth and Fibonacci Numbers

"I find" said 'e, "things very much as / 'ow I've always found, / For mostly they goes up and down or / else goes round and round"

P. R. CHALMERS

The Golden Ruler

Loves me? Loves me not? Which daisy will give the right answer? (Laurence Pringle/Photo Researchers)

. . . loves me . . . loves me not . . . loves me . . .

The answer to this, one of the more celebrated questions posed by man (or woman) depends of course on whether the number of petals on the daisy is even or odd (even means no, odd means yes). But aren't the number of petals on a daisy random? (The answer is no.) Are there any rules that govern the way petals grow on a daisy? (The answer is yes.) Moreover, what is really surprising is that while these rules are not fully understood, it is quite clear that there is some unusual and interesting mathematics lurking behind them.

One of the purposes of this chapter is to acquaint the reader with the mathematical pieces of an exotic puzzle in nature. Why do field daisies almost always have 34 petals, while African daisies almost always have 55 or 89 petals? [Note that knowing this is of great value in romance— she (or he) will almost always love us if we pick the right variety of daisy!]

It would be exciting but misleading to claim that this chapter is about the mathematical aspects of falling in love. Stacking the deck in the game of "loves me, loves me not" is hardly sufficient reason to devote a whole chapter to this topic. The issues at stake in this chapter are a little more general: First, why do flowers, pinecones, sunflowers, and other spiral-shaped objects in the plant world exhibit an affinity for certain kinds of numbers, and second, what is the overall mathematical structure of spiral growth in nature? While mathematics alone cannot answer all these questions, it does shed some light on them.

FIBONACCI NUMBERS

Figure 9-1

Listed in Fig. 9-1 is a very remarkable group of numbers called **Fibonacci numbers**. They are named after the Italian mathematician Leonardo of Pisa (better known by the nickname Fibonacci).[1] The remarkable frequency with which Fibonacci numbers occur among certain plants and flowers appears to be related to the equally remarkable mathematical properties of these numbers. We will discuss some of these now, and again in the next few sections.

First, the list of Fibonacci numbers is infinite. (This is what the three

Fibonacci (c.1170–1250) was the son of a merchant and as a young man he traveled extensively with his father. Through his travels and studies in northern Africa, he became acquainted with the Arabic system of numeration and algebra, which he introduced to Christian Europe in his book *Liber Abaci* ("The Book of the Abacus"), published in 1202. Although he is best remembered for the discovery of Fibonacci numbers, they were only a minor part of his book and of his contributions to history.

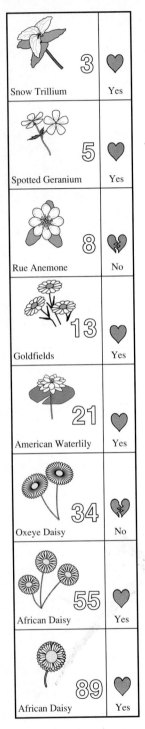

Snow Trillium	Yes
Spotted Geranium	Yes
Rue Anemone	No
Goldfields	Yes
American Waterlily	Yes
Oxeye Daisy	No
African Daisy	Yes
African Daisy	Yes

dots following the 89 indicate.) The list is also ordered, meaning that there is a first Fibonacci number (1), a second (1), a third (2), . . ., a seventh (13), . . ., a tenth (55), an eleventh (89), etc. Some logical questions: What is the 12th Fibonacci number? What is the 100th Fibonacci number? What is the Nth Fibonacci number? (This is the mathematical equivalent of asking for the description of a typical Fibonacci number at some location, which we call N, in the list.) Before we attempt to answer these questions, let's introduce some convenient notation. We will use F to denote Fibonacci numbers in general, and a numerical subscript below the F to denote the specific number to which we are referring. Thus,

$$F_1 = 1, F_2 = 1, \ldots, F_6 = 8, F_7 = 13, \ldots, F_{10} = 55, F_{11} = 89,$$

and F_N = the Nth Fibonacci number.

Note that while this notation is just a convenient form of bookkeeping, we must be careful how we keep our books. For example,

$$F_{6+1} = F_7 = 13, \text{ while } F_6 + 1 = 8 + 1 = 9.$$

Likewise, F_{N+1} is not the same as $F_N + 1$.

Thus we can see that the action takes place on two levels: Downstairs (at the subscript level) all our work deals with positions, and upstairs (at the Fibonacci level) our calculations refer to the actual Fibonacci numbers themselves. Let's say, for example, that we want to say "Add two consecutive Fibonacci numbers" in our new notation. If we write $F_1 + F_2$, we are really saying, "Add the first two Fibonacci numbers," and if we write $F_{27} + F_{28}$, we are just saying, "Add the 27th and the 28th Fibonacci numbers." One way to write the more general statement "Add two consecutive Fibonacci numbers" is $F_N + F_{N+1}$ (or, equivalently, $F_{N-1} + F_N$).

Let's go back to our original listing of the Fibonacci numbers shown in Fig. 9-1.

F_1	F_2	F_3	F_4	F_5	F_6	F_7	F_8	F_9	F_{10}	F_{11}	F_{12}
1	1	2	3	5	8	13	21	34	55	89	?

Figure 9-2

There should be no great mystery as to what the F_{12} position in Fig. 9-2 should contain. The pattern seems clearly to suggest that each Fibonacci number (from F_3 on) is the sum of the preceding two Fibonacci numbers. Thus, $F_{12} = 89 + 55 = 144$. Now, how about F_{100}? Presumably, this would be easy if we only knew F_{99} and F_{98}, which we don't. In fact, at this point we don't know F_{99}, F_{98}, F_{97}, . . ., F_{14}, or F_{13}. On the other hand, it is clear that if we set our minds to it we could slowly but surely march up this Fibonacci ladder one rung at a time: $F_{13} = 144 + 89 = 233$, $F_{14} = 233 + 144 = 377$, etc. Let's cheat a little bit and say that we got $F_{97} = 83621143489848422977$ and $F_{98} = 135301852344706746049$ from a friend. We can now move up to

$$F_{99} = 135301852344706746049 + 83621143489848422977$$

$$= 218922995834555169026,$$

and finally,

$$F_{100} = 218922995834555169026 + 135301852344706746049$$

$$= 354224848179261915075.$$

Next, let's tackle F_N. We know that this Fibonacci number, wherever it may be (as long as it is past F_2), is the sum of its two predecessors. In our notation, we have

$$F_N = F_{N-1} + F_{N-2}.$$

Of course the above rule cannot be applied to F_1 (which has no predecessors) or F_2 (which has only one predecessor), so to complete our description we add these two special cases by writing $F_1 = 1$ and $F_2 = 1$. The combination of facts

$$F_1 = 1, F_2 = 1, \text{ and } F_N = F_{N-1} + F_{N-2}$$

gives a complete description of everything we need to know to compute Fibonacci numbers. It is, in essence, their definition.

SEQUENCES

Any ordered infinite list of numbers is called a **sequence**. The ordered list of positive integers

$$1, 1, 2, 3, 5, 8, 13, 21, 34, 55, \ldots$$

is a sequence called (for obvious reasons) the **Fibonacci sequence**. The individual numbers in a sequence are called **terms**, and every sequence has a first term, a second term, etc.

Mathematically, sequences can be described using the notation we introduced in the preceding section: Each term is represented by a letter (which denotes the name of the sequence) and a numerical subscript (which identifies the position of the term in the sequence). Here is a metaphor for our notation: If we think of the terms of the sequence as houses all lined up on an infinitely long street, then we can think of the letter on top as the name of the street and the subscripts below as the addresses on this street. The numerical value of the term is the resident at that address.

Example 1.   1, 3, 5, 7, 9, 11, 13, 15, ...

This is a well-known sequence consisting of the odd positive integers. If we use O as a name for this sequence, then we have

$$O_1 = 1, O_2 = 3, O_3 = 5, \ldots, O_{100} = 199 \text{ (check it out!), etc.}$$

(Pursuing our metaphor, the mathematical statement $O_{100} = 199$ can be thought of as saying that the number 199 resides at the 100th location in the sequence O.) Here are two different mathematical descriptions of the sequence:

(a) $O_1 = 1, O_N = O_{N-1} + 2.$
(b) $O_N = 2N - 1.$ ∎

Example 2.   5, 12, 19, 26, 33, ...

Here is a sequence that starts with 5 and then adds 7 to each successive term. If we use A as the name of this sequence, we can then describe it by

(a) $A_1 = 5, A_N = A_{N-1} + 7.$

What is the 100th term of this sequence? This is not as hard as it may seem. We start at 5 and at each step add 7. After 99 such hops (that's the number we need to get from the first location to the 100th location), we have

$$A_{100} = 5 + 99 \times 7 = 698.$$

The same approach works for the generic term A_N in this sequence. It yields

$$A_N = 5 + 7(N-1) = 5 + 7N - 7 = 7N - 2.$$

This results in an alternative way to define the sequence A:

(b) $A_N = 7N - 2.$ ∎

A sequence like A in which each term (after the first) is obtained by adding a fixed amount to the preceding term is called an **arithmetic sequence**. We will discuss arithmetic sequences in more detail in Chapter 10.

Example 3   **(Egyptian Fractions).** The terms of a sequence do not have to be integers. Consider the sequence

$$1, \tfrac{1}{2}, \tfrac{1}{3}, \tfrac{1}{4}, \tfrac{1}{5}, \ldots$$

Fractions with numerator 1 are sometimes known as Egyptian fractions, so we call this sequence E. While it is not obvious, it is possible to verify (Exercise 13) that this sequence can be defined by

(a) $E_1 = 1$ and $E_N = \dfrac{E_{N-1}}{1 + E_{N-1}}.$

On the other hand, a much more direct way to define the sequence E is

(b) $E_N = \dfrac{1}{N}$ ∎

Example 4. $3, 6, 12, 24, 48, 96, \ldots.$

It is clear that in this sequence each term is twice the preceding one, with a starting term equal to 3. If we call this sequence G, we can describe it by

(a) $G_1 = 3$ and $G_N = 2 \times G_{N-1}.$

Alternatively, to get to G_N we start at 3 and take $N-1$ steps, each one of which is a multiplication by 2. This leads to

(b) $G_N = 3 \times 2^{N-1}$

A sequence like G in which every term (after the first) is obtained by multiplying the preceding term by a fixed number is called a **geometric sequence**, and we will discuss such sequences in more detail in Chapter 10.   ∎

While each of the sequences in Examples 1 through 4 is nice and orderly (and therefore the kind we would be comfortable dealing with), the reader should be aware that many sequences are not so well behaved and can be much more difficult to work with. We will stay away from these as much as we can.

Sequences are a useful mathematical tool for studying both natural and artificial phenomena, especially when we want to study dynamic events (things that change over time). Such situations will come up repeatedly in this and future chapters. Let's illustrate what we mean by a very simple example.

Example 5. **(Fibonacci's Rabbits).** In his famous book *Liber Abaci*, Fibonacci raised the following problem: A pair of baby rabbits (one female and one male) are placed inside a large, fenced area for the purposes of breeding. The rules of this rabbit breeding game are as follows:

1. It takes 1 month for a newborn baby rabbit to become a mature rabbit.

2. In 1 month, a mature pair (male and female) of rabbits will produce one pair (male and female) of baby rabbits.

How many pairs of rabbits will there be in 1 year?

While these rules are a gross oversimplification of the way rabbits breed in the real world, they give us a workable start. Over the long term, other factors such as the natural death of older rabbits would have to be taken into account, and additional rules would have to be introduced. By the way, this is an example of a problem in population growth, a theme we will take up in full in the next chapter.

Table 9-1 shows the growth of the rabbit population.

We can see that the short-term growth of this rabbit population follows

	Start	After 1 Month	After 2 Months	After 3 Months	After 4 Months	After 5 Months	After 6 Months	
Mature pairs	0	1	1	2	3	5	8	...
Baby pairs	1	0	1	1	2	3	5	...
Total pairs	1	1	2	3	5	8	13	...

Table 9-1

the Fibonacci sequence.[2] The number of pairs of rabbits after 6 months, for example, is F_7; the number of pairs of rabbits after month 7 is F_8; and thus, the number of pairs of rabbits after month 12 (after 1 year has passed) is $F_{13} = 233$. ■

Recursive and Explicit Descriptions of Sequences

We have seen in previous examples that there is more than one way to describe a sequence. One way is what we may call *empirical*: We list a few of the beginning terms of the sequence, add a " . . . ," and leave it to the reader to draw his or her own conclusions. Mathematically speaking, this is not a very satisfactory description. Given the sequence 1, 3, 5, 7, 9, 11, 13, . . ., a true nonconformist could argue that the next term in the sequence should be 35, and we would be hard put to prove that such an answer is wrong.

A second kind of description is the one we used to describe the Fibonacci sequence, namely,

$$F_1 = 1, F_2 = 1, \text{ and } F_N = F_{N-1} + F_{N-2} .$$

A description of this type is called a **recursive definition**—the terms of the sequence are defined by means of one or more preceding terms. This allows us to work our way up the sequence one term at a time: Known terms are used to calculate new terms, new terms become known terms and are used to calculate further new terms, etc. We will call this type of feedback looping process a **recursive process**. This is a concept of fundamental importance, and it will come up again and again in this section of the book.

Recursive processes run on their own once they get going, but they do need a jump start. That's the role of the **seeds**. In the recursive definition of the Fibonacci sequence, the seeds are $F_1 = 1$ and $F_2 = 1$ and the **recursive rule** is $F_N = F_{N-1} + F_{N-2}$. In Example 1, the odd positive integers are given by the seed $O_1 = 1$ and the recursive rule $O_N = O_{N-1} + 2$. Note that a change in the seed would produce an entirely different sequence. Thus, the recursive definition $V_1 = 2$, $V_N = V_{N-1} + 2$ produces the even positive integers (Exercise 15).

Let's look at some additional recursive definitions of sequences:

Example 6. $T_1 = 1, T_2 = 1, T_3 = 2, T_N = T_{N-1} + T_{N-2} + T_{N-3}$
The above describes the sequence 1, 1, 2, 4, 7, 13, 24, 44, 81,. . . .

---

[2] It was in fact this rabbit problem that motivated Fibonacci to "discover" the Fibonacci numbers in 1202.

Here we need three seeds because the recursive rule needs three terms to get going. ∎

Example 7. $K_1 = -3$, $K_2 = 4$, $K_N = 2K_{N-1} - 5K_{N-2}$

The above describes the sequence -3, 4, 23, 26, -63,. . . . An explanation might be in order here. How did we get $K_3 = 23$? Using the recursive rule, we have

$$K_3 = 2K_2 - 5K_1 = 2 \times 4 - 5 \times (-3) = 8 + 15 = 23.$$

Likewise,

$$K_4 = 2K_3 - 5K_2 = 46 - 20 = 26, \text{ etc.} \qquad ∎$$

In contrast to the recursive definition of a sequence, we have the **explicit** definition. This is a direct form of description—no preceding terms are needed to calculate any of the terms. We say in Example 1(b) that the sequence of odd positive integers can be described by the rule $O_N = 2N - 1$. This allows us to calculate any term in the sequence in a very direct way. For example, the 378th odd number is $O_{378} = 2 \times 378 - 1 = 755$.

For each of the sequences in Examples 1 through 4 we have given both recursive and explicit definitions. While it is generally true that when both descriptions are available the explicit one is preferable, there are cases where the explicit definition either is not known or is so complicated that we may have second thoughts about using it. Consider the Fibonacci sequence. We know about its recursive definition—it is simplicity itself. How about an explicit definition? Well, it's not pretty, but here it is:

$$F_N = \frac{\left(\dfrac{1 + \sqrt{5}}{2}\right)^N - \left(\dfrac{1 - \sqrt{5}}{2}\right)^N}{\sqrt{5}}.$$

This complicated-looking formula is known as **Binet's formula**.[3] Clearly, this is an explicit definition and no preceding terms are involved. At the same time, Binet's formula is not one that can be easily used without a computer or a good calculator.

Example 8.   As a final example of a sequence, we will consider the sequence R of ratios between successive Fibonacci numbers, in other

[3] The formula was actually published by Leonhard Euler in 1765. The Frenchman Jacques Binet rediscovered it in 1843 and got all the credit.

words, the sequence defined by

$$R_N = \frac{F_{N+1}}{F_N}.$$

Table 9-2 illustrates what we mean.

N	1	2	3	4	5	6	7	8	9	10	11	12	13
F_N	1	1	2	3	5	8	13	21	34	55	89	144	233
R_N	$\frac{1}{1}$	$\frac{2}{1}$	$\frac{3}{2}$	$\frac{5}{3}$	$\frac{8}{5}$	$\frac{13}{8}$	$\frac{21}{13}$	$\frac{34}{21}$	$\frac{55}{34}$	$\frac{89}{55}$	$\frac{144}{89}$	$\frac{233}{144}$	

Table 9-2

When we use a calculator to express the terms of R as decimals, we get (rounded off to five decimal places)

N	1	2	3	4	5	6	7	8	9	10	11	12
R_N	1	2	1.5	1.66667	1.6	1.625	1.61538	1.61905	1.61765	1.61818	1.61798	1.61806

Table 9-3

We leave it to the reader to verify with a calculator that the further we move into the sequence R, the less change there is from one term to the next (see Exercise 10). This point can best be driven home by going to extremes, such as evaluating

$$R_{98} = \frac{F_{99}}{F_{98}} = \frac{218922995834555169026}{135301852344706746049}$$

and

$$R_{99} = \frac{F_{100}}{F_{99}} = \frac{354224848179261915075}{218922995834555169026}.$$

If we carry our division to 40 decimal places, we get

$$R_{98} \approx 1.6180339887498948482045868343656381177203$$

and

$$R_{99} \approx 1.6180339887498948482045868343656381177202.$$

While these two numbers are not identical, they match up in their first 39 decimal places so that the difference between them is practically insignificant. ■

Implicit here is one of the most important facts about Fibonacci numbers: The further we go into the Fibonacci sequence, the closer the ratios of two successive Fibonacci numbers get to each other and to a specific constant. The value of this constant, approximated to three decimal places is 1.618. Its exact value is given by the expression $(1 + \sqrt{5})/2$. This number, called the **golden ratio**, will play a critical role in the rest of this chapter. In the best mathematical tradition of recognizing especially important and distinguished numbers with their own symbol, it is denoted by the Greek letter Φ (phi).

To summarize: When two "large" consecutive Fibonacci numbers are divided (F_{N+1}/F_N) the ratio is almost identical to a very special number—the golden ratio $\Phi = (1 + \sqrt{5})/2$—and the larger the numbers, the closer this ratio is to Φ. An approximate value of Φ is 1.618 but it should be noted that Φ (just like π) is a number with an infinite, nonrepeating decimal expansion. For example, to the first 40 decimal places Φ is given by

$$1.6180339887498948482045868343656381177203.$$

GNOMONS

So far we have discussed some exotic mathematical concepts such as Fibonacci numbers and the golden ratio Φ, but not a word about the theme of the chapter—spiral growth in nature. The connection between these ideas is made through an even more exotic piece of mathematics, the concept of a gnomon.

As far as we can tell, gnomons were first discussed by Aristotle.[4] In geometry, a **gnomon** to a figure A is another figure which, when suitably attached to A results in a new figure A' which is similar (in the geometric sense) to A.

Before we go into gnomons let's have a very quick review of the concept of *geometric similarity*. We know from high school geometry that two objects are similar if one is a scaled version of the other. Thus, two objects are similar when they represent the same picture at different scales. When we take a photo to the photo lab and have it blown up we are in fact dealing with the notion of similarity. Likewise, a slide projector takes a slide and blows it up onto a screen—once again we are dealing with similarity.

---

[4] Aristotle and the Pythagorean school of Greek geometry were fascinated by gnomons and ascribed mystical properties to them. (One must remember that there was a thin line separating religion and geometry in ancient Greece.)

Here is a list of some very basic facts from elementary geometry that we will use in this chapter.

- Two triangles are similar if their sides are proportional. Alternatively, two triangles are similar if the sizes of their respective angles are the same.

- Two squares are always similar.

- Two rectangles are similar if their sides are proportional, that is,

$$\frac{\text{long side 1}}{\text{short side 1}} = \frac{\text{long side 2}}{\text{short side 2}}$$

- Two circles are always similar.

- Two circular rings are similar if their inner and outer radii are proportional, that is,

$$\frac{\text{outer radius 1}}{\text{inner radius 1}} = \frac{\text{outer radius 2}}{\text{inner radius 2}}$$

We are now ready to take on gnomons.

Example 9. Consider square A in Fig. 9-3(a). The L-shaped figure in Fig. 9-3(b) is a gnomon to square A, because when suitably attached to square A [Fig. 9-3(c)], it results in a figure similar to A—the square A'.

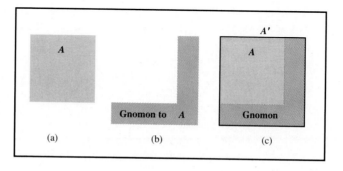

Figure 9-3

Note that the wording is *not* reversible: The square A is not a gnomon to the L-shaped figure since there is no way to combine the two to form a similar L-shaped figure. ∎

Example 10. The O-ring in Fig. 9-4(b) (with inner radius r and outer radius R) is a gnomon to the circle of radius r [Fig. 9-4(a)], since when

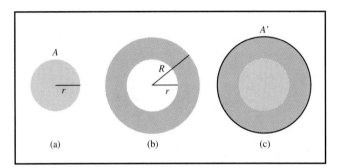

Figure 9-4

we attach the O-ring to the original circle A, we get a new circle A' (and as we know, two circles are always similar). ◼

Example 11. Suppose A is an O-ring with outer radius r [Fig. 9-5(a)]. Consider an O-ring with inner radius r and outer radius R [Fig. 9-5(b)].

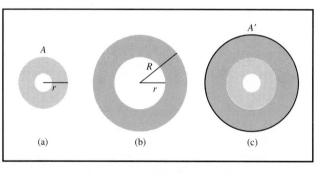

Figure 9-5

Figure 9-5 Red

Is this latter O-ring a gnomon to A? The answer is no, because while we do get another O-ring A' when we attach the two O-rings together [Fig. 9-5(c)], A' is not similar to A. What would a gnomon to A look like? (For an answer, see Exercise 22.) ◼

Example 12. Suppose that we have an arbitrary rectangle A of height h and base b [Fig. 9-6(a)]. The L-shaped object shown in Fig. 9-6(b) is a gnomon to rectangle A as long as the ratios b/h and y/x are equal. In this

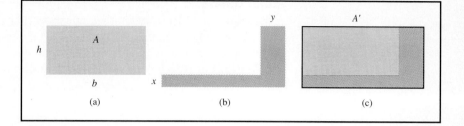

Figure 9-6

case, they can be combined to form a rectangle A' similar to A (see Exercise 28).   ■

Example 13.   Say ABC is an arbitrary triangle with interior angles of α degrees, β degrees, and γ degrees, and suppose α is larger than β [Fig. 9-7(a)]. We can now construct a triangle ABD such that side AD is a

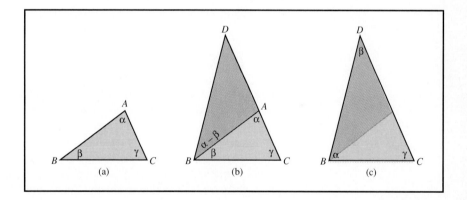

Figure 9-7

continuation of side AC and the interior angle at B is $\alpha - \beta$ degrees [Fig. 9-7(b)]. The triangle ABD is a gnomon to the original triangle ABC because the resulting triangle BDC [Fig. 9-7(c)] has angles the same size as those of the original triangle and is therefore similar to it.   ■

Example 13 illustrates a particularly nice situation much appreciated by Pythagoras and his followers:[5] gnomons for triangles that are themselves triangles. A particularly interesting case of this construction occurs when $\alpha = 72°$, $\beta = 36°$, and $\gamma = 72°$ (see Exercise 29).

[5] The original idea is attributed to Hero of Alexandria, a disciple of Pythagoras.

Example 14.   Here is another question that fascinated Greek geometers: Can a figure be its own gnomon? The answer is yes. Consider the rectangle A with sides l and s (standing for long and short) as shown in Fig. 9-8(a). For A to have itself as a gnomon, it must be the case that the

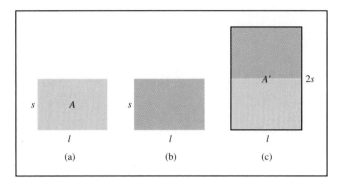

Figure 9-8

rectangle A' of sides l and $2s$ [Fig. 9-8(c)] is similar to A. Here the longer side has length $2s$, and the shorter side has length l. This means that

$$\frac{l}{s} = \frac{2s}{l} \quad \text{or} \quad l^2 = 2s^2; \quad \text{i.e., } l = s\sqrt{2}.$$

In short, if the longer side of a rectangle is $\sqrt{2}$ (≈ 1.414) times the shorter side, the rectangle is its own gnomon.   ■

Example 15.   In this example we raise a slightly more difficult but important question: Under what circumstances does a rectangle have a square gnomon? Consider again the rectangle A with sides l and s [Fig. 9-9(a)]. The only possible hope we could have for a square gnomon is by

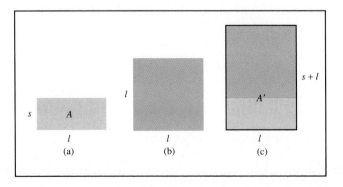

Figure 9-9

using a square of side l, as in Fig. 9-9(b). (A square of side s could never work. The resulting rectangle would have a short side of length s and a long side of length $l + s$ and couldn't possibly be similar to A).

For the resulting rectangle A' [Fig. 9-9(c)] to be similar to A, we must have

$$\frac{l}{s} = \frac{s + l}{l}.$$

Breaking up the right-hand side of this equation gives

$$\frac{l}{s} = \frac{s}{l} + \frac{l}{l} = \frac{s}{l} + 1.$$

We could solve this equation for l in terms of s (or s in terms of l), but a slightly more convenient strategy is to solve the equation for the ratio l/s. If we call $l/s = x$, then $s/l = 1/x$ and

$$\frac{l}{s} = \frac{s}{l} + 1$$

becomes

$$x = \frac{1}{x} + 1,$$

which is equivalent to

$$x^2 = 1 + x$$

and therefore to

$$x^2 - x - 1 = 0.$$

The quadratic formula will now bail us out. (That's the one that says $ax^2 + bx + c = 0$ has solutions

$$x = \frac{-b \pm \sqrt{b^2 - 4ac}}{2a}.)$$

Using it, we get the solutions

$$x = \frac{1 + \sqrt{5}}{2} \approx 1.618 \qquad \text{or} \qquad x = \frac{1 - \sqrt{5}}{2} \approx -0.618.$$

The second solution is impossible (both *l* and *s* are the lengths of the sides of a rectangle so $x = l/s$ cannot be negative!), which means we have a single possible answer: $x = l/s = (1 + \sqrt{5})/2 \approx 1.618$. *A rectangle has a square gnomon if and only if the ratio of the longer side to the shorter side is the golden ratio* $\Phi = (1 + \sqrt{5})/2 \approx 1.618$. ∎

Example 16. This is an example that comes from nature. It also gives us a hint as to why we are so interested in gnomons. Figure 9-10 shows a photograph of a cross section of a well-known seashell—that of a chambered nautilus. The spiral-shaped shell of the chambered nautilus is a wonder of natural architecture and a classic example of nature's ability to combine function with beauty. The nautilus builds its shell in stages, each stage consisting of the addition of a chamber to the already existing shell. At every stage of its growth, the shape of the chambered nautilus shell remains the same—the beautiful and distinctive spiral shown in Fig. 9-10. In essence, we can interpret this as a case in which each new chamber added to the shell is a gnomon to the already existing set of chambers. We will come back to this example soon.

Figure 9-10

∎

THE GOLDEN RATIO

With the possible exception of π, there is no other number that plays a more important role in our physical world than the golden ratio Φ. The ancient Greeks ascribed to it mystical characteristics and called it the *divine proportion*. The great astronomer Johannes Kepler wrote

Geometry has two great treasures: one is the theorem of Pythagoras; the other, the division of a line into [the golden] *ratio. The first we may compare to a measure of gold; the second we may name a precious jewel.*

Many books have been written about the role of the golden ratio in geometry, art, architecture, music, and nature. One of the first of these

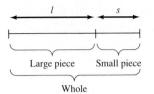

Figure 9-11

was *De Divina Proportione* (*The Divine Proportion*) written in 1509 by Fra Luca Pacioli and beautifully illustrated by Leonardo da Vinci. Among the modern classics devoted to the golden ratio are references 1, 5, and 6.

What is it that sets the number Φ apart from other numbers? The answer is that Φ is blessed with a special and unique virtue: *It strikes a perfect balance between the large and the small.* Let's explain. Suppose we have a rod we want to break into two pieces (one large and one small) (Fig. 9-11) in some sort of perfect balance, that is, in such a way that the large piece is not too large and the small piece is not too small. In essence what this means is that the proportion between the large piece l and the small piece s should be the same as the proportion between the whole piece $(l + s)$ and the large piece l. In other words,

$$\frac{l}{s} = \frac{l + s}{l}.$$

This is exactly the same problem we solved in Example 15, and the solution is given by

$$\frac{l}{s} = \frac{1 + \sqrt{5}}{2} = \Phi \qquad \text{or equivalently} \qquad l = s\Phi.$$

The perfect balance between large and small represented by the golden ratio is particularly striking in a rectangle. A rectangle with sides of length l and s such that their ratio is the golden ratio is called a **golden rectangle**. We were first introduced to golden rectangles in Example 15 where we saw that it is precisely this kind of rectangle that has square gnomons. Figure 9-12 shows some approximate golden rectangles.

Note that the rectangles in Figs 9-12(c) and (d) have sides whose lengths are consecutive Fibonacci numbers. We will call such rectangles **Fibonacci rectangles**. We know from Example 8 that Fibonacci rectangles are only approximations of golden rectangles, but that by the time we get to

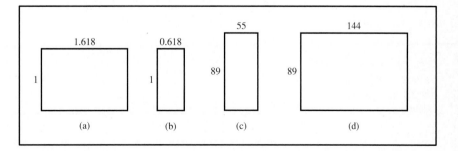

Figure 9-12

rectangles such as the ones in Figs 9-12(c) and (d), these approximations are quite good.

■ The Golden Ratio in History

While examples of the use of the golden ratio in architecture can be traced as far back as biblical times (the ancient Pyramid of Cheops is said to have proportions based on Φ), the Greeks were the first to systematically incorporate the golden ratio into their art and architecture. The famous Greek sculptor Phidias consistently used the golden ratio to obtain the

Horsemen. **One of the friezes on the Parthenon attributed to the Greek sculptor Phidias (c. 440 B.C.). Phidias used the golden ratio to proportion not only the friezes but the Parthenon itself. (British Museum)**

best proportions for his sculptures.[6] Another famous example of the way the golden ratio was used by the Greeks is in the proportions of the Parthenon, as illustrated in Fig. 9-13.

The golden ratio was also extensively used by Renaissance artists, Leonardo da Vinci, Botticelli, and others. Detailed accounts of this can be found in reference 5.

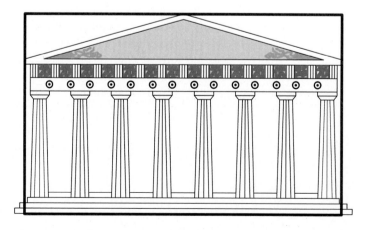

Figure 9-13 The face of the Parthenon fits almost exactly in a golden rectangle.

[6] In fact the choice of Φ to represent the golden ratio comes from Phidias' name.

In 1876 the famous German psychologist Gustav Fechner performed some experiments trying to establish whether certain proportions were more naturally pleasing to people than others. In one of his experiments he let subjects choose from an assortment of rectangles of various proportions the one they found the most aesthetically pleasing to the eye. Table 9-4 summarizes the results of Fechner's experiment. We can see that the golden rectangle was the overwhelming favorite and that over 75% of the subjects chose rectangles with ratios within 10% of the golden ratio.

$R = \dfrac{\text{longer side}}{\text{shorter side}}$	1	1.2	1.25	1.33	1.45	**1.49**	$\Phi \approx 1.618$	**1.75**	2	2.5
Percent of subjects preferring a rectangle with proportions R	3%	0.2%	2%	2.5%	7.7%	**20.6%**	**35%**	**20%**	7.5%	1.5%

Table 9-4.   Fechner's data

Fechner's experiments confirmed the intuitions of Greek sculptors and architects about the special aesthetic value of the golden ratio. Modern merchandisers have taken advantage of this by packaging products in boxes that are of aesthetically pleasing proportions. (Many cereal boxes, for example, have proportions close to the golden ratio. This is supposed to encourage "impulse" buying.)

From Greek temples to Renaissance art to cereal boxes, one number has stood out among all others in the search for beauty and balance—the golden ratio. Not surprisingly, nature itself discovered this number long before humans did.

GNOMONIC GROWTH

We started this chapter with a question about daisies which we now reconsider in a slightly more general form: What are the rules that govern the geometric arrangement of leaves on a stem or of petals on a flower?[7]

The starting point for our discussion is the observation that there is a surprisingly large and varied group of natural objects that have a special affinity for Fibonacci numbers. Figure 9-14, shown on the next page, illustrates some typical examples of this. While there is no generally accepted explanation for this affinity, there is one common element in all the natural objects shown in Fig. 9-14: the nature of their growth.

Natural organisms grow in essentially two different ways. The more

---

[7] The study of the arrangement and distribution of paristiches (leaves, stalks, petals, bracts, etc.) on plants has a technical name: *phyllotaxis*.

**Figure 9-14 (a) Daisies:
Numbers of petals are
most frequently 8, 13, 21,
34, 55, or 89. (b) Pine-
cones: bracts spiral in sets
of 8 and 13 rows. (c)
Sunflowers: Florets spiral
out from the center in sets
of 55 rows and 89 rows. (d)
Pineapples: Scales spiral in
sets of 8, 13, and 21 rows.**

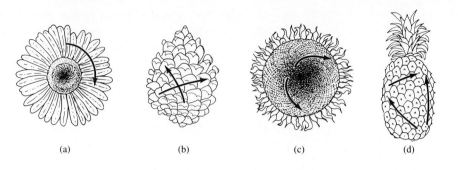

(a)   (b)   (c)   (d)

common type of growth (and the one we are most familiar with) is the
growth exhibited by humans, animals, and many plants. This can be called
all-around growth, in which all living parts of the organism grow simultaneously (although not necessarily at the same rate.) One of the characteristics of this type of growth is that there is no obvious way to distinguish between the newer and the older parts of the organism. In fact,
the distinction between new and old parts does not make much sense.
The historical record (so to speak) of the organism's growth is lost. By
the time the child becomes an adult, no identifiable traces of the child (as
an organism) remain—that's why we need photographs!

Contrast this with the kind of growth exemplified by the shell of the
chambered nautilus, a ram's horn, the trunk of a redwood tree, or the
inflorescence of a daisy. This is a kind of growth that we may informally
call *growth at one end* or *asymmetric growth.* In this type of growth the
organism has a part added to it (either by its own or outside forces) in
such a way that the old organism together with the added part form the
new organism. At any stage of the growth process we can see not only

**A sunflower loves Fibonacci
numbers: 34 petals on the
outside; 55 and 89 rows of
florets on the inside.
(Wendell/Photo
Researchers)**

the present but the organism's entire past. All the previous stages of its growth are the building blocks that make up the present structure.

A second important fact is that in most instances organisms that grow in this fashion do so in a way that preserves their overall shape; in other words, they remain similar to themselves. This is where gnomons come into the picture: Regardless of how the new growth comes about, its shape is a gnomon of the entire organism. We will call this kind of growth process **gnomonic growth**.

We will illustrate the process of gnomonic growth with some examples.

Example 17. We know from Example 10 that the gnomon to a circle is an O-ring with an inner radius equal to the radius of the circle. We can thus have circular gnomonic growth (Fig. 9-15). Rings added one layer at a time to a starting circular structure preserve the circular shape throughout the structure's growth. When carried to three dimensions, this is a good model for the way the trunk of a redwood tree grows. ■

Figure 9-15 Circular gnomonic growth.

Example 18. We have already seen in Example 16 that in the shell of a chambered nautilus, each chamber is a gnomon for the entire shell. The gnomonic growth of the shell proceeds in essence as follows: Starting with the shell of a baby chambered nautilus (which is a tiny spiral similar in all respects to the adult spiral shape), the animal builds a chamber (by producing a special secretion around its body that calcifies and hardens). The resulting, slightly enlarged spiral shell is similar to the original one. The process then repeats itself in a *recursive* way: A new chamber is added (it is a gnomon to the shell that is similar but slightly larger than the first one), resulting in a new enlarged spiral. This process continues ad infinitum or until the animal reaches maturity, whichever comes first. ■

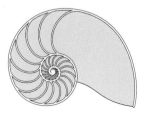

Example 19. This is a fictitious example. Imagine, if you will, a family of small extraterrestrial beings that wants to build itself a structure in which to live. Let's say, for the sake of argument, that the perfect floor plan for this family is an isosceles triangle [Fig. 9-16(a)]. As the family grows, the original triangle becomes too small. They need to move into a new shelter just like the original one only bigger. They can do this by adding a triangle that is a gnomon to the original triangle (see Example 13). This results in the structure shown in Fig. 9-16(b). As the process of growth continues, we get something like the structure shown in Fig. 9-16(c). The successive stages of growth are labeled 1, 2, 3, 4, and 5. This means that by the fifth growth period the structure is the entire triangle shown in Fig. 9-16(c).

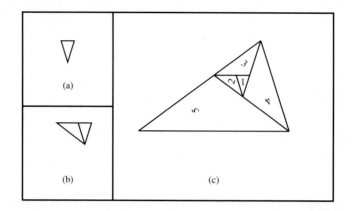

Figure 9-16

If we track the trajectory of any point (for simplicity say a vertex) throughout this process of gnomonic growth, we see a very special kind of spiral shape (Fig. 9-17) called a **logarithmic spiral**. The logarithmic spiral

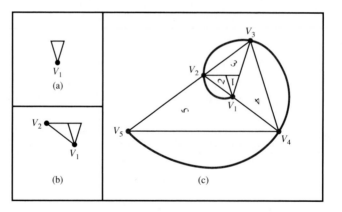

Figure 9-17

is characteristic of gnomonic growth in nature and can be seen in many seashells, animal horns, etc. ∎

Example 20.   Suppose that we start with a Fibonacci rectangle such as the one shown in Fig. 9-18(a). We know from our discussion of Fibonacci numbers and the golden ratio that for all practical purposes this is a golden rectangle. As we discovered in Example 15, it has a square gnomon [Fig. 9-18(b)]. As the process of gnomonic growth continues, we get a nested sequence of Fibonacci rectangles [Fig. 9-18(c)]. Once again,

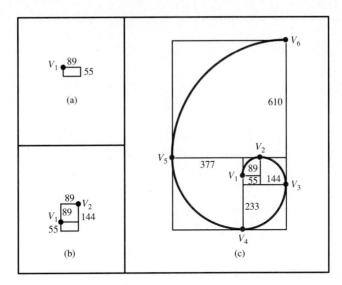

Figure 9-18

if we track the trajectory of any point throughout the growth process, we recognize a logarithmic spiral.   ■

CONCLUSION

Fibonacci numbers, gnomons, the golden ratio, logarithmic spirals—these are the mathematical elements nature uses to produce some of its complex, beautiful forms, among them the daisy, the sunflower, the chambered nautilus, etc. This chapter was not an attempt to explain exactly how and why these relationships between nature and mathematics occur (to the best of our knowledge there are several theories but no confirmed facts) but rather to point out that a relationship does indeed exist.

A technical account of some of the competing theories that attempt to explain the surprising connections between mathematics and spiral growth can be found in references 3, 7, and 9.

KEY CONCEPTS

arithmetic sequence
Binet's formula
explicit definition (of a sequence)
Fibonacci number
Fibonacci rectangle
Fibonacci sequence
geometric sequence
gnomon
gnomonic growth
golden ratio

golden rectangle
recursive definition (of a
  sequence)
recursive process
recursive rule
seeds
sequence
similarity
terms (of a sequence)

EXERCISES

 Walking

1. Suppose F is the Fibonacci sequence.
   (a) Find F_{16}.
   (b) Given $F_{36} = 14930352$ and $F_{37} = 24157817$, find F_{35} and F_{38}.

2. Suppose F is the Fibonacci sequence.
   (a) Find F_{15}.
   (b) Given $F_{31} = 1346269$ and $F_{33} = 3524578$, find F_{32}.

3. Let A_N be the sequence explicitly defined by $A = 3N + 2$.
   (a) Write out the first 10 terms of A.
   (b) What is A_{500}?
   (c) Is 3053 one of the terms of A? If so, which one. If not, why not?

4. Let B be the sequence explicitly defined by $B_N = (3N + 2)/(2N + 1)$.
   (a) Write out the first five terms of B.
   (b) What is B_{23}?
   (c) Explain why 3851/2286 cannot be a term of B.

5. Let C be the sequence explicitly defined by $C_N = 1 + 2(-1)^N$.
   (a) Write out the first five terms of C.
   (b) What is C_{317}?
   (c) Find $C_1 + C_2 + \cdots + C_{100}$.

6. Let D be the sequence explicitly defined by $D_N = 2 + (-1)^{N+1}$.
   (a) Write out the first five terms of D.
   (b) What is D_{317}?
   (c) Find $D_1 + D_2 + \cdots + D_{100}$.

7. Let A be the sequence recursively defined by $A_N = 2A_{N-1} + 4$, with $A_1 = 0$.
   (a) Write out the first 10 terms of A.
   (b) Find A_{N+1} in terms of A_N.
   (c) Is 21,519 one of the terms of A? Why?

8. Let B be the sequence recursively defined by $B_N = 2B_{N-1} + 3$, with $B_1 = 1$.
   (a) Write out the first 10 terms of B.
   (b) Find B_{N+1} in terms of B_N.
   (c) Is 43,214 one of the terms of B? Why?

9. Let C be the sequence recursively defined by $C_N = 2C_{N-1} + 3C_{N-2}$, with $C_1 = 1$ and $C_2 = -1$. Write out the first five terms of C.

10. Suppose F is the Fibonacci sequence and let R be the sequence of ratios between successive Fibonacci numbers, i.e., $R_N = F_{N+1}/F_N$ (see Example 8).

(a) Find R_{13}, R_{14}, and R_{15} rounded off to seven decimal places. (Use a calculator.)

(b) Without doing any calculations, guess the value of R_{16} rounded off to five decimal places. Verify the answer with a calculator.

(c) Explain why if we take $F_{16} \times 1.61803$ and round to the nearest integer we get F_{17}.

11. Let A be the sequence recursively defined by $A_N = A_{N-1} - 2A_{N-2} + 3A_{N-3}$, with $A_1 = 3$, $A_2 = 0$, and $A_3 = 1$. Find the first 10 terms of A.

12. Let B be the sequence recursively defined by $B_N = B_{N-1} - 2B_{N-2} - B_{N-3}$, with $B_1 = 1$, $B_2 = 0$, and $B_3 = -1$. Find the first 10 terms of B.

13. Consider the sequence E recursively defined by $E_1 = 1$, and
$E_N = E_{N-1}/(1 + E_{N-1})$.
(a) Find E_2, E_3, E_4, and E_5.
(b) Suppose that $\frac{1}{213}$ is a term in the sequence. Find the next term in the sequence. (Write your answer in simplified form.)

14. Consider the sequence E recursively defined by $E_1 = 1$, and
$E_N = E_{N-1}/(1 + 2E_{N-1})$.
(a) Find E_2, E_3, E_4, and E_5.
(b) Suppose that $\frac{1}{400}$ is a term in the sequence. Find the next term in the sequence. (Write your answer in simplified form.)

15. Consider the sequence V recursively defined by $V_1 = 2$, and
$V_N = V_{N-1} + 2$.
(a) Find the first six terms of this sequence.
(b) Find V_{100}.
(c) Find an explicit definition of V_N.

16. Consider the sequence W recursively defined by $W_1 = 1$, and
$W_N = 3W_{N-1} - 2$.
(a) Find the first six terms of this sequence.
(b) Find W_{100}.
(c) Find an explicit definition of W_N.

17. Find the values of x and y so that the shaded figure is a gnomon to the given triangle.

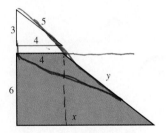

18. Find the values of x and y so that the shaded triangle is a gnomon to the given triangle ABC.

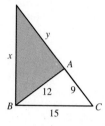

19. Find the value of x so that the shaded figure is a gnomon to the given rectangle.

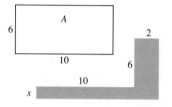

20. Find the length c of the shaded rectangle so that it is a gnomon to the given rectangle with sides 3 and 9.

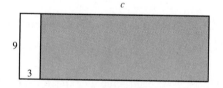

■ **Jogging**

21. Find x, y, and z so that the shaded figure has an area that is eight times the area of the given triangle and at the same time is a gnomon to the given triangle.

22. (a) Which of the following O-rings is similar to I? Explain your answer.

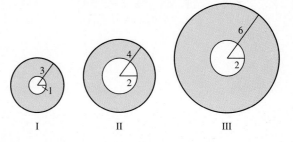

I   II   III

    (b) Explain why an O-ring cannot have a gnomon.

23. Find x and y so that the shaded figure has an area of 75 and at the same time is a gnomon to the given rectangle.

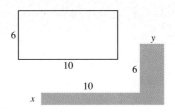

24. Under what conditions is a triangle its own gnomon?

25. Explain why the reciprocal of the golden ratio is exactly 1 less than the golden ratio (i.e., $1/\Phi = \Phi - 1$).

26. In the following figure $ABCD$ is a square and the three triangles I, II, and III have equal areas. Show that x/y is the golden ratio.

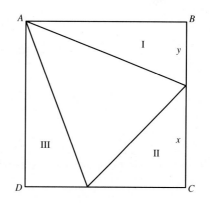

27. Let F be the Fibonacci sequence.
    (a) Compute: $F_1 + F_2 + F_3$, $F_1 + F_2 + F_3 + F_4$, $F_1 + F_2 + F_3 + F_4 + F_5$, etc. (continue until you see the pattern).
    (b) From the pattern found in part (a), state a formula for $F_1 + F_2 + \cdots + F_N$.

28. Show that the L-shaped object in the following figure is a gnomon for rectangle A as long as the ratios b/h and y/x are equal.

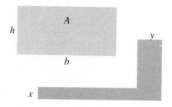

29. Triangle ABC has angles of $72°$, $36°$, and $72°$ as shown in the following figure. Let M be the point on side BC so that AM and AC have equal lengths. The line segment AM divides the triangle into two triangles (I and II) as shown in the figure. Explain why triangle II is a gnomon to triangle I.

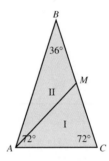

30. (a) Use the explicit description of the Fibonacci sequence

$$F_N = \frac{\left(\dfrac{1 + \sqrt{5}}{2}\right)^N - \left(\dfrac{1 - \sqrt{5}}{2}\right)^N}{\sqrt{5}}$$

and a good calculator to compute F_{10}.

(b) Compare your answer in part (a) with the known value $F_{10} = 55$. Can you explain any discrepancy?

31. Let

$$a = \frac{1 + \sqrt{5}}{2} \quad \text{and} \quad b = \frac{1 - \sqrt{5}}{2}.$$

Without using a calculator, expand and simplify
(a) $a + b$
(b) ab
(c) $a^2 + b^2$
(d) $a^3 + b^3$.

32. Consider the following recursively defined sequence A: $A_1 = 5$, $A_2 = 5$, $A_N = A_{N-1} + A_{N-2}$.

(a) Find A_3, A_4, and $A_{5.}$.

(b) Find a numerical relation between F_N (the Nth term of the Fibonacci sequence) and A_N.

33. Consider the following recursively defined sequence A: $A_1 = 5$, $A_2 = 8$, $A_N = A_{N-1} + A_{N-2}$.

(a) Find A_3, A_4, and $A_{5.}$.

(b) Find a numerical relation between F_N (the Nth term of the Fibonacci sequence) and A_N.

34. In the following figure, suppose that $ABCD$ is a golden rectangle and that $ABEF$ is a square. Show that $ECDF$ is also a golden rectangle.

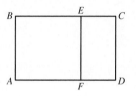

35. Consider the following equation relating various terms of the Fibonacci sequence.

$$F_{N+2}^2 - F_{N+1}^2 = F_N \cdot F_{N+3}.$$

(a) Verify that the equation is true for $N = 1, 2, 3$.

(b) Using the algebraic identity $A^2 - B^2 = (A - B)(A + B)$, show that the equation is true for every positive integer N.

■ Running

36. Show that the sum of any 10 consecutive Fibonacci numbers is a multiple of 11.

37. Show that the equation $(F_N \, F_{N+3})^2 + (2F_{N+1} \, F_{N+2})^2 = (F_{2N+3})^2$ holds for every positive integer N. (F is the Fibonacci sequence.)

38. Show that the geometric sequence G explicitly defined in terms of the golden ratio by

$$G_N = \left(\frac{1 + \sqrt{5}}{2}\right)^N$$

satisfies the same recursive relation as the Fibonacci sequence; that is,

$$G_N = G_{N-1} + G_{N-2}.$$

39. **The Arithmetic-Geometric Mean Inequality.** Let A and B be two sequences defined recursively by

$$B_N = \frac{A_{N-1} + B_{N-1}}{2}$$

and

$$A_N = \sqrt{A_{N-1}B_{N-1}},$$

with $A_1 = 1$ and $B_1 = 9$.

(a) Write out the first four terms of each of the sequences. (Use a calculator if you need to.)

(b) Try to prove that $A_N \leq B_N$. [*Hint*: $(A_{N-1} - B_{N-1})^2 \geq 0$.]

(c) Use part (b) to show that $A_N \geq A_{N-1}$.

40. (a) The triangle *ABC* in the following figure has angles of 36°, 72°, and 72° and sides of length 1, *x*, and *x*. (Triangles such as these are called **golden triangles**.) Show that $x = \Phi$.

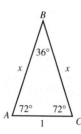

(b) The regular pentagon in the figure has sides of length 1. Show that the length of any one of its diagonals is Φ.

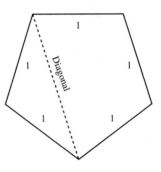

41. During the time of the Greeks the star pentagram was a symbol of the Brotherhood of Pythagoras. A typical diagonal of the large outside regular pentagon is broken up into three segments of lengths *x*, *y*, and *z* as shown in the following figure.

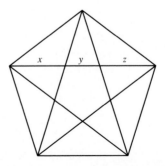

(a) Show that $\dfrac{x}{y} = \Phi,\quad \dfrac{x + y}{z} = \Phi,\quad \dfrac{x + y + z}{x + y} = \Phi.$

(b) Show that if $y = 1$, then $x = \Phi,\quad x + y = \Phi^2$, and $x + y + z = \Phi^3$.

42. **The puzzle of the missing area.** Consider a square 8 units on a side and cut into four pieces as shown in the following figure.

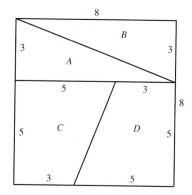

If we rearrange the pieces into a rectangle as shown in the next figure, we see that although the square has area $8 \times 8 = 64$, the rectangle has area $13 \times 5 = 65$.

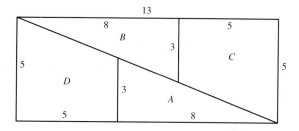

(a) Draw similar figures using other (larger) Fibonacci numbers. How does the area of the square compare with the area of the rectangle?

(b) Explain the discrepancies in the areas.

(c) Consider the figures shown on the next page.

What are the conditions on a and b so that this puzzle is not a puzzle, that is, the areas are the same?

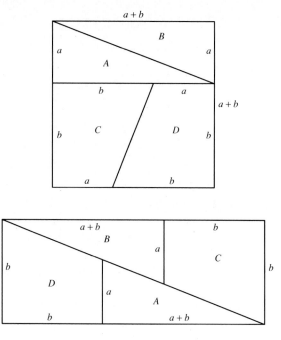

REFERENCES AND FURTHER READINGS

1. Cook, Theodore, *The Curves of Life*. New York: Dover Publications, Inc., 1979.

2. Coxeter, H. S. M., *Introduction to Geometry*. New York: John Wiley & Sons, Inc., 1961, chap. 11.

3. Erickson, R. O., "The Geometry of Phyllotaxis," in *The Growth and Functioning of Leaves*, ed. J. E. Dale and F. L. Milthrope. New York: Cambridge University Press, 1983.

4. Gardner, Martin, "About Phi, an Irrational Number that Has Some Remarkable Geometrical Expressions," *Scientific American*, 201 (August 1959), 128–134.

5. Ghyka, Matila, *The Geometry of Art and Life*. New York: Dover Publications, Inc., 1977.

6. Huntley, H. E., *The Divine Proportion: A Study in Mathematical Beauty*. New York: Dover Publications, Inc., 1970.

7. Jean, R. V., *Mathematical Approach to Pattern Form in Plant Growth*. New York: John Wiley & Sons, Inc., 1984.

8. Kappraff, Jay, *Connections: The Geometric Bridge Between Art and Science*. New York: McGraw-Hill Book Company, 1991.

9. Thompson, D'Arcy, *On Growth and Form*. New York: Macmillan Publishing Co., Inc., 1942, chaps. 11, 13, and 14.

10 The Growth of Populations

There Is Strength in Numbers

In Example 5, Chapter 9, we discussed for the first time a simple problem about population growth: What happens to a family of rabbits if it is allowed to breed without restraint. We will devote this entire chapter to the study of problems of a similar kind.

The rise and fall of populations of various kinds is a very general problem, and one of fundamental importance to all of us. We all know that rain forests are dwindling, condors are almost extinct, there are too many cars on the road, and we are being buried (metaphorically speaking, of course) in mountains of garbage. All these issues fall under the heading of problems in population growth.

The breadth of meaning the expression "population growth" has acquired in modern times stems from the generous way in which both the word "population" and the word "growth" are used. The Latin root of the word "population" is *populus* (which means "people"), so that in its original interpretation the word refers to human populations. This original scope has been expanded, however, so that now the word can be applied to any collection of objects (animate or inanimate) that changes over time and about which we want to make a numerical or quantitative statement.

The many faces of "population growth." (Clockwise from top left Carl Frank/Photo Researchers, Carl B. Koford/Photo Researchers, Eugene Gordon, AP/Wide World Photos)

Thus, we speak of populations of fish, insects, tires, aluminum cans, dollars, and occasionally even of people.

Second, we normally think of the word "growth" as being applied to things that get bigger, but in this chapter we will ascribe a slightly more technical meaning to it: "Growth" can mean decay (i.e., getting smaller) as well as real growth (i.e., getting bigger). This is convenient, because often we don't know ahead of time what is going to happen when studying a population. Is it going to go up or down? By allowing "growth" to mean up or down, we do not need to concern ourselves with making this distinction.

THE DYNAMICS OF POPULATION GROWTH

The growth of a population is a **dynamical process**, meaning that it represents values that change over time. Mathematicians distinguish two kinds of situations: continuous growth and discrete growth. In **continuous growth** the dynamics of change are in effect all the time—every hour, every minute, every second, there is growth. A typical example of this kind of growth is represented by money left in an account that is drawing interest on a continuous basis (yes, there are banks that offer such accounts). We will not study this type of growth in this chapter.

The second type of growth, **discrete growth**, is the most common and natural way by which populations change. We can think of it as a *stop-and-go* type of situation: For a while nothing happens, then there is a sudden change in the population (we will call this a **transition**), then for a while nothing happens again, then another transition takes place, and so on. Of course the "for a while" (i.e., the period between transitions) can be 100 years, an hour, or a millionth of a second. To us the length of time between transitions will not make a difference. The human population of our planet is an example of what we mean: Nothing happens until someone is born or someone dies, at which point there is a change (+ 1 or − 1); then, again there is no change until the next birth or death. Since, however, someone is born every fraction of a second and someone dies slightly less often, it is somewhat tempting to think that the world's human population is for all practical purposes changing in a continuous way. On the other hand, the laws of growth affecting the world's population may be only quantitatively different from the laws of growth affecting the population of Hinsdale County, Colorado (population 408),[1] where a change in the population may not come about for months or even years.

The basic problem of population growth is to figure out what happens to a given population over time. Sometimes we talk about a specific period of time ("The Hispanic population of the United States will grow by 30% before the end of the century."[2]), and sometimes we may talk about the long-term behavior of the population ("The wild rhino population is heading for extinction."[3]). In either case, the most basic way to deal with the question of growth of a particular population is to find the rules that govern the transitions. We will call these the **transition rules**. After all, if we have a way to figure out how the population changes each time there is a transition, then (with a little help from mathematics) we can usually figure out how the population changes after many transitions. In this sense, the growth of a population over time can be conveniently thought of as a

[1] Source: *The World Almanac and Book of Facts*, 1990.

[2] U.S. Bureau of the Census, *Statistical Abstract of the United States: 1990*. 110th edition, Washington D.C., 1990, 14.

[3] See reference 8.

sequence of numbers, something we discussed in depth in the preceding chapter (we had our motives!). We will use P to represent a **population sequence** and we will let P_1 represent the starting population, P_2 the population after the first transition, P_3 the population after two transitions, P_{174} the population after 173 transitions, and so on.

Let's review Example 5 in Chapter 9 (Fibonacci's rabbits) in the context of population sequences.

Example 1 (Fibonacci's Rabbits Revisited).   Recall that in this example we discussed a family of rabbits that grew by the following rules:

1.   It takes 1 month for a newborn baby rabbit to become a mature rabbit.

2.   In 1 month, a mature pair of rabbits will produce one pair (male and female) of baby rabbits.

We first note that according to these rules, all the action takes place at the end of a monthly period (in other words, the period between transitions is 1 month.) From our detailed discussion of this problem in Chapter 9, we know that the transition rules are $P_N = P_{N-1} + P_{N-2}$. (Notice that we have changed from F to P because we are talking about populations, but this is still the same old Fibonacci sequence.) We can now rephrase the original question, What is the size of the population of rabbits (in pairs) after 1 year? to read, What is the size of the population after 12 transitions? The answer is $P_{13} = F_{13} = 233$ pairs (male and female) of rabbits. ∎

Before we take up the study of specific types of population growth, we will make one final comment of a general nature. Even the most superficial look at the assumptions of the rabbit growth example indicates that they are absurdly simplistic. Rabbits do not breed in assembly-like fashion (once a month, one male-female pair producing one male-female pair, etc.) In fact, if we think about it, it would be almost impossible to find a set of rules that accurately describes the exact way that rabbits breed. First, there are too many variables involved (health, the environment, rabbit love, etc.), some of which we understand and some of which we don't. Second, even if we knew and understood all the variables involved, the set of rules would be so complicated and there would be so many of them that the logistics of expressing them mathematically and then applying them in the form of transition rules to get the population sequence would get out of hand.

The question then becomes, What good does it do to use absurdly simplistic rules that are far from reflecting the reality of how a population makes transitions in its growth? The answer is surprising: We can make excellent predictions about the growth of a population over time even when we don't have a completely realistic set of transition rules. The secret is to capture the variables that are really influential in determining how the population grows, put them into a few transition rules that describe how the variables interact, and forget about the small things. This, of course, is easier said than done. In essence, it is what population biologists and mathematical ecologists do for a living, and it is as much an art as it is a science.

For the rest of this chapter we will scratch the surface of this kind of activity. We will discuss three of the most basic models of population growth: linear growth, exponential growth, and logistic growth.

LINEAR GROWTH

Example 2. A newly opened garbage dump has collected 8000 tons of garbage since its opening. It is projected that the dump will collect 120 tons of garbage per month over the next several years. At this rate how much garbage will the dump have collected 5 years from now?

In this example, the population we are studying is the garbage collected by the particular dump in question. Since we are looking at monthly averages, by definition the transitions occur once a month. (This, of course, is only the way we look at it on paper; in reality, the transitions occur every time a garbage truck dumps a load, but for our purposes that's just a stinking detail.) The essential fact about this population growth problem is that the monthly garbage at the dump grows by a constant amount of 120 tons a month.

Our starting population (P_1) is 8000 tons. Thus, we have the following sequence P:

$$P_1 = 8000, P_2 = 8120, P_3 = 8240, P_4 = 8360, \ldots.$$

In 5 years we will have had 60 transitions, each of which will represent an increase of 120 tons. In short, the answer to our question is

$$P_{61} = 8000 + 60(120) = 15{,}200 \text{ tons.} \qquad \blacksquare$$

Example 2 is typical of a general situation called **linear growth**. In a linear growth situation there is a starting population P_1 and a transition rule in which a constant amount d is added during each transition. The recursive description of the sequence P is

$$P_N = P_{N-1} + d, \text{ with starting population } P_1.$$

The sequence P also has a simple explicit description, given by

$$P_{N+1} = P_1 + Nd.$$

(After all, we start at P_1 and make N transitions, each one representing an increase by the amount d.)

Any sequence in which each term is obtained by adding a fixed amount d to the preceding term is called an **arithmetic sequence**. The number d is called the **common difference** (two consecutive terms differ by d). Our first encounter with arithmetic sequences was in Example 2, Chapter 9.

■ **Plotting Population Growth**

A very convenient way to describe population growth is by means of a **plot** or **graph**. The horizontal axis usually represents time (with the tick marks generally corresponding to the transitions), and the vertical axis usually represents the size of the population. Because we have complete freedom in choosing both the horizontal and vertical scales, plots can be misleading. Consider Figs. 10-1(a) and (b). Which population grows faster?

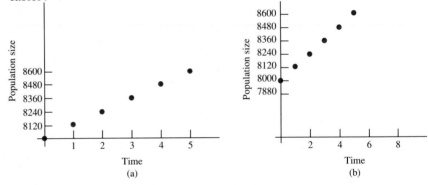

Figure 10-1

Actually, both plots represent the growth of the same population: the garbage problem in Example 2. These plots illustrate why linear growth is called linear growth—no matter how we plot it, the values of the population *line up* (in a straight line).

Example 3.   Jane Doe is a company that manufactures tractors. The company has decided to start up a new plant. On the first of each month (for a period of 6 years), a module of equipment that will produce three tractors per month will be installed in this plant. What is the total number of tractors that the company will produce over the 6 years the program is in effect ?

In this problem, the total production of each module conforms to a linear growth pattern, but the number of months each module works is different. Let's make a list:

- Module installed the 1st month works for 72 months, producing $3 \times 72 = 216$ tractors.

- Module installed the 2nd month works for 71 months, producing $3 \times 71 = 213$ tractors.

- Module installed the 3rd month works for 70 months, producing $3 \times 70 = 210$ tractors.

.

.

.

- Module installed the 72nd month works for 1 month, producing 3 tractors.

The total number of tractors produced at the end of 72 months is

$$216 + 213 + 210 + \cdots + 3.$$

This sum is the sum of consecutive terms of the arithmetic sequence 3, 6, 9, . . ., 213, 216. We could, of course, add these numbers up, with or without a calculator, but that's dull. Let's take a slightly more elegant tack. Let's write our total twice, once forward and once backward:

$$\text{Total} = 216 + 213 + 210 + \cdots + 6 + 3.$$
$$\text{Total} = 3 + 6 + 9 + \cdots + 213 + 216.$$

If we add each term in the first row to each term in the second row, we get

$$2 \times \text{Total} = 219 + 219 + 219 + \cdots + 219 + 219.$$

Since there are 72 such terms, we end up with

$$2 \times \text{Total} = 219 \times 72,$$

and therefore

$$\text{Total} = \frac{219 \times 72}{2} = 7884. \qquad \blacksquare$$

The same approach works with *any* arithmetic sequence: We can add up any number of consecutive terms in a very convenient way. The formula is

---

**Adding N Consecutive Terms
of an Arithmetic Sequence**

$$\text{First term} + \cdots + \text{last term} = \frac{(\text{first term} + \text{last term}) \times N}{2}.$$

Example 4. $\underbrace{5 + 12 + 19 + 26 + 33 + \cdots}_{132 \text{ terms}} = ?$

Here we are adding 132 consecutive terms of an arithmetic sequence P. The first term is $P_1 = 5$; the common difference is $d = 7$. We need to find the 132nd term P_{132}. We already know how to do this: $P_{132} = 5 + 131 \times 7 = 922$. We can now apply the formula:

$$5 + 12 + 19 + 26 + 33 + \cdots + 922 = \frac{(5 + 922) \times 132}{2} = 61,182.$$

■

Example 5. $4 + 13 + 22 + 31 + 40 + \cdots + 922 = ?$

Here we are adding the terms of an arithmetic sequence with $P_1 = 4$ and $d = 9$. To apply the formula we need to find the number of terms N. To find N we set up an equation: $922 = 4 + 9(N - 1)$. From it we get $9(N - 1) = 918$, and therefore $N - 1 = 102$ and $N = 103$. It follows that

$$4 + 13 + 22 + 31 + 40 + \cdots + 922 = \frac{(4 + 922) \times 103}{2} = 47,689.$$

■

**EXPONENTIAL
GROWTH**

Before we start our discussion of exponential growth in earnest, let's look at a couple of preliminary examples.

Example 6. A firm manufactures an item at a cost of C dollars. The item is marked up 10% and sold to a distributor. The distributor then marks the item up 20% (based on the price he paid) and sells the item to a retailer. The retailer marks the price up 50% and sells the item to the public. By what percent has the item been marked up over its original cost?

- Original cost of item: C

- Price to distributor after 10% mark up (D): D = 110% of C = $(1.1)C$

- Price to retailer after 20% mark up (R): R = 120% of D = $(1.2)D$ = $(1.2)(1.1)C$ = $(1.32)C$

- Price to the public after 50% markup (P): P = 150% of R = $(1.5)R$ = $(1.5)(1.32)C$ = $(1.98)C$.

Therefore, the markup over the original cost is 98%. ■

Example 7. A retailer buys an item for C dollars and marks it up 80%. He then puts the item on sale for 40% off the marked price. What is the net percentage mark up on this item?

- Original cost of item: C

- Price after 80% mark up (P): P = 180% of C = $(1.8)C$

- Sale price after 40% discount (S): S = 60% of P = $(0.6)P$ = $(0.6)(1.8)C$ = $(1.08)C$

The net markup is 8%. ■

We are now ready to tackle exponential growth.

Example 8. $1000 is deposited in a retirement account that pays 10% *annual* interest (i.e., interest is paid once a year at the end of the year). How much money is there in the account after 25 years if the interest is left in the account?

This example is a typical case of a situation involving **exponential growth**: The money draws interest; then the money plus the interest draw interest; and so on. While we will concentrate on examples from the world of finance, exponential growth represents a common pattern of growth in many other areas as well. The essence of exponential growth is *recursive multiplication*: Each transition consists of a multiplication by some fixed amount called the **growth rate**. We will illustrate these ideas by means of Example 8. Table 10-1 will help us get started.

	Account Balance at Beginning of Year	Interest Earned for the Year	Account Balance at End of Year
Year 1	$1000	$100	$1100
Year 2	1100	110	1210
Year 3	1210	121	1331
⋮	⋮	⋮	⋮
Year 24	???	???	???
Year 25	???	???	???

Table 10-1

The critical observation is that the account balance at the end of year 1 is obtained by adding the **principal** ($1000) and the interest earned for the year (10% of $1000). We know that this is equivalent to taking 110% of $1000—in other words $1000 × 1.1. Likewise the account balance at the end of year 2 is

(Account balance at beginning of year 2) × (1.1) = $1000 × $(1.1)^2$.
　　　　　　　　　　　　　　　　　　　　　$1000 × (1.1)

It isn't hard to see that each transition (which occurs at the end of each year) corresponds to taking 110% of the money at the start of that year. This, of course, is the same as a multiplication by 1.1. In our terminology, we will say that the rate of growth in Example 8 is 1.1 (not 10% as is often mistakenly thought). The answer to the question raised in Example 8 is now easy: The account balance at the end of year 25 is

$1000 × $(1.1)^{25}$ = $10,834.70.

In this example the period between transitions is 1 year; the population (the money in the account) is given by the sequence

$P_1 = \$1000, P_2 = \$1100, P_3 = \$1210, \ldots, P_{26} = \$10,834.70,$

and in general, at the end of N years we have a balance of

$P_{N+1} = \$1000 \times (1.1)^N.$

Figure 10-2 plots the growth of the money in the account for the first 8 years. ■

A sequence of the kind described in Example 8, where every term is obtained by multiplying the preceding term by a fixed amount, is called

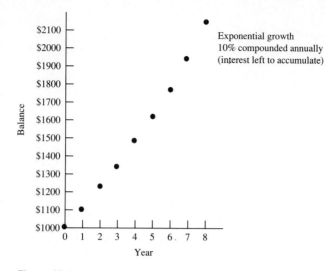

Exponential growth
10% compounded annually
(interest left to accumulate)

Figure 10-2

a **geometric sequence**. In any exponential growth problem, the size of the population is always a term in some geometric sequence, and it can therefore be expressed as

$$P_{N+1} = P_N \times (\text{growth rate})$$

or, alternatively,

$$P_{N+1} = P_N \times (\text{growth rate})^N.$$

A common misconception is that in an exponential growth situation the population always gets bigger. The truth of the matter is that this need not be the case.

Example 9.   A population grows exponentially with an annual growth rate of 0.3 and starting population $P_1 = 1,000,000$. What is the size of the population at the end of 6 years?

In this case, we have $P_7 = 1,000,000 \times (0.3)^6$. We don't need a calculator to compute this number:

$$P_7 = 10^6 \times \frac{3^6}{10^6} = 3^6 = 729.$$

Figure 10-3 plots the growth of this population for the first 6 years, and we can clearly see that it is heading toward extinction. ∎

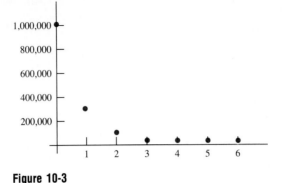

Figure 10-3

It is often convenient to distinguish between exponential growth situations in which populations get bigger (as in Example 8) and exponential growth situations in which populations actually decrease (as in Example 9). The latter situation is commonly referred to as **exponential decay**. The difference, of course, is in the growth rate: *If it is positive but less than 1, we have decay; if it is more than 1, we have real growth.*

■ Putting Your Money Where Your Math Is

Let's return to the world of savings accounts. In Example 8 we discussed a special case of a general situation: A certain sum of money P_1 (called the *principal*) is deposited in an account that draws interest at an annual interest rate r (the interest is paid once a year at the end of the year). How much money is there in the account at the end of N years assuming that the interest is left in the account? We now know that the answer is given by the formula

$$P_{N+1} = P_1 \times (\text{growth rate})^N.$$

How do we compute the growth rate? When the annual interest was 10% (Example 8), we computed the growth rate as 1.1 (110%). If the annual interest had been 12%, the growth rate would have been 1.12 (112%), and if the annual interest had been $6\frac{3}{4}\%$, the annual growth rate would have been 1.0675 (106.75%). In general, if we write the annual interest rate as a decimal r (rather than a percent), then the growth rate will be $1 + r$. Placing this in the preceding formula gives

$$P_{N+1} = P_1(1 + r)^N.$$

Example 10. If we deposit $367.51 in a savings account yielding an annual interest of $9\frac{1}{2}\%$ a year and we leave both the principal and the interest in the account, how much money will there be in the account at the end of 7 years?

Here $P_1 = 367.51$, $r = 0.095$, and $N = 7$. The answer is

$$P_8 = 367.51 \times (1.095)^7 = 693.69.$$ ■

Example 11. Let's now consider a variation of Example 8. Suppose we find a bank that pays 10% annual interest compounded *monthly*. If we deposit $1000 (and leave the interest in the account), how much money will there be in the account at the end of 5 years?

This problem is still a problem in exponential growth. The big difference now is that the period between transitions is a month (instead of a year). At the end of 5 years we will have gone through 60 transitions, so in this example we are after P_{61}. Just as before (remember we are dealing with a geometric sequence), we have

$$P_{61} = \$1000 \times (\text{growth rate})^{60},$$

and just as before, it all boils down to finding the growth rate. Since the period between transitions is a month, we can't use the 10% annual interest to compute the growth rate. Instead, we must divide the annual interest by 12, which gives the **periodic interest**

$$\frac{10\%}{12} = 0.83333 \ldots \%.$$

The growth rate is then $100.83333 \ldots \%$ which, written in decimal form is $1.0083333 \ldots$, and therefore

$$P_{61} = \$1000 \times (1.008333 \ldots)^{60} \approx \$1645.30.$$

Just out of curiosity, what would happen if we left the money in the account for 25 years? Everything is the same as in our last computation, except that the number of transitions is $25 \times 12 = 300$. It follows that after 25 years the amount of money in the account will be

$$P_{301} = \$1000 \times (1.008333 \ldots)^{300} \approx \$12{,}056.94.$$ ■

Example 12. Now suppose that we find a bank that pays 10% annual interest compounded *daily*. If we deposit the same $1000 for 5 years (just as we did in Example 11), how much will we have at the end of the 5 years?

In this case the period between transitions is 1 day. The total number of transitions in 5 years is $365 \times 5 = 1825$, and the periodic interest is 10%/365, which we write as $0.10/365 \approx 0.00027397$. The growth rate for this exponential growth problem is

$$\left(1 + \frac{0.10}{365}\right) \approx 1.00027397,$$

so that the final answer is

$$1000 \left(1 + \frac{0.10}{365}\right)^{365 \times 5} \approx 1000(1.00027397)^{1825} \approx 1648.60. \quad \blacksquare$$

Examples 11 and 12 illustrate the general rules for growth under compound interest after N transitions. They are, in short:

■ $P_{N+1} = P_1 \times$ (growth rate)^N.

■ Growth rate $= 1 +$ periodic interest.

■ Periodic interest $= \dfrac{\text{annual interest}}{\text{number of periods in 1 year}}$.

Example 13.   (Shopping for a Bank). You have an undisclosed amount of money to invest. Bank A offers savings accounts that pay 10% annual interest compounded yearly. Bank B offers accounts that pay 9.75% annual interest compounded monthly. Bank C offers accounts that pay 9.5% annual interest compounded daily. Which bank offers the best deal?

Note that the problem does not indicate the amount of money we invest or the length of time we plan to leave the money in the account. The answer to the problem depends only on the annual interest and the compounding period. The way to compare these different accounts is to use a common yardstick, for example, How much does $1 grow in 1 year?

With bank A, in 1 year $1 becomes $1.10.

With bank B, in one year $1 becomes

$$\$\left(1 + \frac{0.0975}{12}\right)^{12} \approx \$1.102,$$

And with bank C, in one year \$1 becomes

$$\$\left(1 + \frac{0.095}{365}\right)^{365} \approx \$1.0996.$$

We can now see that bank B offers the best deal.

These same calculations are described by banks in a slightly different form called the annual yield. The **annual yield** is the percentage increase that the account will produce in 1 year. In Example 13, the annual yield for bank A is 10%, for bank B, 10.2%, and for bank C, 9.96%. These numbers can be read off directly from the preceding calculations. ■

One of the tidbits we learned in Example 3 is that there is a straightforward formula that allows us to add up the beginning terms of any arithmetic sequence. We would like to be able to do something similar with a geometric sequence. The basic fact that we need is given by the following formula:

**Adding Consecutive Terms
of a Geometric Sequence**

$$1 + r + r^2 + \cdots + r^N = \frac{r^{N+1} - 1}{r - 1}.$$

(For an explanation, see Exercise 31.) Notice that the above formula only tells us how to find the sum of the beginning terms of a geometric sequence that starts with a 1. However, some very basic algebra will allow us to do the rest.

Example 14. Find
$$8 + 8 \times 3 + 8 \times 3^2 + 8 \times 3^3 + \cdots + 8 \times 3^{13}.$$

This is the sum of the terms of a geometric sequence with first term 8 and growth rate equal to 3. If we factor out the common factor 8, we have

$$8 \times (1 + 3 + 3^2 + 3^3 + \cdots + 3^{13}).$$

According to our formula, the sum inside the parentheses is

$$\frac{3^{14} - 1}{3 - 1},$$

and the whole thing is therefore

$$8 \times \left(\frac{3^{14} - 1}{3 - 1}\right) = 19{,}131{,}872. \qquad \blacksquare$$

(The reader is encouraged to try Exercises 17 and 18 at this point.)

Example 15. A mother decides to set up a college trust fund for her newborn child. The plan is to deposit $100 a month for the next 18 years (i.e., 216 months) in a savings account that pays 6% annual interest compounded monthly. How much money will there be in the account at the end of 18 years?

This is a problem of exponential growth with a twist: Each $100 deposit grows at the same rate of 1.005 [growth rate $= 1 + (0.06/12) = 1.005$] but the number of periods is different for each deposit. Let's make a list:

- ■ 1st deposit of $100 draws interest compounded for 216 months, producing $100(1.005)^{216}$.

- ■ 2nd deposit of $100 draws interest compounded for 215 months, producing $100(1.005)^{215}$.

- ■ 3rd deposit of $100 draws interest compounded for 214 months, producing $100(1.005)^{214}$.

$$\cdot$$
$$\cdot$$
$$\cdot$$

- ■ 216th deposit of $100 draws interest for 1 month, producing $100(1.005)^{1}$.

The total amount in the account at the end of 18 years will be

$$100(1.005)^{216} + \cdots + 100(1.005).$$

This is the sum of the beginning terms of a geometric sequence (written backward). We use the same approach we used in Example 14 and get

$$\$100(1.005)\left[\frac{(1.005)^{216} - 1)}{1.005 - 1}\right] \approx \$38{,}929.$$

Eighteen years from now this amount of money might cover tuition for about one semester! ■

LOGISTIC GROWTH

When dealing with animal populations, the two models we have studied so far are mostly inadequate. As we now know, *linear growth* represents the case in which there is a fixed amount of growth during each period between transitions. This model might work with inanimate objects (garbage, production goods, sales figures, etc.) but fails completely when there is an element of breeding that must be taken into account. *Exponential growth*, on the other hand, represents the case in which there is a constant rate of growth, which is appropriate when there is unrestrained breeding (as with money left to compound in a bank account). In population biology, however, it is generally the case that the rate of growth of an animal population is not always the same. It depends on the relative sizes of other interacting populations (predators, prey, etc.) and, even more importantly, on the relative size of the population itself. When the relative size of the population is small (we will define more precisely what we mean by this soon) and there is plenty of room to grow, then the rate of growth is high. As the population gets larger, there is less room to grow and the growth rate starts to taper off. Sometimes the population gets too large for its own good, leading toward decay and possibly to extinction.

A well-known experiment with rats in a cage illustrates some of these ideas. Put a few rats in a cage with plenty of food. If the cage is big enough, the rats will start breeding in an unrestrained fashion, and for a while the growth of the rat population will follow an exponential growth model. As the cage gets more crowded, the rate of growth will slow down dramatically. The force that regulates this slowdown is competition for the resources that are essential for growth: food, sex, and space. Eventually, the competition gets so keen that the rats start killing each other off—it is their own quick fix for dealing with the overcrowding problem. Often when the population gets back down to an acceptable level, the cannibalism stops. Sometimes nature's growth-regulating mechanism may get out of kilter—the killing frenzy may not quite stop in time, and the rats will wipe each other out in a kind of rodent nuclear holocaust.

The above scenario applies (with variations) to almost every situation in which there is a limited environment for a population to live in. Population biologists call such an environment the **habitat**. The habitat might be a cage (as in the example of the rats), a lake (as for a population of fish), a garden (as for a population of snails), and of course the planet itself which is everyone's habitat.

Of the many mathematical models that attempt to deal with the basic principle of a variable growth rate in a fixed habitat, the simplest is the **logistic growth model**. To put it in a very informal way, the key idea in the logistic growth model is that the rate of growth of the population is directly proportional to the amount of elbow room available in the population's habitat. Thus, lots of elbow room means a high growth rate, little elbow room means a low growth rate (possibly less than 1 which, as we know, means that the population is actually going down), and finally if it ever gets to be the case that the habitat is completely saturated, the population will die out.

There are two equivalent ways we can describe the situation mathematically. If C is some constant that describes the total saturation point of the habitat (population biologists call C the **carrying capacity** of the habitat), then for a population of size P_N we can say that the amount of elbow room is the difference between the carrying capacity and the population size, namely, $(C - P_N)$. Then, if the growth rate is proportional to the amount of elbow room (as described above) we have

$$\text{Growth rate for period } N = R(C - P_N)$$

(where R is a constant of proportionality that depends only on the particular population we are studying). Using the fact that (population at period N) × (growth rate for period N) = population for period $(N + 1)$, we get the following transition rule for the logistic growth model:

$$P_{N+1} = R(C - P_N)P_N .$$

There are two constants in the above transition rule: R which depends on the animal population we are studying, and C which depends on the habitat.

A computationally more convenient way to describe exactly the same thing is to put everything in relative terms: The maximum of the population is 1 (i.e., 100% of the habitat is taken up by the population); the minimum is 0 (i.e., the population is extinct); and every possible population size P_N is represented by some fraction between 0 and 1 which we will denote by p_N (to distinguish it from P_N). The relative amount of elbow room then is $(1 - p_N)$, and the transition rules for the logistic model can be rewritten in the form of the following equation, called the **logistic equation**[4]:

$$p_{N+1} = r(1 - p_N)p_N .$$

In this equation the value p_N represents the fraction of the habitat's carrying capacity taken up by P_N ($p_N = P_N/C$), and the constant r depends

---

[4] This equation is sometimes known as the *Verhulst equation* after the Frenchman P. F. Verhulst who proposed it in the late nineteenth century.

on both the original growth rate R and the habitat's carrying capacity C. We will call r the **growth parameter**.

Because it looks at population growth using a single common yardstick (the fraction of its habitat's carrying capacity taken up by the population), the second description is particularly convenient when making growth comparisons between populations and is the one preferred by ecologists and population biologists. We will stick to it ourselves. In the examples that follow we will look at the growth pattern of an imaginary population under the logistic growth model. In each case, all we need to get started is the original population p_1 (given as a fraction of the habitat's carrying capacity) and the value of the growth parameter r. The logistic equation rule and recursion will do the rest. Needless to say, a good calculator is essential.

Example 16. Suppose that we have a pond we wish to stock with a particular type of fish—let's say for the sake of argument, a particular kind of trout. We are told that the growth parameter for this breed of trout is $r = 2.5$. Let's suppose, moreover, that our starting population is $p_1 = 0.2$ (remember, this means that we stocked the pond to 20% of its carrying capacity). Now let's see what the logistic growth model predicts. After one breeding season[5] we have

(The population of the pond has doubled. Things are looking good!)

$$p_2 = 2.5 \times (1 - 0.2) \times (0.2) = 0.4.$$

After the next breeding season we have

$$p_3 = 2.5 \times (1 - 0.4) \times (0.4) = 0.6.$$

(The growth rate has slowed down a bit.)

By the third generation, we have

$$p_4 = 2.5 \times (1 - 0.6) \times (0.6) = 0.6.$$

Let's try one more.

$$p_5 = 2.5 \times (1 - 0.6) \times (0.6) = 0.6.$$

It is quite clear that the trout population has now stabilized at 60% of the pond's carrying capacity, and unless some external change is made, it will remain at the same level for all future generations. This may be

---

[5] In animal populations, the transitions usually correspond to breeding seasons.

somewhat disappointing for the grower, but it is a perfect situation for the trout. We have here an example of a population that is in stable balance with its environment. ∎

Example 17. Suppose that we have the same pond and the same trout as in Example 16 (in other words, we still have $r = 2.5$), but we wonder what would happen if we stocked the pond differently—let's say we started with $p_1 = 0.3$. We now have

$$p_2 = 2.5 \times (1 - 0.3) \times (0.3) = 0.525$$

$$p_3 = 2.5 \times (1 - 0.525) \times (0.525) = 0.6234$$

(Things are looking good!)

(A bit of a disappointment!)

$$p_4 = 2.5 \times (1 - 0.6234) \times (0.6234) = 0.5869$$

$$p_5 = 2.5 \times (1 - 0.5869) \times (0.5869) = 0.6061.$$

(Up again!)

$$p_6 = 2.5 \times (1 - 0.6061) \times (0.6061) = 0.5969.$$

(And down.)

$$p_7 = 2.5 \times (1 - 0.5969) \times (0.5969) = 0.6015.$$

We leave it to the reader to verify that the population of the pond is once again moving closer and closer to the value 0.6 but in an oscillating (up, down, up down, . . .) manner. ∎

Example 18. What happens in Example 17 if $p_1 = 0.7$? After the first generation the population behaves identically with that in Example 17. This follows from the fact that in both cases we get the same value for p_2:

$$p_2 = 2.5 \times 0.3 \times 0.7 = 2.5 \times 0.7 \times 0.3 = 0.525.$$

A useful general rule about logistic growth can be spotted here: If we replace p_1 with its complement $(1 - p_1)$, then after the first generation the populations will behave identically. ∎

Example 19. Let's consider a different population of fish—let's say catfish—and let's say that the growth parameter is $r = 3.1$. We start again as in Example 16 with $p_1 = 0.2$. For the sake of brevity we will write the values of the population in sequence form and leave the calculations to the reader.

$p_1 = 0.2,$ $\qquad p_2 = 0.496,$ $\qquad p_3 = 0.775,$ $\qquad p_4 = 0.541,$

$p_5 = 0.770,$ $\qquad p_6 = 0.549,$ $\qquad p_7 = 0.767,$ $\qquad p_8 = 0.553,$

$p_9 = 0.766,$ $\qquad p_{10} = 0.555,$ $\qquad p_{11} = 0.766,$ $\qquad p_{12} = 0.556,$

$p_{13} = 0.765,$ $\qquad p_{14} = 0.557,$ $\qquad p_{15} = 0.765,$ $\qquad p_{16} = 0.557, \ldots$

An interesting pattern emerges here: The population of the pond seems to have settled into a two-period cycle alternating between a high population period at 0.765 (famine) and a low population period at 0.557 (feast). There are many animal populations whose growth is cyclical—a lean period followed by a boom period followed by a lean period, etc.

■

Example 20. Next let's try a population of beetles. For this particular type of beetle we are told that $r = 3.5$. Let's assume that $p_1 = 0.56$ and use the logistic equation to analyze the growth pattern of this beetle population. We leave it to the reader to verify these numbers and fill in the missing details (a calculator is all that is needed).

$p_1 = 0.560,$ $\qquad p_2 = 0.862,$ $\qquad p_3 = 0.415,$ $\qquad p_4 = 0.850,$

$p_5 = 0.446,$ $\qquad p_6 = 0.865,$ $\qquad \ldots,$ $\qquad p_{21} = 0.497,$

$p_{22} = 0.875,$ $\qquad p_{23} = 0.383,$ $\qquad p_{24} = 0.827,$ $\qquad p_{25} = 0.500,$

$p_{26} = 0.875,$ $\qquad \ldots.$

It took a while, but we can now see a pattern: Since $p_{26} = p_{22}$, the population will repeat itself in a four-period cycle ($p_{27} = p_{23}, p_{28} = p_{24}, p_{29} = p_{25}, p_{30} = p_{26} = p_{22}, \ldots$). There are many populations that follow more complicated cyclical patterns of growth: Locusts are the most notorious example.

■

Example 21. Our last and most remarkable example is given by the case $r = 4$. Let's start with $p_1 = 0.2$. The first 20 values of the population sequence are given by

$p_1 = 0.2000,$ $\qquad p_2 = 0.640,$ $\qquad p_3 = 0.9216,$ $\qquad p_4 = 0.2890,$

$p_5 = 0.8219,$ $\qquad p_6 = 0.5854,$ $\qquad p_7 = 0.9708,$ $\qquad p_8 = 0.1133,$

$p_9 = 0.4020,$ $\qquad p_{10} = 0.9616,$ $\qquad p_{11} = 0.1478,$ $\qquad p_{12} = 0.5039,$

$p_{13} = 0.9999,$ $\qquad p_{14} = 0.0002,$ $\qquad p_{15} = 0.0010,$ $\qquad p_{16} = 0.0039,$

$p_{17} = 0.0157,$ $\qquad p_{18} = 0.0617,$ $\qquad p_{19} = 0.2317,$ $\qquad p_{20} = 0.7121.$

Figure 10-4 shows the behavior of the population for the first 20 generations. The reader is encouraged to chart this population for a few additional generations. The surprise here is the absence of any predictable pattern. Even though the population sequence is governed by a very pre-

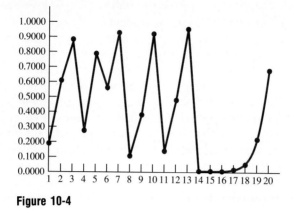

Figure 10-4

cise rule (the logistic equation), to an outside observer the pattern of growth appears to be quite erratic and seemingly random. ■

The behavior of populations under the logistic growth model exhibits many interesting surprises. In addition to the exercises at the end of this chapter, the reader is encouraged to experiment on his or her own in a manner similar to the work we did in the preceding examples. (Choose a p_1 between 0 and 1, choose an r between 0 and 4, and fire up both your calculator and your imagination!)

CONCLUSION

In this chapter we studied three simple models that describe the way that populations grow.

In the *linear model* of population growth, the population sequence is described by an arithmetic sequence, and at each transition period the population grows by a constant amount. Linear growth is most common with populations consisting of inanimate objects.

In the *exponential model* of population growth, the population is described by a geometric sequence. Here in each transition period the population is multiplied by a constant amount called the growth rate. Exponential growth is typical of situations in which there is unrestrained breeding. Money drawing interest in a bank account is one such example.

The *logistic model* of population growth represents situations in which the rate of growth of the population varies from one season to the next depending on the amount of space available in the population's habitat. Many animal populations are governed by the logistic model or simple variations of it.

It goes without saying that most serious studies of population growth involve models with much more complicated mathematical descriptions, but to us that is neither here nor there. Ultimately, the details are not as important as the overall picture: A realization that mathematics can be useful even in its most simplistic forms to describe and predict the rise and fall of populations in many fields—from the artificial world of industry and finance to the natural world of population biology and animal ecology.

KEY CONCEPTS

annual yield
arithmetic sequence
carrying capacity
common difference
continuous growth
discrete growth
exponential decay
exponential growth
geometric sequence
growth parameter

growth rate
habitat
linear growth
logistic equation
logistic growth
periodic interest
population sequence
transition
transition rule

EXERCISES

■ Walking

1. You have a coupon worth 15% off any item in the store (including sale items). The particular item you want is on sale at 30% off the marked price of $100. The store policy allows you to use your coupon before the 30% discount or after the 30% discount (i.e., you can take 15% off the marked price first and then take 30% off the resulting price, or you can take 30% off the marked price first and then take 15% off the resulting price).
   (a) What is the dollar amount of the discount in each case?
   (b) What is the total percentage discount in each case?
   (c) Suppose the article cost P dollars (instead of $100). What is the percentage discount in each case?

2. You have $1000 to invest in one of two competing banks (bank A or bank B) both of which are paying 10% annual interest on deposits left for 1 year. Bank A is offering a 5% bonus credited to your account at the time of the initial deposit, provided the funds are left in the account for a year. Bank B is offering a 5% bonus paid on your account balance at the end of the year after the interest has been credited to your account.
   (a) How much money would you have at the end of the year if you invested in bank A? In bank B?
   (b) What is the total percentage gain (interest plus bonus) at the end of the year for each of the two banks?
   (c) Suppose you invested P dollars (instead of $1000). What is the total percentage gain (interest plus bonus) at the end of the year for each bank?

3. A membership store gives a 10% discount on all purchases to its members. If the store marks each item up 50% (based on its cost), what is the markup actually realized by the store when an item is sold to a member?

4. A hardware store gives a 10% discount to all its charge account customers. In addition, items purchased in large quantities (e.g. a box of 100 bolts, a case of air-conditioning filters) are priced at 10% off the individual item cost. What percent does a charge account customer save off the individual item cost when she purchases items in large quantities?

5. The amount of $3250 is deposited in a savings account that draws 9% annual interest with interest credited to the account at the end of each year. Assuming no withdrawals are made, how much money will be in the account after 4 years?

6. The amount of $1237.50 is deposited in a savings account that draws $8\frac{1}{4}$% annual interest with interest credited to the account at the end of each year. Assuming no withdrawals are made, how much money will be in the account after 3 years?

7. (a) The amount of $5000 is deposited in a savings account that pays 12% annual interest compounded monthly. Assuming no withdrawals are made, how much money will be in the account after 5 years?
   (b) What is the annual yield on this account?

8. (a) The amount of $874.83 is deposited in a savings account that pays $7\frac{3}{4}$% annual interest compounded daily. Assuming no withdrawals, how much money will be in the account after 2 years?
   (b) What is the annual yield on this account?

9. You have some money to invest. The Great Bulldog Bank offers accounts that pay 6% annual interest. The First Northern Bank offers accounts that pay 5.75% annual interest compounded monthly. The Bank of Wonderland offers 5.5% annual interest compounded daily. What is the annual yield for each bank?

10. Complete the following table:

Annual interest rate	Compounded	Annual yield
12%	Yearly	12%
12%	Semiannually	?
12%	Quarterly	?
12%	Monthly	?
12%	Daily	?

11. You decide to open a Christmas Club account at a bank that pays 6% annual interest compounded monthly. You deposit $100 on the first of January and on the first of each succeeding month through November. How much will you have in your account on the first of December?

12. You decide to save money to buy a car by opening a special account at a

bank that pays 8% annual interest compounded monthly. You deposit $300 on the first of each month for 36 months. How much will you have in your account at the end of the thirty-sixth month?

13. Consider the arithmetic sequence P with $P_1 = 23$ and common difference $d = 7$.

arithmatic sequence

(a) Find P_{100}.

(b) Find P_N.

(c) Find $P_1 + P_2 + \cdots + P_{1000}$.

(d) Find $P_{50} + P_{51} + \cdots + P_{1000}$.

14. Consider the arithmetic sequence P with first four terms 7, 11, 15, 19.

(a) Find P_{100}.

(b) Find P_N.

(c) Find $P_1 + P_2 + \cdots + P_{1000}$.

(d) Find $P_{101} + P_{102} + \cdots + P_{1000}$.

15. The city of Lightsville currently has 137 street lights. As part of an urban renewal program the city council has decided to install and have operational two additional street lights at the end of each week for the next 52 weeks. Each street light costs $1 to operate for 1 week.

(a) How many street lights will the city have at the end of 38 weeks?

(b) How many street lights will the city have at the end of N weeks? ($N \le$ 52)

(c) What is the cost of operating the original 137 lights for 52 weeks?

(d) What is the additional cost for operating the newly installed lights for the 52-week period during which they are being installed?

16. A manufacturer currently has on hand 387 widgets. During the next 2 years, the manufacturer will be increasing his inventory by 37 widgets per week. Each widget costs 10 cents a week to store.

(a) How many widgets will the manufacturer have on hand 21 weeks from today?

(b) How many widgets will the manufacturer have on hand N weeks from today? ($N \le 104$)

(c) What is the cost of storing the original 387 widgets for 2 years?

(d) What is the additional cost for storing the increased inventory of widgets for the next 2 years?

17. Consider the geometric sequence P with first four terms 1, 3, 9, 27.

(a) Find P_{100}.

(b) Find P_N.

(b) Find $P_1 + P_2 + \cdots + P_{100}$.

(c) Find $P_{50} + P_{51} + \cdots + P_{100}$.

18. Consider the geometric sequence P with $P_1 = 3$ and growth rate $r = 2$.

(a) Find P_{100}.

(b) Find P_N.

(b) Find $P_1 + P_2 + \cdots + P_{100}$.

(c) Find $P_{50} + P_{51} + \cdots + P_{100}$.

19. Suppose that we have a confined colony of squirrels that grows according to the logistic growth model with growth parameter $r = \frac{29}{10}$. Suppose also that the starting population is $p_1 = \frac{4}{29}$.

(a) What is p_2?

(b) What is the population after the second transition?

(c) What does the logistic growth model predict in the long term for this population?

(d) What other starting population would result in the same population sequence after the first transition?

20. Suppose we know that under the logistic growth model the growth parameter for a particular colony of birds is $r = 2.5$. Suppose also that the starting population is $p_1 = 0.8$.

(a) What is p_2?

(b) What is the population after the second transition?

(c) What does the logistic growth model predict in the long term for this population?

(d) What other starting population would result in the same population sequence after the first transition?

■ Jogging

21. How much should a retailer mark up her goods so that when she has a 25% off sale, the resulting prices will still reflect a 50% mark up (on her cost)?

22. What annual interest rate compounded semiannually gives an annual yield of 21%?

23. Before Annie set off for college, Daddy Warbucks offered her a choice between the following two incentive programs:

■ *Option 1.* A $100 reward for every A she gets in a college course.

■ Option 2. One cent for her first A, 2 cents for the second A, 4 cents for the third A, 8 cents for the fourth A, etc.

Annie chose option 1. After getting a total of 30 A's in her college career, Annie is happy with her reward of $100 × 30 = $3000. Unfortunately, Annie did not get an A in math. Help her figure out how much she would have made had she chosen option 2.

24. Consider a population that grows according to the logistic growth model with growth parameter r ($r > 1$). Show that for $p_1 = (r - 1)/r$ the population is constant.

25. Suppose the habitat of a population of snails has a carrying capacity of $C = 20,000$ and the current population size is 5000. Suppose also that the growth parameter for this particular type of snail is $r = 3.0$. What does the logistic growth model predict for this population after four transition periods?

26. A certain type of moth has a growth parameter of $r = 3.2$. The starting population is $p_1 = 0.27$. Use the logistic equation to study the long term behavior of this population.

27. $1 + 5 + 3 + 8 + 5 + 11 + 7 + 14 + \cdots + 99 + 152 =$

28. $1 + 1 + 2 + \frac{1}{2} + 4 + \frac{1}{4} + 8 + \frac{1}{8} + \cdots + 4096 + \frac{1}{4096} =$

Exercises 29 and 30 refer to the following situation. If B dollars are borrowed at a periodic interest rate i and N equal periodic payments are to be made, then the periodic payment p is given by the formula

$$p = \frac{Bi(1 + i)^N}{(1 + i)^N - 1}.$$

29. **(a)** You buy a house for $120,000 with $20,000 down and finance the balance over 30 years at 9% annual interest (with equal monthly payments). What is your monthly payment?
    (b) What is the monthly payment if the loan described in part (a) is financed over 40 years instead of 30 years?

30. You decide that you can afford a $1000-per-month house payment. The current going interest rate on 30-year home loans is 11%. How much money can you borrow at this rate so that your payment will not exceed $1000 per month?

■ **Running**

31. **(a)** Show that
    $$1 + r + r^2 + \cdots + r^{100} = (r^{101} - 1)/(r - 1).$$
    [*Hint:* Do the multiplication
    $$(1 + r + r^2 + \cdots + r^{100})(r - 1)$$
    and see what you get!]
    (b) Show that
    $$1 + r + r^2 + \cdots + r^N = (r^{N+1} - 1)/(r - 1).$$
    (c) Show that
    $$1 + r + r^2 + \cdots + r^N = (1 - r^{N+1})/(1 - r).$$

32. Show that the sum of the first N terms of an arithmetic sequence with first term c and common difference d is
    $$\frac{N}{2} [2c + (N - 1)d].$$

33. Suppose $r > 3$. Using the logistic growth model, find a population p_1 such that $p_1 = p_3 = p_5 = \ldots$, but $p_1 \neq p_2$. [*Hint:* Exercise 24 shows that $p_1 = (r - 1)/r$ would give $p_1 = p_2 = p_3 = \ldots$].

34. You are purchasing a home for $120,000 and are shopping for a loan. You have a total of $31,000 to put down, including the closing costs of $1000 and any loan fee that might be charged. Bank *A* offers a 10% annual interest loan amortized over 30 years with 360 equal monthly payments. There is no loan fee. Bank *B* offers a 9.5% annual interest loan amortized over 30 years with 360 equal monthly payments. There is a 3% loan fee (i.e., a one-time up-front charge of 3% of the loan). Which loan is better?

35. A friend of yours sold his car to a college student and took a personal note (cosigned by the student's rich uncle) for $1200 with no interest, payable at

$100 per month for 12 months. Your friend immediately approaches you and offers to sell you this note. How much should you pay for the note if you want an annual yield of 12% on your investment?

REFERENCES AND FURTHER READINGS

1. Gleick, James, *Chaos: Making a New Science*. New York: Viking Penguin, Inc., 1987, chap. 3.

2. Hoppensteadt, Frank C., *Mathematical Theories of Population: Demographics, Genetics and Epidemics*. Philadelphia: Society for Industrial and Applied Mathematics, 1975.

3. Kingsland, Sharon E., *Modeling Nature: Episodes in the History of Population Ecology*. Chicago: University of Chicago Press, 1985.

4. May, Robert M., "Simple Mathematical Models with Very Complicated Dynamics," *Nature*, 261 (1976), 459–467.

5. May, Robert M., "Biological Populations with Nonoverlapping Generations: Stable Points, Stable Cycles and Chaos," *Science*, 186 (1974), 645–647.

6. May, Robert M., and George F. Oster, "Bifurcations and Dynamic Complexity in Simple Ecological Models," *American Naturalist*, 110 (1976), 573–599.

7. Smith, J. Maynard, *Mathematical Ideas in Biology*. Cambridge: Cambridge University Press, 1968.

8. Stevens, William K., "Experts Race to Save Dwindling Rhinos," *New York Times*, May 7, 1991, B5.

11

Symmetry of Motion

Mirror, Mirror, off the Wall

It is said that Eskimos have over 50 different words for ice. Ice is a pervasive and fundamental element in the Eskimo's world. In this sense, we would expect mathematics to have at least as many words to describe the notion of symmetry. Unfortunately, just the opposite is the case. In mathematics, "symmetry" is a rather general all-encompassing term, and even in common usage it's a word that has different meanings for different people. (Before reading on, the reader might want to stop and write down his or her own definition of symmetry.)

Regardless of how we define it, in its everyday use symmetry is some form of attribute we give to a physical object, and in this sense symmetry is a distinctly geometric concept. We should mention in passing that this need not be the only possible setting for the concept of symmetry. In a more general way symmetry is a notion that applies equally well to music, literature, and many areas of mathematics besides geometry.

Be that as it may, in this chapter we will only discuss traditional symmetry in the setting of physical objects. We will refer to this kind of symmetry by the more specific term of symmetry of motion. In Chapter 12 we will discuss a nontraditional but still geometric type of symmetry which we will call symmetry of scale.

* Hermann Weyl, *Symmetry* (Princeton, N.J.: Princeton University Press, 1952).

Day and Night by M. C. Escher. (© 1938 M. C. Escher/Cordon Art, Baarn, Holland) "At the left, the white silhouettes merge and form a dark sky; to the right, the black ones melt together and form a nocturnal background. The two landscapes are each other's mirror image and flow together through the fields, from which, once again, the birds take shape."—M. C. Escher

Hand with Reflecting Sphere by M. C. Escher. (© 1935 M. C. Escher/ Cordon Art, Baarn, Holland) ". . . (a) spherical mirror, resting on (my) *left* hand. But as print is the reverse of the original stone drawing, it is my *right* hand that you see depicted. (Being left-handed, I needed my left hand to make the drawing.)"—M. C. Escher

SYMMETRY OF MOTION

One of the foremost popularizers of symmetry in our times was the Dutch artist M. C. Escher. The popularity of his work proved that the notion of symmetry, when creatively packaged, can be of interest to everyone.

Let's start with a somewhat informal definition: We will say that an object has **symmetry** if it looks exactly the same when seen from two or more different vantage points. Imagine a tiny, nearsighted ant standing at one of the vertices of triangle I (Fig. 11-1) and looking inward. The ant would see exactly the same thing whether standing at vertex A, B, or C.

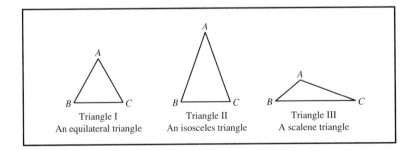

Triangle I
An equilateral triangle

Triangle II
An isosceles triangle

Triangle III
A scalene triangle

Figure 11-1

In fact, if the vertices were not labeled and without any other frame of reference, the ant would be unable to distinguish one position from the other. In triangle II, the ant would see the same thing when standing at *B* or *C* but not when standing at *A*. In triangle III, the ant would see a different thing at each vertex. Informally, we might say that triangle I has more symmetry than triangle II, which in turn has more symmetry than triangle III.

All of this is fine, but still rather vague. Our goal in this chapter is to develop a systematic set of tools for analyzing *how much* and *what kind* of symmetry a given object has.

A different, but equivalent, way to think about symmetry of motion is that rather than moving the observer, we can move the object itself. To say, for example, that triangle II in Fig. 11-1 looks exactly the same to a tiny ant standing at vertex *B* as it does to the same ant standing at vertex *C* is the same as saying that there must be some way to move triangle II so that vertex *B* is where vertex *C* used to be and yet the triangle itself is in exactly the same place as it was before. Thus, the term **symmetry of motion**.

Having agreed that we will restrict our attention to symmetry as it applies to physical objects, let's make one further restriction and agree to focus our attention on two-dimensional (i.e., planar) objects and shapes. Needless to say, three-dimensional objects and shapes are equally deserving of our attention, but it just happens that the analysis is a little more complicated. Moreover, we can gain most of the important insights by looking at the two-dimensional case.

RIGID MOTIONS

The first step in a detailed discussion of symmetry involves the concept of a **rigid motion**. When we take an object and move it from some starting position to some ending position without altering its shape or size, that's a rigid motion. When in the process of moving the object we stretch it, tear it, or generally alter its shape or size, that's *not* a rigid motion. While in general a rigid motion might be a complicated sequence of moves, we will discover that the net result isn't. Take a quarter sitting on top of a dresser. In the morning we might pick it up, put it in a pocket, drive around town with it, take it out of the pocket and flip it into the air, put it back in a different pocket, go home, take it out of the pocket, and lay it on the same dresser again. While the actual trip taken by the quarter was a complicated one, the end result is all that we care about: The quarter started somewhere on top of the dresser and ended somewhere else on the dresser. We could have accomplished the whole thing by a much simpler action such as sliding the quarter a certain distance, turning it around its center, and flipping it over (if the starting and final positions had opposite faces up).

The example of the quarter illustrates a little known but fundamental fact. The *net effect* of any rigid motion, no matter how complicated the actual motion is, can be boiled down to just one of a few possibilities. If we consider the case of two-dimensional objects that start and end in the same plane, the list of possibilities is only four: reflection, rotation, translation, and glide reflection. For convenience, we will call these four rigid motions the **basic rigid motions in the plane.**[1] We will discuss each of them separately.

REFLECTIONS

A **reflection** is often called a *mirror reflection*. It is the kind of motion that moves an object into a new position that is a mirror image of the starting position. The line corresponding to the position of the mirror is called the **axis** of the reflection, and any reflection can be completely described by specifying its axis. Figures 11-2 through 11-4 show examples of reflections. For convenience the original figure is shown in black, and its reflected or mirror image in red. Note that there is no prohibition against the axis of reflection cutting through the object itself, as in Fig. 11-4.

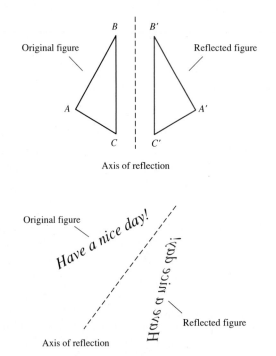

Figure 11-2

Figure 11-3

[1] The reader may be interested to know that in the case of three-dimensional objects moving in space, the number of basic rigid motions is only six (reflection, rotation, translation, and glide reflection, plus two other motions called rotary reflection and screw displacement).

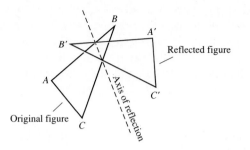

Figure 11-4

We will find it convenient to have a more mathematical description of reflection, so here it is: Given the axis of the reflection, we can tell exactly where any point P will end up under the reflection by drawing a line through P perpendicular to the axis and finding the point P' that is on this line and at the same distance as P from the axis.

This process can also be reversed, so that if we know the starting and ending positions (P and P', respectively) of any point under the reflection, we can find the axis. It is the perpendicular bisector of the segment joining the two points (Fig. 11-5). A useful consequence of this is that if we are given a point P and its ending position P' under the reflection (usually called the **image** of P), then we can completely specify the reflection.

An important characteristic of a reflection is that it reverses all the traditional frames of reference one uses for orientation. As illustrated in Fig. 11-6, in a reflection left is interchanged with right, and clockwise with counterclockwise. We will say that reflection is an **improper** rigid motion to indicate the fact that it reverses the left-right and clockwise-counterclockwise orientations.

Another important fact about reflections is the following: If we apply the same reflection twice, every point ends up exactly where it started. In other words, the net effect of applying the same reflection twice is the same as not having moved the object at all. This leads us to an interesting

Figure 11-5

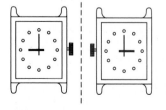

Figure 11-6

semantic question: Should not moving an object at all be considered in itself a rigid motion? On the one hand, it seems rather absurd to say yes. If we are talking about motion, then there should be some kind of movement, however small. On the other hand, we are equally compelled to argue that the result of combining two (or more) consecutive rigid motions should itself be a rigid motion. If this is the case, then combining two consecutive reflections on the same axis (which produces the same result as no motion at all) should be a rigid motion. We will opt for the latter alternative because it is the mathematically correct way to look at things. We will formally agree therefore that not moving an object at all is itself a rigid motion of the object, and we will call it the **identity motion**.

ROTATIONS

The second type of rigid motion we will consider is a **rotation**. A rotation is described by giving the **center** of the rotation and the **angle** or amount of the rotation. It should be noted that in two dimensions there is no axis of rotation. Figures 11-7 through 11-9 show examples of rotations. Again we show the original position in black and the final position in red.

A few comments about the examples above are in order. In all three examples we have specified the angle of rotation in degrees. This is strictly a matter of personal choice—some people prefer degrees, other radians. We will be consistent and stick with degrees, while at the same time reminding the reader that one can always change degrees to radians, or radians to degrees, by using the following relations:

$$\text{Radians} = \frac{\text{degrees}}{180} \times \pi \quad ; \quad \text{Degrees} = \frac{\text{radians}}{\pi} \times 180.$$

Our second observation starts with the well-known fact that a rotation by 360° is the same as no motion at all—in other words, it is the identity motion. This has several useful consequences. First, any rotation by an angle that is more than 360° is equivalent to another rotation with the

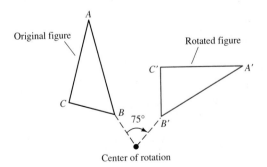

Figure 11-7

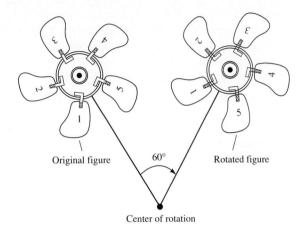

Original figure   60°   Rotated figure

Center of rotation

Figure 11-8

same center by an angle that is between 0° and 360°. All we have to do is divide the angle by 360 and take the remainder. For example, as a rigid motion, a clockwise rotation by 759° is the same as a clockwise rotation by an angle of 39°, because 759 divided by 360 gives a quotient of 2 and a remainder of 39. Second, any rotation that is specified in a clockwise orientation can just as well be specified in a counterclockwise orientation. In Fig. 11-8 for example, we gave the angle of rotation as 60° clockwise, but we could have just as well said 300° counterclockwise. While it is possible to eliminate the need to specify the orientation (clockwise or counterclockwise) of the angle of rotation by choosing one or the other and sticking to it, we prefer the freedom of choosing the orientation that is most convenient.

One rotation for which we do not need to specify the orientation is a rotation by 180°. Here clockwise or counterclockwise produces the same

Original figure

Rotated figure

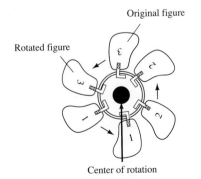

Center of rotation

Figure 11-9

result. The fact that two consecutive 180° rotations produce the same result as two consecutive reflections on the same axis gives rise to the frequent misconception that a 180° rotation is the same as a reflection. A moment's reflection (no pun intended) and/or Fig. 11-10 can make it clear that this is not the case.

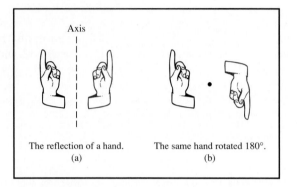

The reflection of a hand.   The same hand rotated 180°.
(a)   (b)

Figure 11-10

A second way to argue that a 180° rotation cannot be the same as a reflection is to note that a rotation is a rigid motion that leaves all original orientations (left, right, clockwise, counterclockwise) unchanged. Any rigid motion that does this is called a **proper** rigid motion. We have already observed that a reflection is an *improper* rigid motion.

Our final comment about rotations is that, unlike reflections, they cannot be specified by giving a single point P and its ending position P'. There are infinitely many rotations that move P to P'. Any point on the perpendicular bisector of the segment PP' can be the center of a possible rotation moving P to P' [Fig. 11-11(a)]. A second pair Q, Q' will allow us to nail down the rotation: The center is the point where the perpendicular bisectors of PP' and QQ' meet [Fig. 11-11(b)]. In the special case

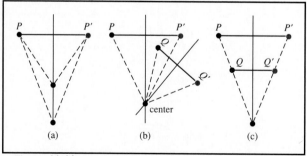

(a)   (b)   (c)

Figure 11-11

where PP' and QQ' are parallel, the center of rotation is the intersection of PQ and $P'Q'$ [Fig. 11-11(c)].

TRANSLATIONS

A **translation** is essentially a slide of the object in the plane. A translation is completely specified by the direction and amount of the slide. These two pieces of information are combined in the form of a **vector**. A vector can be represented by an arrow giving its direction and length. As long as the arrow points in the proper direction and has the right length, its actual placement is immaterial. Figures 11-12 and 11-13 show examples of translations. As usual, the original position is in black and the final position is in red. In both figures, any one of the arrows labeled ①, ②, or ③ (and for that matter infinitely many others) represent the same vector.

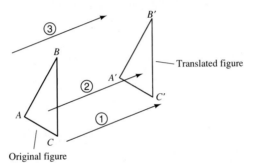

Figure 11-12

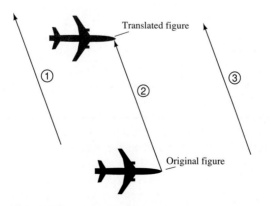

Figure 11-13

Translations, unlike reflections, are *proper* rigid motions of the plane: They do not change the left-right or clockwise-counterclockwise orientations. There is one characteristic that translations do share with reflections: A translation can be completely specified by giving a point P and its final position P'. The arrow joining P to P' gives the vector for the translation.

GLIDE REFLECTIONS

A **glide reflection**, as the name suggests, is a rigid motion consisting of a translation (the glide part) followed by a reflection. The axis of the reflection *must* be parallel to the direction of the translation. The wording "translation followed by a reflection" is somewhat misleading: We can just as well do the reflection first and the translation second, and the end result will be the same. Figures 11-14 and 11-15 show examples of glide reflections with the starting position shown in black, the ending position in red, and the intermediate position in gray.

Figure 11-14

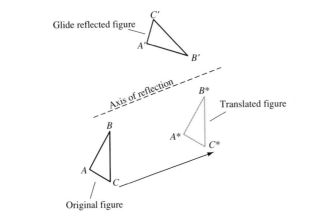

A glide reflection is an *improper* rigid motion—it changes left-right and clockwise-counterclockwise orientations. We can thank the reflection part of the glide reflection for that.

A glide reflection cannot be determined by just one point P and its image P'. As with a rotation, another point Q and its image Q' are needed. Given the two pairs P, P' and Q, Q', the axis of the reflection can be found by joining the midpoints of the segments PP' and QQ' [Fig. 11-16(b)]. This follows from the fact that in any glide reflection the midpoint between a point and its image belongs to the axis of the reflection [Fig. 11-16(a)]. Once the axis of reflection is known, the vector of the translation can be determined by locating the intermediate point $P*$ which is the image of P' under the reflection [Fig. 11-16(c)].

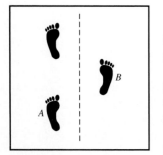

Figure 11-15 Footprints: A glide reflection moves footprint *A* to footprint *B*.

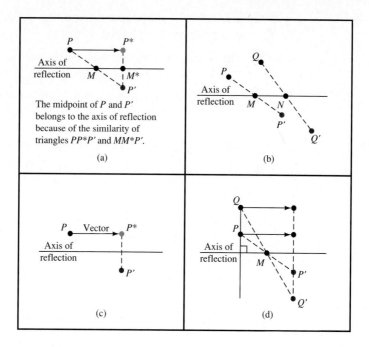

Figure 11-16

In the unlikely event that the midpoints of PP' and QQ' are the same point M, then the line passing through P and Q must be perpendicular to the axis of reflection [Fig. 11-16(d)], so the axis of reflection is obtained by taking a line perpendicular to the line PQ and passing through the common midpoint M.

SYMMETRY OF MOTION REVISITED

We have already discussed the fact that in rigid motions, unlike sight-seeing trips, *how* we get to the final destination is irrelevant. If rigid motion A moves every point in the plane to exactly the same point as rigid motion B, then we say that A and B are **equivalent** rigid motions. We have already mentioned that the result of applying two consecutive reflections with the same axis is a rigid motion equivalent to the identity rigid motion. Likewise, we have also noted that the result of applying two consecutive 180° rotations with the same center is also equivalent to the identity rigid motion. Here are a few additional facts about combining rigid motions:

■   The result of applying two consecutive rotations with the same center is equivalent to a single rotation with the same center (Exercise 21).

■ The result of applying the same glide reflection twice is equivalent to a translation (Exercise 22).

■ The result of applying two consecutive reflections with parallel axes is equivalent to a translation (Exercise 23).

■ The result of applying two consecutive reflections with intersecting axes is a rotation (Exercise 24). A particularly interesting special case of this is when the axes of reflection are perpendicular to each other in which case we get a 180° rotation.

■ The result of applying two consecutive translations is another translation (Exercise 25).

We can play this game for quite a while, and it gets increasingly complicated, especially when we start mixing different types of rigid motions. What, for example, is the result of applying a rotation, followed by a translation, followed by a reflection with the axis not passing through the center of the first rotation, followed by blah, blah, blah? The remarkable thing is that whatever it is, it has to be equivalent to one of the four basic rigid motions: a reflection, a rotation, a translation, or a glide reflection.

We are finally in a position to give a formal definition of **symmetry of motion**: A symmetry of motion of an object is any rigid motion that moves the object back onto itself. It is important to note that this does not necessarily force the rigid motion to be the identity motion. Individual parts of the object may be moved to different starting and ending positions, even while the whole object is moved back into itself. Of course, the identity motion is itself a symmetry.

Based on the fact that there are only four basic types of rigid motion in the plane, we can classify the symmetries of planar objects accordingly.

■ Reflection Symmetry

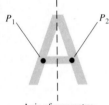

Axis of symmetry

Figure 11-17

Any object that falls back onto itself after a reflection is said to have **reflection symmetry**, more commonly referred to as **bilateral symmetry**. The axis of the reflection is called the **axis of symmetry**. In our illustrations we will use a dashed line to indicate an axis of symmetry. Figure 11-17 shows an example of reflection symmetry. Note that while the object looks exactly the same after the reflection, the pair of points P_1 and P_2 interchange positions.

The equilateral triangle shown in Fig. 11-18(a) has three different reflection symmetries—the axes are the perpendicular bisectors of the three sides. On the other hand, the scalene triangle shown in Fig. 11-18(b) has no reflection symmetries. A circle has infinitely many reflection symmetries (any line passing through the center is an axis of symmetry [Fig. 11-18(c)]).

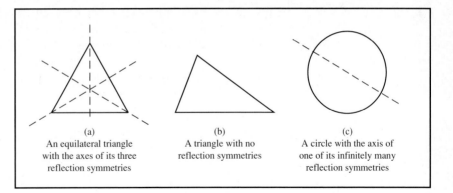

(a)
An equilateral triangle
with the axes of its three
reflection symmetries

(b)
A triangle with no
reflection symmetries

(c)
A circle with the axis of
one of its infinitely many
reflection symmetries

Figure 11-18

■ Patterns

In a border pattern a theme
repeats itself in just one
direction (be it horizontal,
vertical, or wound up in a
roll). (Joshua Sheldon)

Let's take a brief commercial break from our study of symmetry to discuss patterns. A **pattern** is an abstraction representing an infinite figure with infinitely repeating themes. In two dimensions, the most common occurrences of patterns are on wallpaper, printed fabrics, friezes (the decorations marking the ceiling lines of older buildings), ribbons, etc.

For planar patterns we will distinguish two different situations. On the one hand, we have the case exemplified by ribbons, borders, and friezes in which the repetition of the theme occurs in only one direction. We will call these **border patterns** (Fig. 11-19).

Note that the pattern can repeat itself in any direction. For us, it is typographically convenient to make the direction of the pattern horizontal (it wastes the least amount of page space), and we will do so from now on. Thus, when we say "horizontal direction" we mean the direction of

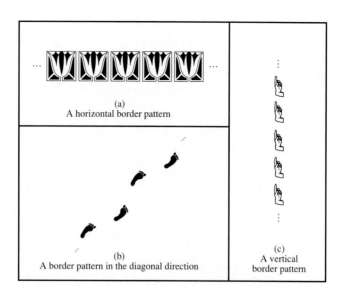

(a)
A horizontal border pattern

(b)
A border pattern in the diagonal direction

(c)
A vertical
border pattern

Figure 11-19

the pattern, and it follows that when we say "vertical direction" we mean the direction perpendicular to the direction of the pattern.

When it comes to reflection symmetries, a border pattern can have

■ No reflection symmetries at all [Fig. 11-20(a)]

■ Horizontal (i.e., in the direction of the pattern) reflection symmetry only [Fig. 11-20(b)]

■ Vertical (i.e., perpendicular to the direction of the pattern) reflection symmetry only [Fig. 11-20(c)]. (Note that here there are infinitely many different choices for the axis of reflection.)

■ Horizontal and vertical reflection symmetries [Fig. 11-20(d)].

Figure 11-20

The second type of pattern we will consider is one that takes up the entire plane. Such is the situation with wallpaper and printed fabrics, and we call these patterns **wallpaper patterns**. In wallpaper patterns the repetition of the theme occurs in more than one direction.

In a wallpaper pattern a theme repeats itself in two nonparallel directions filling up the entire plane.

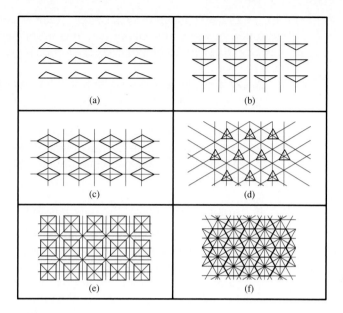

Figure 11-21

The possible reflection symmetries of wallpaper patterns are a little more complicated than we would expect. The possibilities are

- No reflections [Fig. 11-21(a)]
- Reflections in one direction only [Fig. 11-21(b)]
- Reflections in two directions only [Fig. 11-21(c)]
- Reflections in three directions only [Fig. 11-21(d)]
- Reflections in four directions only [Fig. 11-21(e)]
- Reflections in six directions only [Fig. 11-21(f)].

There are no other possibilities. Note that particularly conspicuous in its absence is the case of reflections in exactly five different directions.

Now let's return to our main theme, the analysis of symmetries of motion.

■ Rotation Symmetry

An object that falls back onto itself after a rotation by a certain angle is said to have **rotation symmetry**. We already know that a 360° rotation is the identity motion and that any rotation by an angle more than 360° can be reduced to one by an angle less than 360°, so we can assume that in rotation symmetry the angle of rotation is less than 360°. When an object

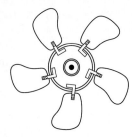

Figure 11-22 A five-blade fan has 72°, 144°, 216°, 288°, and 360° rotation symmetry about the center. It is enough to specify the 72°.

has rotation symmetry by an angle, then it automatically has rotation symmetry by any multiple of that angle. For this reason, we usually just specify the *smallest* positive angle for which the object has rotation symmetry (Fig. 11-22).

A circle is the one two-dimensional shape that has infinitely many rotation symmetries about its center, and any angle, no matter how small, is an appropriate angle of rotation. While the fan shown in Fig. 11-22 has only rotation symmetry (no reflections), it is often the case that rotation symmetry is accompanied by reflection symmetry as in Fig. 11-23.

What kinds of rotation symmetry are possible in patterns? The answers are a bit surprising:

■ In a *border pattern*, the only possible angle of rotation is 180°.

■ In a *wallpaper pattern*, the only possible angles of rotation are 60°, 90°, 120°, and 180°. These correspond to what are known respectively as sixfold [Fig. 11-24(d)], fourfold [Fig. 11-24(b)], threefold [Fig. 11-24(c)], and twofold [Fig. 11-24(a)] rotation symmetry.

Figure 11-23

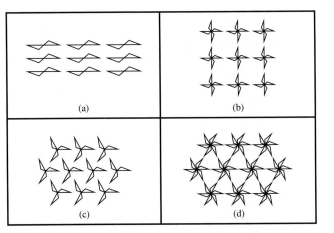

Figure 11-24

It is worth noting that in wallpaper patterns there is no fivefold rotation symmetry (angle of rotation = 72°). An interesting consequence of this is that while we can tile our bathroom floor with tiles that are triangles, squares, or regular hexagons, we can't tile a floor (without leaving any gaps, that is) with a tile that is a regular pentagon.

■ **Translation Symmetry**

Any figure that falls back onto itself after a translation is applied to it is said to have **translation symmetry**. The first thing to note is that a finite figure cannot possibly have translation symmetry—*only patterns* can have this type of symmetry. In fact, we can define patterns as exactly those figures that have translation symmetry. Border patterns are those that have translation symmetry in only one direction (the direction of the pattern) [Fig. 11-25(a)]; wallpaper patterns have translation symmetry in more than one direction [Fig. 11-25(b)].

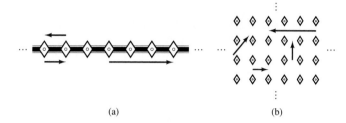

(a)   (b)

Figure 11-25 (a) Three (of the infinitely many) translation symmetries are indicated by the arrows. The translations are always in the horizontal direction. (b) Four (of the infinitely many) translation symmetries are indicated by the arrows.

■ Glide Reflection Symmetry

A glide reflection can also move a figure onto itself, in which case it is a symmetry. Since there is a translation involved, the figure cannot be finite. If a figure has glide reflection symmetry then it must be a pattern. If a pattern has a glide reflection symmetry, then applying the same glide reflection twice results in a translation symmetry. The direction of the glide and the direction of the translation are the same, and the amount of translation is double the amount of the glide (see Exercise 22). Figure 11-26 shows a border pattern with glide reflection symmetry. The point A moves to the position A', and the point B moves to the position B' under the glide reflection. Applying the glide reflection again moves A' to A'' and B' to B''. The combined sequence moving A to A'' and B to B'' is a translation symmetry.

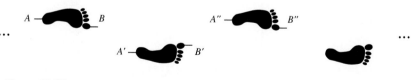

Figure 11-26

Consider now the pattern in Fig. 11-27. It has both translation symmetry and horizontal reflection symmetry, and these symmetries can be combined into a glide reflection that moves the pattern back onto itself. In this case, however, the glide reflection is a symmetry by default, in contrast to the situation in Fig. 11-26 where the glide reflection is a sym-

$$\cdots \quad \mathbb{D} \; \mathbb{D} \; \mathbb{D} \; \mathbb{D} \; \mathbb{D} \; \mathbb{D} \; \mathbb{D} \quad \cdots$$

Figure 11-27

metry in its own right. We will find it convenient to draw a distinction between these two situations. From now on when we say that a pattern has a glide reflection symmetry, we will mean only the situation exemplified by Fig. 11-26.

With wallpaper patterns, the possibilities for glide reflection symmetries are

- No glide reflections [Fig. 11-28(a)]
- Glide reflections in one direction only [Fig. 11-28(b)]
- Glide reflections in two directions only [Fig. 11-28(c)]
- Glide reflections in three directions only [Fig. 11-28(d)]
- Glide reflections in four directions only [Fig. 11-28(e)]
- Glide reflections in six directions only [Fig. 11-28(f)].

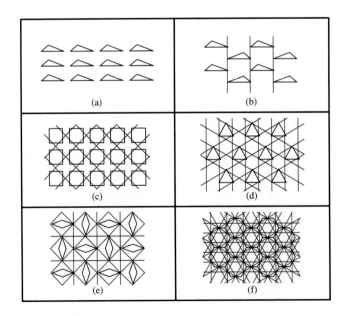

Figure 11-28

CONCLUSION: CLASSIFYING PATTERNS

Patterns play an important role in many areas of science. Archaeologists can recognize cultures and cultural influences from the types of patterns showing up in pottery, wall paintings, textiles, etc. Chemists and crystallographers are interested in the patterns formed by crystals, because they say a great deal about the chemical structure and properties of a particular material.

One of the fundamental tools used in any analysis of patterns is the classification of patterns according to their symmetries. Take, for example, wallpaper patterns. There are infinitely many different possible wallpaper designs. (In fact, it often seems that there are that many in a single wallpaper store!) Two very different looking wallpaper designs, however, can be obtained by just replacing one particular theme (say a doll) by another (say a flower). In this sense, we might say that Figs. 11-29(a) and (b) represent two different designs based on the same pattern type. In fact, we can see that mathematically what really makes the pattern special is not the doll or the flower but the fact that it has a very well-defined set of symmetries: translations in more than one direction, a reflection in only one direction, and a glide reflection in only one direction.

This particular set of symmetries describes a **pattern type**, and any pattern having exactly these symmetries (no more and no less) is said to be of the same type. Thus, classifying patterns means spelling out the various combinations of symmetries that can come together to form a pattern type. When confronted with this problem, a frequent response is Can't we put together any combination of symmetries we want to form a pattern type? (sort of like picking symmetries out of a buffet line). While

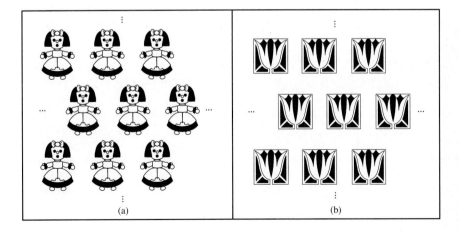

(a)   (b)

Figure 11-29

the answer (no) is only mildly surprising, the actual details almost always come as a real shock:

- ■ **Fact 1.** Any *border* pattern is of one of only *7* different possible pattern types.

- ■ **Fact 2.** Any *wallpaper* pattern is of one of only *17* different possible pattern types.

A complete and accurate proof of these facts (especially fact 2) requires some very abstract, sophisticated mathematical ideas. In fact, while it is believed that these facts were informally understood as far back as the time of the Egyptians (who used all 17 of the wallpaper patterns in their ornamental designs), a rigorous mathematical proof of fact 2 was given only in 1924 by the Hungarian mathematician George Polya.

In Appendix 1 we illustrate and briefly discuss the 7 pattern types for border patterns. The 17 wallpaper pattern types are somewhat more briefly illustrated and described in the table shown in Appendix 2. It is certainly not intended for the student to memorize the information in these appendices but rather that they be used as a handy reference guide by those whose interest has been kindled and who wish to pursue the subject further.

We conclude this chapter with a brief quote from the great mathematician Hermann Weyl:

Symmetry is a vast subject, significant in art and nature. Mathematics lies at its root, and it would be hard to find a better one on which to demonstrate the working of the mathematical intellect.

KEY CONCEPTS

axis (of reflection)	reflection
basic rigid motions in the plane	reflection symmetry
border pattern	rigid motion
equivalent rigid motion	rotation
glide reflection	rotation symmetry
glide reflection symmetry	symmetry of motion
image (of a point)	translation
improper rigid motion	translation symmetry
pattern	vector (of translation)
pattern type	wallpaper pattern
proper rigid motion	

EXERCISES

■ **Walking**

1. Given a reflection that sends the point P to the point P' as shown in the figure, find

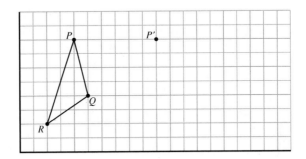

    (a) the axis of reflection
    (b) Q' (the image of Q) under the reflection
    (c) the image of triangle PQR under the reflection.

2. Given a reflection with the axis of reflection as shown in the figure, find

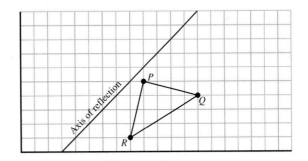

    (a) P' (the image of P) under the reflection
    (b) the image of triangle PQR under the reflection.

3. In each of the following give an answer between 0° and 360°.
    (a) A clockwise rotation by an angle of 500° is equivalent to a clockwise rotation by an angle of _____.
    (b) A clockwise rotation by an angle of 3681° is equivalent to a counterclockwise rotation by an angle of _____.

4. In each of the following give an answer between 0° and 360°.
   (a) A clockwise rotation by an angle of 500° is equivalent to a counterclockwise rotation by an angle of _____.
   (b) A clockwise rotation by an angle of 3681° is equivalent to a clockwise rotation by an angle of _____.

5. Given a rotation that sends the point B to the point B' and the point C to the point C' as shown in the figure, find

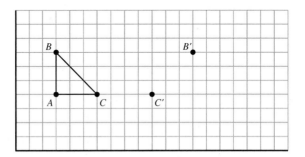

   (a) the center of the rotation
   (b) the image of triangle ABC under the rotation.

6. Given a rotation that sends the point B to the point B' and the point C to the point C' as shown in the figure, find

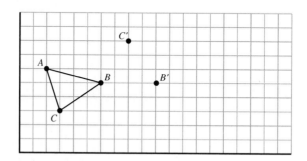

   (a) the center of the rotation
   (b) the image of triangle ABC under the rotation.

7. Given a translation that sends E to E' as shown in the figure, find

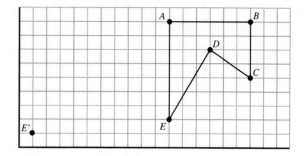

(a) the image A' of A under the translation

(b) the image of the figure $ABCDE$ under the translation.

8. Given a translation that sends Q to Q' as shown in the figure, find

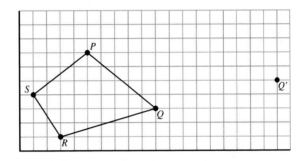

(a) the image P' of P under the translation

(b) the image of the figure $PQRS$ under the translation.

9. Given a glide reflection that sends the point B to the point B' and the point D to the point D' as shown in the figure, find

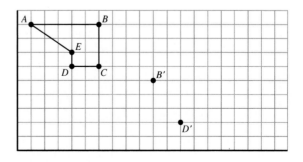

(a) the axis of the glide reflection

(b) A' (the image of A) under the glide reflection

(c) the image of figure $ABCDE$ under the glide reflection.

10. Given a glide reflection that sends the point A to the point A' and the point C to the point C' as shown in the figure, find

   (a) the axis of the glide reflection
   (b) B' (the image of B) under the glide reflection
   (c) the image of figure $ABCD$ under the glide reflection.

11. For each of the following symbols describe all the symmetries. (You don't have to mention the identity symmetry.)
    (a) A
    (b) D
    (c) +
    (d) Z
    (e) Q.

12. For each of the following shapes describe all the symmetries (other than the identity symmetry).
    (a) equilateral triangle
    (b) square
    (c) rectangle
    (d) parallelogram (not a rectangle)
    (e) circle.

13. Give an example of a capital letter of the alphabet that has
    (a) horizontal reflection only
    (b) vertical reflection only
    (c) horizontal and vertical reflection
    (d) 180° rotation but no reflection
    (e) no symmetries other than the identity symmetry.

14. List all the digits (0, 1, 2, 3, 4, 5, 6, 7, 8, 9) that have
    (a) horizontal reflection
    (b) vertical reflection
    (c) horizontal and vertical reflections
    (d) 180° rotation.

15. (a) Give an example of a figure that has a 120° rotation symmetry and three different reflection symmetries.

(b) Give an example of a figure that has a 120° rotation symmetry and *no* reflection symmetries.

16. (a) Give an example of a figure that has 45° rotation symmetry and reflection symmetries.

(b) Give an example of a figure that has 45° rotation symmetry and *no* reflection symmetries.

17. Describe all the symmetries of each border pattern.

(a)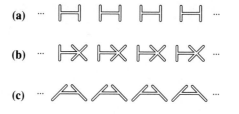

(b)

(c) ⋯

18. Describe all the symmetries of each border pattern.

(a)

(b)

(c) ⋯

19. Describe all the symmetries of each border pattern.

(a) ⋯ A A A A A A A A ⋯

(b) ⋯ D D D D D D D D ⋯

20. Describe all the symmetries of each border pattern.

(a) ⋯ S S S S S S S ⋯

(b) ⋯ L L L L L L L L ⋯

■ Jogging

21. (a) Rotation 1 has center *C* and a clockwise angle of 30°, and rotation 2 has the same center *C* and a clockwise angle of 50°. Show that the result of applying rotation 1 followed by rotation 2 is equivalent to a rotation with center *C* and a clockwise angle of 80°.

(b) Show that if rotation 1 has center C and clockwise angle α and rotation 2 has center C and clockwise angle β, then the result of applying rotation 1 followed by rotation 2 is equivalent to a rotation with center C and angle $\alpha + \beta$.

(c) Show that for the two rotations described in part (b) the result of applying rotation 1 followed by rotation 2 is equivalent to applying rotation 2 followed by rotation 1.

22. (a) Given a glide reflection with axis and vector as shown, find the image Q'' of the point Q when the glide reflection is applied twice.

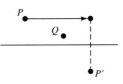

(b) Show that the result of applying the same glide reflection twice is equivalent to a translation. Describe the direction and amount of the translation in terms of the direction and amount of the original glide.

23. Reflection 1 has axis l_1; reflection 2 has axis l_2; l_1 and l_2 are parallel, and the distance between them is d.

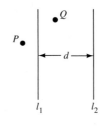

(a) Find the images of P and Q when we apply reflection 1 followed by reflection 2.

(b) Show that the result of applying reflection 1 followed by reflection 2 is a translation. Describe the direction and amount of the translation.

(c) Show that the result of applying reflection 1 followed by reflection 2 is *not* the same as the result of applying reflection 2 first followed by reflection 1. Describe the difference.

24. Reflection 1 has axis l_1; reflection 2 has axis l_2; l_1 and l_2 intersect at C. The angle between l_1 and l_2 as shown in the figure is α.

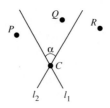

(a) Find the images of P, Q, and R when we apply reflection 1 followed by reflection 2.

(b) Show that the result of applying reflection 1 followed by reflection 2 is a rotation with center C. Give the clockwise angle of rotation.

(c) Show that the result of applying reflection 2 first followed by reflection 1 is a different rotation from the one found in part (b). Describe the difference.

25. Translation 1 moves point P to P'; translation 2 moves point Q to point Q' as shown in the figure.

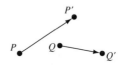

(a) Find the images of P and Q when we apply translation 1 followed by translation 2.

(b) Find the images of P and Q when we apply translation 2 followed by translation 1.

(c) Show that the result of applying translation 1 followed by translation 2 is a translation. Give a geometric description of the vector of the translation.

26. (a) Find all the reflection symmetries of a regular pentagon.
    (b) Find all the reflection symmetries of a regular hexagon.
    (c) Describe all the reflection symmetries of a regular polygon with N sides.

27. Describe all the symmetries of each border pattern.

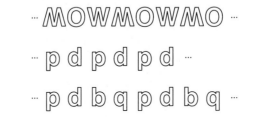

28. For each of the following sets of symmetries, give an example that has exactly those symmetries (no more and no less). Do not use any of the patterns in Exercise 27. You can use letters of the alphabet, numbers, or symbols to create the patterns.
    (a) Translations only
    (b) Translations and vertical reflections
    (c) Translations and horizontal reflections
    (d) Translations and 180° rotations
    (e) Translations and glide reflections
    (f) Translations, vertical reflections, glide reflections, and 180° rotations
    (g) Translations, vertical reflections, horizontal reflections, and 180° rotations.

29. A rigid motion moves the triangle PQR into the triangle $P'Q'R'$ as shown in the figure.

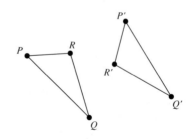

    (a) Explain why the rigid motion cannot possibly be a rotation or a translation. (*Hint*: Is the rigid motion proper or improper?)

    (b) Explain why the rigid motion cannot possibly be a reflection.

    (c) What kind of rigid motion is it then?

30. A *palindrome* is a word that is the same when read forward or backward. MOM is a palindrome, and so is ANNA. (For simplicity we will assume all letters are capitals.)

    (a) Explain why if a word has vertical reflection symmetry, then it must be a palindrome.

    (b) Give an example of a palindrome (other than ANNA) that doesn't have vertical reflection symmetry.

    (c) If a palindrome has vertical reflection symmetry, what can you say about the symmetries of the individual letters in the word?

■ Running

31. Suppose that a rigid motion moves points P to P', Q to Q', and R to R' (as in the figure). We do not know what kind of rigid motion it is.

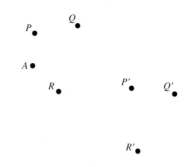

    (a) For an arbitrary point A in the plane, find the image A'. (*Hint*: $AP = A'P'$, $AQ = A'Q'$, and $AR = A'R'$.)

    (b) Describe a general procedure for finding the image of a point A under some rigid motion when we know three points P, Q, and R (not on a straight line) and their images P', Q', and R'.

32. Find all the symmetries of the following wallpaper pattern.

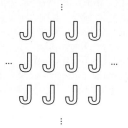

33. Find all the symmetries of the following wallpaper pattern.

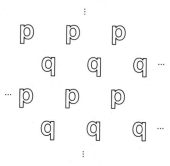

34. Find all the symmetries of the following wallpaper pattern.

35. Find all the symmetries of the following wallpaper pattern.

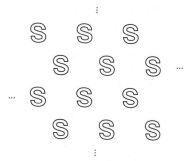

APPENDIX 1:
The Seven Border
Pattern Types

In this appendix we illustrate and describe the seven different pattern types that are possible with border patterns. There are several different ways to organize this classification. We will adopt one that is consistent with the standard notation used in crystallography.

Let's start with the fact that every border pattern has translations in the direction of the pattern. As we have already observed, when dealing with patterns, translations come with the territory. We will now divide the possibilities into two categories.

- **Category** *m*. The pattern has vertical reflections (remember that this really means reflections in a direction perpendicular to the direction of the pattern)

- **Category 1.** The pattern has no vertical reflections.

In category *m* there are three possibilities.

- *m*1. Translations, vertical reflections and nothing else

- *mg*. Translations, vertical reflections, plus glide reflections, and 180° rotations.

- *mm*. Translations, vertical reflections, plus horizontal reflections, and 180° rotations.

In category 1 we have the other four possible pattern types.

- **11.** Translations and nothing else

- **1*m*.** Translations plus horizontal reflections

- **12.** Translations plus 180° rotations

- **1*g*.** Translations plus glide reflections.

The following table summarizes the seven pattern types with their symmetries.

Type	Translation	Reflection Along Direction of Pattern	Reflection Along Perpendicular to Pattern	180° Rotation	Glide Reflection	Example
11	✓					
1m	✓	✓				
m1	✓		✓			
12	✓			✓		
1g	✓				✓	
mg	✓		✓	✓	✓	
mm	✓	✓	✓			

Why are these the only possibilities? Without going into a complete proof, a clue as to what happens can be found by looking at the impossible combination of symmetries consisting of just translations, vertical reflections, and horizontal reflections. Such a combination is impossible because the vertical reflection combined with the horizontal reflection automatically forces a 180° rotation (and therefore forces the pattern to be of type *mm*). In a similar fashion we can show that any other combination not on the list is impossible. (Some of them are a little bit harder to show than the one we picked.)

APPENDIX 2: The Seventeen Wallpaper Pattern Types

In this appendix we will just show the 17 wallpaper pattern types with the standard notation used in crystallography.

Type	Translation	Rotations				Reflections (Number of Directions)					Glide Reflections (Number of Directions)					Example
		60°	90°	120°	180°	1	2	3	4	6	1	2	3	4	6	
p2gg	✓				✓	✓					✓					
p211	✓				✓							✓				
p2mg	✓				✓											
p4mm	✓		✓		✓				✓			✓				
p4gm	✓		✓		✓	✓								✓		
p4	✓		✓		✓											

Type	Translation	Rotations				Reflections (Number of Directions)					Glide Reflections (Number of Directions)					Example
		60°	90°	120°	180°	1	2	3	4	6	1	2	3	4	6	
*c*1*m*1	✓					✓					✓					
*p*1*m*1	✓					✓										
*p*1*g*1	✓										✓					
*p*1	✓															
*p*2*mm*	✓				✓		✓									
*c*2*mm*	✓				✓		✓					✓				

Type	Translation	Rotations				Reflections (Number of Directions)					Glide Reflections (Number of Directions)					Example
		60°	90°	120°	180°	1	2	3	4	6	1	2	3	4	6	
p3m1	✓			✓				✓					✓			
p31m	✓			✓				✓					✓			
p3	✓			✓												
p6mm	✓	✓		✓	✓					✓					✓	
p6	✓	✓		✓	✓											

**REFERENCES
AND FURTHER
READINGS**

1. Bunch, Bryan, *Reality's Mirror: Exploring the Mathematics of Symmetry*. New York: John Wiley & Sons, Inc., 1989.

2. Coxeter, H. S. M., *Introduction to Geometry* (2d ed.). New York: John Wiley & Sons, Inc., 1967.

3. Crowe, Donald W., "Symmetry, Rigid Motions and Patterns," *UMAP Journal*, 8 (1987), 206–236.

4. Gardner, Martin, *The New Ambidextrous Universe: Symmetry and Asymmetry from Mirror Reflections to Superstrings* (3rd ed.). New York: W. H. Freeman & Co., 1990.

5. Grunbaum, Branko, and G. C. Shephard, *Tilings and Patterns: An Introduction*. New York: W. H. Freeman & Co., 1989.

6. Hofstadter, Douglas R., *Gödel, Escher, Bach: An Eternal Golden Braid*. New York: Vintage Books, 1980.

7. Martin, George E., *Transformation Geometry: An Introduction to Symmetry*. New York: Springer Publishing Co., Inc., 1982.

8. Rose, Bruce, and Robert D. Stafford, "An Elementary Course in Mathematical Symmetry," *American Mathematical Monthly*, 88 (1981), 59–64.

9. Schattsneider, Doris, *Visions of Symmetry: Notebooks, Periodic Drawings, and Related Work of M. C. Escher*. New York: W. H. Freeman & Co., 1990.

10. Shubnikov, A. V., and V. A. Kopstik, *Symmetry in Science and Art*. New York: Plenum Publishing Corp., 1974.

11. Weyl, Hermann, *Symmetry*. Princeton N.J.: Princeton University Press, 1952.

12. Yale, Paul B., *Geometry and Symmetry*. San Francisco: Holden-Day, 1968.

12 Symmetry of Scale and Fractals

Fractally Speaking

In Chapter 11 we discussed the concept of *symmetry of motion*—the characteristic of shapes, objects, and patterns that look exactly the same after they have been moved in various ways. Symmetry of motion represents the conventional way that people think about symmetry, and in common usage symmetry of motion is referred to as just plain symmetry, as if it were the only kind. In fact, it isn't.

In this chapter we will introduce a different, nontraditional, somewhat more exotic type of symmetry called *symmetry of scale*, and we will use symmetry of scale to discuss some remarkable geometric objects called *fractals*. The fantastic geometric shapes we will come across in our study of symmetry of scale are unlike anything we could imagine based on our experience with Euclidean (high school) geometry.

We will motivate our definition of symmetry of scale with some preliminary examples.

THE KOCH SNOWFLAKE

The Koch snowflake is a remarkable geometric object first discovered by the Swedish mathematician Helge von Koch in 1904. The construction of the Koch snowflake proceeds in steps as follows:

■ **Start.** Start with a solid black equilateral triangle of arbitrary size. For simplicity we will assume that the sides of the triangle are of length 1 and call the area of the triangle A.

■ **Step 1.** Divide each of the sides of the triangle into three equal segments. Attach to the middle segment of each side a small, black equilateral triangle [Fig. 12-1(b)]. The resulting star-shaped object has 12 sides, each of length $\frac{1}{3}$ [Fig. 12-1(c)].

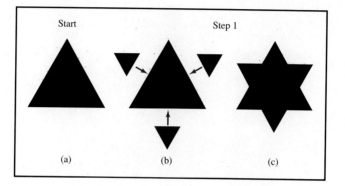

Start   Step 1

(a)   (b)   (c)

Figure 12-1

■ **Step 2.** For each of the 12 sides of the star in Fig. 12-1(c), repeat the process described in step 1: Divide each side into three equal segments and attach a black equilateral triangle to the middle segment. The resulting shape has 48 sides, each of length $\frac{1}{9}$ [Fig. 12-2(a)].

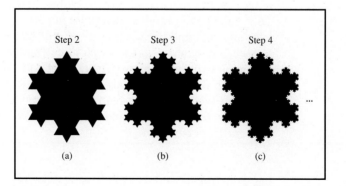

Step 2   Step 3   Step 4

(a)   (b)   (c)

Figure 12-2

■ **Steps 3, 4, etc.** Repeat the process (divide sides into thirds, attach appropriate sized equilateral triangles to the middle thirds) ad infinitum [Fig. 12-2(b), (c), etc.].

Each step in the construction described above takes us closer and closer to the ultimate destiny of this trip: the Koch snowflake itself (Fig. 12-3).

Step ∞

Figure 12-3

The perceptive reader will immediately notice that Fig. 12-3 is not a picture of the real Koch snowflake but that of a useful impostor. The Koch snowflake requires an infinite number of steps in its construction, so a complete and perfect picture of it is impossible. However, the fact that we cannot show a perfect picture of the snowflake should not deter us from drawing good approximations of it (as we did in Fig. 12-3) or from using such drawings to study its mathematical properties. (This is very similar to the situation in high school geometry where we learned a lot about squares, triangles, and circles even when our drawings of them were far from perfect.)

■ **Recursive Processes**

The construction of the Koch snowflake is an example of a **recursive process**, a process in which the same basic rules are applied over and over again with the end product at each step becoming the starting point for the next step (see Fig. 12-4). The concept of a recursive process is not new to us. We saw it in the construction of the Fibonacci and other recursively defined sequences as discussed in Chapter 9, as well as in the transition rules for population growth as discussed in Chapter 10. Here the objects of the process are not numbers but geometric shapes, but other than that the basic principles are quite similar.

The use of recursion allows us to describe the construction of the Koch snowflake in a very efficient form of shorthand we will call a **recursive replacement rule**.

Figure 12-4 Two examples of recursive processes.

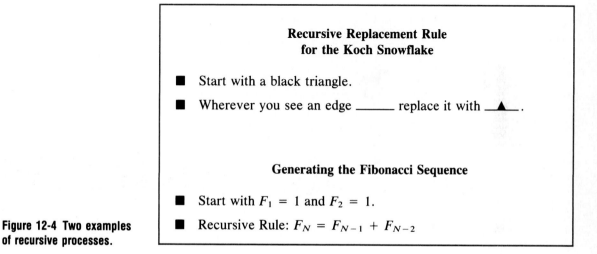

**Recursive Replacement Rule
for the Koch Snowflake**

■ Start with a black triangle.

■ Wherever you see an edge _____ replace it with ___▲___ .

Generating the Fibonacci Sequence

■ Start with $F_1 = 1$ and $F_2 = 1$.

■ Recursive Rule: $F_N = F_{N-1} + F_{N-2}$

When we look at the Koch snowflake from the perspective of Euclidean geometry, we find some very bizarre characteristics. Let's start by asking two typical questions from traditional geometry: What are the perimeter and the area of the Koch snowflake?

■ **Perimeter of the Koch Snowflake**

The boundary of the Koch snowflake will be of particular interest to us. It is commonly known as the **Koch curve** or the **snowflake curve** (see Fig. 12-5).

Figure 12-5 A section of the Koch curve.

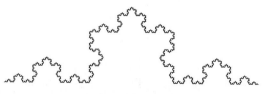

How long is the Koch curve? Figure 12-6 shows what happens when we take the length of the boundary at each of the first few steps of the construction.

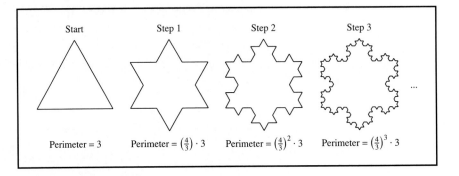

Start   Step 1   Step 2   Step 3

Perimeter = 3   Perimeter = $\left(\frac{4}{3}\right) \cdot 3$ Perimeter = $\left(\frac{4}{3}\right)^2 \cdot 3$ Perimeter = $\left(\frac{4}{3}\right)^3 \cdot 3$

Figure 12-6

We can see from Fig. 12-6 that the perimeters of the figures that result at each successive step of the construction exhibit a pattern of exponential growth with a growth rate of $\frac{4}{3}$. Since this growth rate is more than 1, it follows that the Koch curve must be infinitely long (see Fig. 12-7).

The boundary of the Koch snowflake has infinite length.

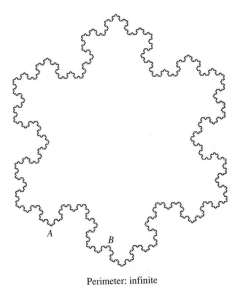

Figure 12-7 A tiny ant trying to get from point *A* to point *B* traveling along the nooks and crannies would be hard put to make much progress.

Perimeter: infinite

■ **Area of the Koch Snowflake**

> *The area of the Koch snowflake is 1.6 times the area of the starting equilateral triangle.*

The preceding statement is a startling fact! It implies that the Koch snowflake has a finite area enclosed by an infinite boundary—a fact that common geometric intuition finds somewhat difficult to accept. Clearly, this is not an everyday, run-of-the-mill geometric object!

To verify that the area of the Koch snowflake is 1.6 times the area of the starting equilateral triangle will take some doing, but the necessary tools are already at our disposal. In what follows, we will give an outline of the argument, leaving the technical details as exercises for the reader. In fact, the reader who wishes to do so may skip the forthcoming explanation without prejudice.

The strategy we will adopt to calculate the area of the Koch snowflake is as follows:

■ Calculate a formula that gives the area of the polygon we get at the Nth step (i.e., a generic step) in the construction.

■ Determine what happens to the formula obtained above as N gets bigger and bigger.

Figure 12-8 (on the next page) shows how the area changes when we go from one step of the construction to the next. From Fig. 12-8 we observe that the area of the polygon at the Nth step of the construction is

$$A + \left(\frac{1}{3}\right)A + \left(\frac{4}{9}\right)\left(\frac{1}{3}\right)A + \left(\frac{4}{9}\right)^2\left(\frac{1}{3}\right)A + \cdots + \left(\frac{4}{9}\right)^{N-1}\left(\frac{1}{3}\right)A$$

(where A is the area of the starting equilateral triangle). Except for the first term, we are looking at the sum of terms of a geometric sequence. Using the formula for adding the consecutive terms of a geometric sequence (Chapter 10), we can simplify the preceding expression to

$$A + \left(\frac{3}{5}\right)A\left[1 - \left(\frac{4}{9}\right)^N\right].$$

We leave the technical details to the reader (Exercise 24).

We are now ready to wrap this up. We only need to figure out what happens to $(\frac{4}{9})^N$ when N gets bigger and bigger. In fact, what happens to any positive number less than 1 when we raise it to higher and higher powers? If you know the answer, then you are finished. If you don't, take a calculator, enter a number between 0 and 1, and multiply it by itself repeatedly. (You will readily convince yourself that the result gets closer

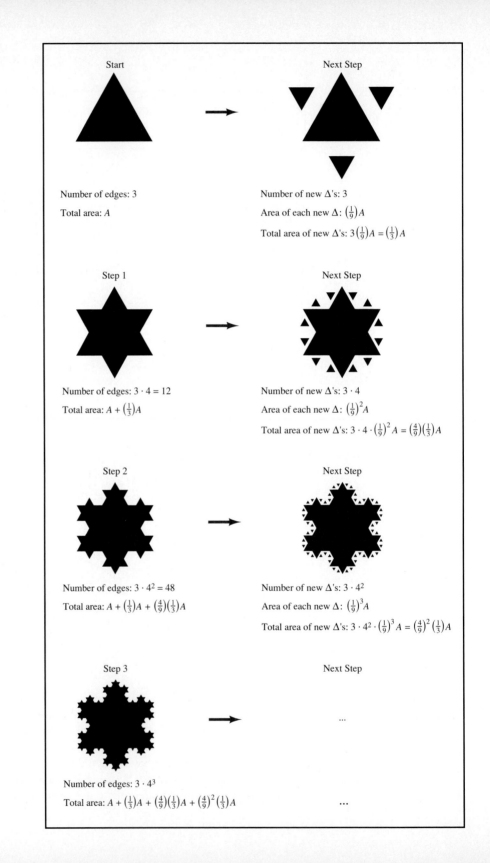

Start

Number of edges: 3

Total area: A

Next Step

Number of new Δ's: 3

Area of each new Δ: $\left(\frac{1}{9}\right)A$

Total area of new Δ's: $3\left(\frac{1}{9}\right)A = \left(\frac{1}{3}\right)A$

Step 1

Number of edges: $3 \cdot 4 = 12$

Total area: $A + \left(\frac{1}{3}\right)A$

Next Step

Number of new Δ's: $3 \cdot 4$

Area of each new Δ: $\left(\frac{1}{9}\right)^2 A$

Total area of new Δ's: $3 \cdot 4 \cdot \left(\frac{1}{9}\right)^2 A = \left(\frac{4}{9}\right)\left(\frac{1}{3}\right)A$

Step 2

Number of edges: $3 \cdot 4^2 = 48$

Total area: $A + \left(\frac{1}{3}\right)A + \left(\frac{4}{9}\right)\left(\frac{1}{3}\right)A$

Next Step

Number of new Δ's: $3 \cdot 4^2$

Area of each new Δ: $\left(\frac{1}{9}\right)^3 A$

Total area of new Δ's: $3 \cdot 4^2 \cdot \left(\frac{1}{9}\right)^3 A = \left(\frac{4}{9}\right)^2 \left(\frac{1}{3}\right)A$

Step 3

Number of edges: $3 \cdot 4^3$

Total area: $A + \left(\frac{1}{3}\right)A + \left(\frac{4}{9}\right)\left(\frac{1}{3}\right)A + \left(\frac{4}{9}\right)^2 \left(\frac{1}{3}\right)A$

Next Step

...

...

Figure 12-8

and closer to 0.) The bottom line is that as N gets bigger and bigger, the expression inside the square brackets gets closer and closer to 1, and therefore the area gets closer and closer to $(1.6)A$.

■ Symmetry of Scale

Let's say, for the sake of argument, that we are not totally sold on the idea of the Koch snowflake and the Koch curve. We demand a closer look! What does the fine detail look like? In Fig. 12-9 we show a section of the Koch curve magnified nine times and then a portion of the magnified curve magnified nine times again. We can continue this magnification process indefinitely, but it won't help. No matter how much we crank up our imaginary microscope, we will continue seeing exactly the same thing!

This remarkable characteristic of the Koch curve is called **symmetry of scale** (or sometimes **self-similarity**). As the name suggests, it is a symmetry between the large and the medium, the medium and the small, the small and the smaller—a symmetry that carries itself across different scales. Thus, we say that the Koch curve has symmetry of scale to indicate that any piece of it can be found again at another level of magnification and that its defining pattern (_/_) occurs at infinitely many scales.

Before we move on, a final point is in order. Suppose we wanted a realistic geometric description of a very jagged piece of coastline (say, something like the Scandinavian fjords). We would be hard put to find among the traditional shapes of Euclidean geometry anything that comes as close to giving the "feel" of that coastline as a piece of the Koch curve. The Koch snowflake and its boundary, the Koch curve, are not mathematical freaks but rather very good mathematical descriptions of the way certain things (such as coastlines) look in nature.

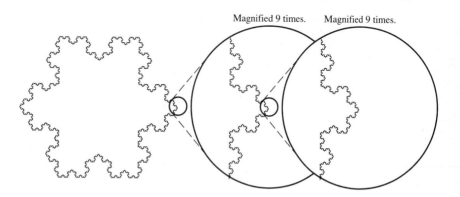

Magnified 9 times.   Magnified 9 times.

Figure 12-9

THE SIERPINSKI GASKET

This construction was first suggested (in a slightly modified form) by the Polish mathematician Waclaw Sierpinski around 1915. It involves another recursive construction like that of the Koch snowflake. It also starts with a black triangle *ABC*, but unlike the Koch snowflake, the starting black triangle does not have to be equilateral—any black triangle will do [Fig. 12-10(a)]. In this construction, instead of *adding* smaller copies of the original triangle, we will *remove* smaller copies of the original triangle as follows:

■   **Start.**   Start with an arbitrary black triangle *ABC*.

■   **Step 1.**   Join the midpoints M_1, M_2, and M_3 of sides *AB*, *AC*, and *BC*, respectively. This gives us four triangles [AM_1M_2, BM_1M_3, CM_2M_3, and the *middle* triangle $M_1M_2M_3$ as shown in Fig. 12-10(b).] Each of the four triangles is similar to the original black triangle (Exercise 1). We now remove the interior of the middle triangle $M_1M_2M_3$, leaving a triangular white hole. If we call the interior of the middle triangle $M_1M_2M_3$ the *heart* of triangle *ABC*, we can use a cruel but convenient metaphor—we have "cut out the heart" of triangle *ABC*!

■   **Step 2.**   Cut out the heart of each of the 3 black triangles in step 1. This leaves us with 9 black triangles (all similar to the original triangle *ABC*) and 4 triangular white holes [Fig 12-10(c)].

■   **Steps 3, 4, etc.**   Repeat the process (cut out the heart of every black triangle) ad infinitum.

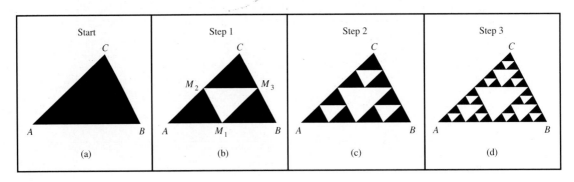

Figure 12-10

After infinitely many steps of this recursive construction one gets a bizarre kind of geometric Swiss cheese called the **Sierpinski gasket** (Fig. 12-11).

Once again we observe that Fig. 12-11 is just an approximation of the Sierpinski gasket. The tiny black triangles that we see are the result of poor eyesight and the inadequacies of printing. The Sierpinski gasket has

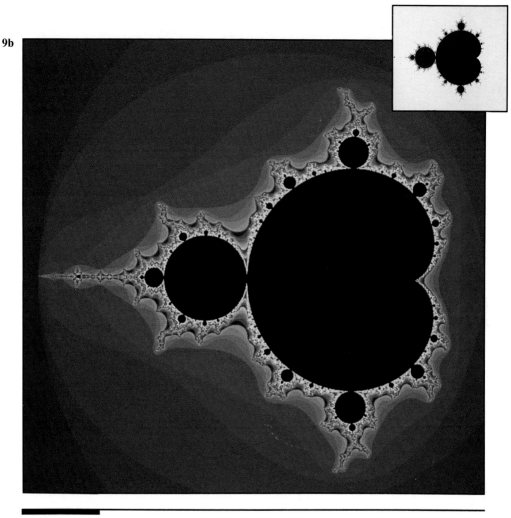

A black and white rendering of the Mandelbrot set (upper insert) comes to life when color is added. What does the picture depict? A cosmic bug? An eclipse in some distant galaxy? A scene from *Star Wars?* Hardly. The Mandelbrot set is a mathematical *fractal*—the result of applying a simple recursive process to the numbers in the complex plane (for details see chapter 12). The plates in the next four pages show the exquisite details and strange beauty of what has been called "the most complex mathematical object known by man." (Computer images by Rollo Silver.)

10a

10b

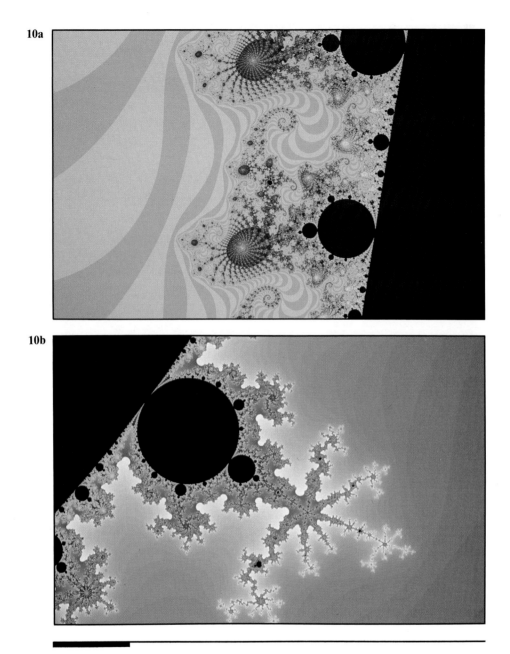

Hundreds of *Mandelbuds* surround the boundary of the Mandelbrot set. Each of the buds, in turn, is surrounded by hundreds of smaller buds and the process repeats itself ad infinitum. They are all similar to each other and to the Mandelbrot set itself. Striking differences show up, however, when we look at the "reefs" around each bud. "Urchins" and "seahorses" (top) are replaced by "starfish" and "snails" (bottom). This blending of infinite repetition and infinite variety is one of the hallmarks of the Mandelbrot set.

11a

11b

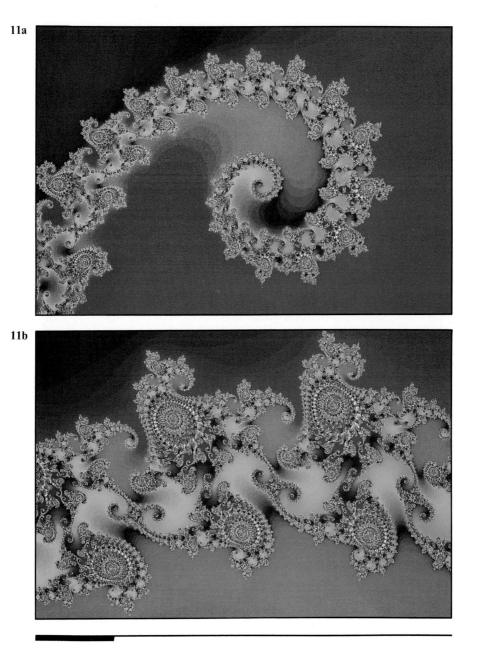

A closer look at a ''seahorse tail'' (top) and a detail of it (bottom). Similar but smaller seahorse tails can be seen throughout the bottom plate sometimes in singles and sometimes in pairs, and if we look deeper, in sets of four (see next page), eight, sixteen, . . .

12a

12b

Old themes and new themes come together. Top: *Seahorse Quadrille*. A foursome of seahorse tails stand guard around a small replica of the original Mandelbrot set. The magnification is approximately 250,000 times the original set (plate 9b). If plate 9b was done at the same scale as this plate, it would be approximately 20 miles wide. Bottom: *Rainbow vortex*. An entirely new scene, deep inside the labyrinth that surrounds the Mandelbrot set.

13a

13b

Left: *The Monkey's Tail.* A new spiral, different in shape but similar in makeup to the seahorse tail (plate 11a), found fishing in deeper waters on the eastern coast of the Mandelbrot set. Right: *The Bird of Paradise.* A closeup of one of the small pieces that make up the monkey's tail. This picture is at a magnification of over 700,000 times the original Mandelbrot set.

14a

14b

A case of life imitating art. Top: *Carolina* by Ken Musgrave and Benoit Mandelbrot). This artificial landscape was created with a computer using fractal constructions not unlike the ones discussed in chapter 12. Bottom: *Sunset over Mt. Annapurna* (by Scott Woolums). A real landscape, photographed in an expedition to the Himalayas. The remarkable resemblance between the two scenes is purely accidental (the photographer had never seen the computer landscape) but it reinforces the fact that fractals are the natural tools for describing nature.

15a

15b

A case of art imitating life. Fractal geometry can be used to create remarkable forgeries of natural scenes. Top: This beautiful scene depicting the rugged Sandia mountains in New Mexico is a fractal forgery created entirely in a computer. Bottom: A blending of man made and natural elements. Hot air balloons (created using elements from traditional euclidean geometry such as spheres, cones and ellipsoids) approach a fractally created island. (Both images were created by John Mareda, Gary Mastin and Peter Wattenberg of Sandia National Laboratories.) The mathematical techniques used to create these images are used not only for fanciful pictures but they are important in many applications, from the training of airplane pilots to the modeling of high energy reactions in particle physics.

16a

16b

The ability to create natural landscapes using fractal geometry (see preceding pages) can be carried one step further to create "otherwordly" landscapes—fantastic but compelling images of imaginary worlds. Top: *Lethe*—the river of forgetfulness and oblivion in Greek and Roman mythology. Drinking its waters caused one to forget all former life. Across the river, a vision of Hades. Bottom: *Blessed State*. A giant moon rising over untroubled waters produces a surreal effect. (Both images created by Ken Musgrave and Benoit Mandelbrot of Yale University.)

Figure 12-11 The Sierpinski gasket.

no solid black triangles! If we were to magnify any of these seemingly black triangles we would see a replica of what we see in Fig 12-11 (Fig. 12-12). We now have a name for this phenomenon: The Sierpinski gasket has symmetry of scale.

Just like the Koch snowflake, the Sierpinski gasket can be described in a very convenient way by a recursive replacement rule.

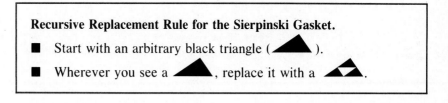

Recursive Replacement Rule for the Sierpinski Gasket.

■ Start with an arbitrary black triangle (▲).

■ Wherever you see a ▲, replace it with a ◥◣ .

We leave the following two facts as exercises to be verified by the reader:

■ The Sierpinski gasket has zero area (Exercise 3).

■ The Sierpinski gasket has an infinitely long boundary (Exercise 4).

(By now, nothing surprises us anymore!)

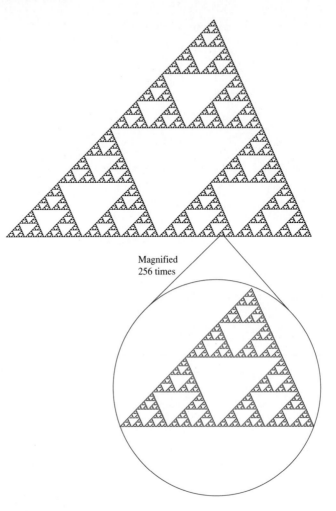

Magnified
256 times

Figure 12-12

THE CHAOS GAME

This example involves an arbitrary triangle ABC and an honest die. To each of the vertices of the triangle we assign two of the six possible outcomes of rolling the die—say, A is assigned the numbers 1 and 2, B is assigned 3 and 4, and C is assigned 5 and 6. (The object is to give each of the three vertices an equal chance of being chosen. Instead of rolling a die, we could just as easily draw the name A, B, or C out of a hat.) We are now ready to play the game.

   Start.   Roll the die. Mark the vertex corresponding to the roll. Say we roll a 5—we then mark vertex C. This is our starting position.

■ **Step 1.** Roll the die again. Say we roll a 2 (i.e., we have picked vertex A). We now mark the point M_1 halfway between the starting position C and the chosen vertex A [Fig. 12-13(a)].

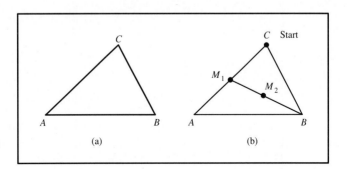

Figure 12-13 The chaos game after three rolls of the die.

■ **Step 2.** Roll the die again. Mark the point M_2 halfway between the previous position M_1 and the chosen vertex. If the roll is 3, for example, mark M_2 halfway between M_1 and B [Fig. 12-13(b)].

■ **Steps 3, 4, etc.** Continue playing the chaos game ad infinitum, each time marking the point halfway between the preceding position and the chosen vertex.

What kind of a picture does the chaos game produce? Figure 12-14(a) shows the results after 50 rolls of the die, just a bunch of scattered dots. Figure 12-14(b) shows the results after 500 rolls, and we begin to detect the vague outlines of something familiar. Figure 12-15 shows the results after 5000 rolls—the pattern is unmistakably a Sierpinski gasket! We won't show the picture of what we get after a billion rolls, but Fig. 12-11 is a very good approximation of it.

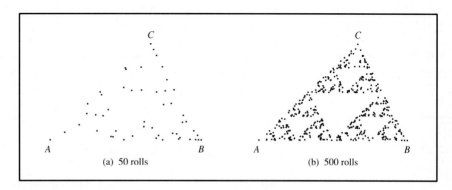

Figure 12-14

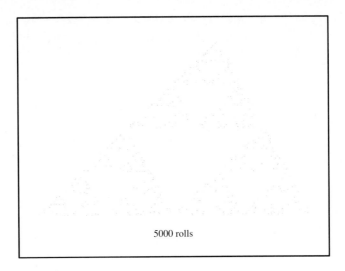

5000 rolls

Figure 12-15

This is a rather surprising turn of events: The chaos game is ruled by the roll of the die and thus by the laws of chance. While our natural expectation is that no predictable pattern should appear, in fact an extremely predictable pattern emerges. There are no ifs or buts about it, the longer you play the chaos game the closer you get to the Sierpinski gasket!

Thus, the chaos game gives us a different way to construct the Sierpinski gasket. While the process is still recursive, an element of chance has been added.

SYMMETRY OF SCALE IN ART AND LITERATURE

The notion of symmetry of scale as exemplified by the Koch snowflake and the Sierpinski gasket is not unique to formal mathematical or geometric constructions. Artists, poets, writers, and even philosophers have used it in their own way and for their own purposes. In this section we will briefly mention just a few such examples.

Figure 12-16 shows a typical artistic expression of symmetry of scale. It is the cover of an issue of "TV Week" in which there is a man sitting in an armchair holding a remote control in his left hand. It so happens that the issue he is holding shows on its cover the same scene: a man sitting in an armchair holding in his right hand an issue of "TV Week," and so on ad infinitum. Just as in our previous examples, there is a recursive construction behind the picture which, in fact, can be described very precisely in geometric terms: Start with the picture of the man with the TV, the remote, the armchair, etc. (everything except the issue of the "TV Week" in his right hand), shrink it down to 35% of the original size, and then apply a rotation and a translation to the new picture so that it

Figure 12-16 (Reproduced by permission of the *Fresno Bee*.)

''slips'' right into the man's right hand in the old picture. If we repeat this process again and again, we get the effect of Fig. 12-16.

As with any other artificial rendition of symmetry of scale, the symmetry of scale in Fig. 12-16 is only illusory: The recursive process is supposed to go on to infinity, but in this case it stops after four steps. (The artist, after all, has to get on with his life.) The message, however, is unmistakable and clear—emotionally all of us accept that the recursion goes on forever even if we know that in reality the process must stop at some point.

More ingenious artistic examples of symmetry of scale can be found in the work of the Dutch artist M. C. Escher and in the literary works of such diverse writers as Lewis Carroll, Aldous Huxley and e.e. cummings. Even philosophers have dealt with the concept of symmetry of scale. The German mathematician and philosopher G. W. Leibniz believed that within each drop of water lived an entire universe containing all the elements of our universe (including of course, more drops of water).

We conclude this section with a poem by Jonathan Swift. The theme is unmistakably symmetry of scale.

So Nat'ralists observe, A Flea
Hath Smaller Fleas that on him prey
and these have smaller Fleas to bite 'em
And so proceed, ad infinitum.

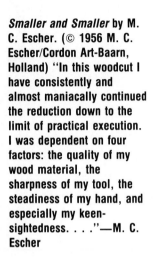

Smaller and Smaller by M. C. Escher. (© 1956 M. C. Escher/Cordon Art-Baarn, Holland) "In this woodcut I have consistently and almost maniacally continued the reduction down to the limit of practical execution. I was dependent on four factors: the quality of my wood material, the sharpness of my tool, the steadiness of my hand, and especially my keen-sightedness. . . ."—M. C. Escher

THE MANDELBROT SET

We now return to a more mathematical example. In fact, the mathematics in this example goes a bit beyond the level of this book (it requires an understanding of *complex numbers*), so we will describe the overall idea in rather general, oversimplified terms. The actual purpose of this example is not to deal with mathematical details but rather to illustrate an important variation of symmetry of scale we haven't seen before. Besides, it gives us an excuse to introduce some incredibly exotic pictures.

The Mandelbrot set is named after the mathematician Benoit Mandelbrot, an IBM Fellow and professor at Yale University. Mandelbrot was the first person to extensively study and fully appreciate the importance of this beautiful and complex mathematical object.

The construction we are going to describe results in an actual geometric shape called the Mandelbrot set. Figure 12-17(a) is a picture of the Mandelbrot set in black and white. To most people, the Mandelbrot set looks like some sort of bug; a cockroach from another planet might be an apt description. The structure of the Mandelbrot set can be described as consisting of a body, a head, and an antenna coming out of the middle of the head. Both the head and the body are surrounded by warts of various sizes. In Fig. 12-17(a) we notice that the larger warts have a structure similar (but not identical) to the original object (body, head, and antennas). When we blow up one of the warts [Fig. 12-17(b)], we can clearly see the

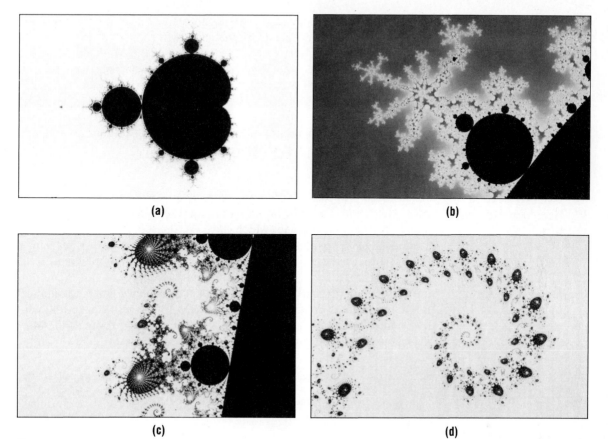

Figure 12-17

repetition of the original overall structure, including the wart's many warts and subwarts. But we can also see new and interesting patterns, such as the swirling clusters appearing on the upper right-hand side and in the lower center of Fig. 12-17(b), as well as the structure resembling a snowflake on the upper left. We also see a few scattered, small, black blots which appear to resemble the original Mandelbrot set itself. Figure 12-17(c) is a close-up of the area around one of the small warts in Fig. 12-17(b). It shows something similar to a jewel-studded brooch shaped like a seahorse's tail in the upper center and a new kind of swirling pattern in the lower center. Smaller versions of both of these patterns can be spotted throughout the picture. In the midst of these new structures we can again detect what appear to be small replicas of the original starting picture. Figure 12-17(d) is a close-up of one of the seahorse tails in Fig. 12-17(c). It is made up of familiar-looking clusters of various sizes as well as new structures we haven't seen before. The process goes on and on and on. At every new level of magnification, familiar and previously unseen formations mingle in a breathtaking dance of infinite repetition and infinite variety—structures repeating themselves at infinitely many different scales but never in exactly the same way.

What sort of symmetry of scale do we find in the Mandelbrot set? In our original examples (the Koch snowflake, the Sierpinski gasket, the cover of "TV Week"), symmetry of scale implied that exact copies of a structure occur across infinitely many scales. This is sometimes described as **exact** symmetry of scale. In the Mandelbrot set, however, there is only an **approximate** symmetry of scale where images repeat themselves across infinitely many scales but are never exactly alike.

■ **Constructing the Mandelbrot Set**

When faced with the beauty and complexity of the Mandelbrot set (see color plates), it is easy to forget its origin: The Mandelbrot set is a completely mathematical object which, just like the Koch snowflake and the Sierpinski gasket, is defined by a relatively simple recursive replacement rule. The recursive process itself works like this: Start by picking a number we will refer to as the **seed** of the recursive process and then apply infinitely many times a recursive rule that at each step squares the number in the preceding step and then adds the seed to the result. The following box summarizes the recursive process, which we will refer to as the **Mandelbrot replacement process**.

Mandelbrot Replacement Process.

■ **Start:** Seed (s).

■ **Recursive step:** Replace x by $x^2 + s$.

The choice of the name "seed" for our starting value gives us a convenient metaphor: Each seed, when planted, produces a different sequence of numbers (the tree). The recursive step is like a rule telling the tree how to grow from one season to the next. Each tree has its own rule for growth, which is encoded in the seed, but the rules for growth for different trees all fall under one general pattern.

Well, let's take a trip to the botanical garden and look at some examples of the Mandelbrot replacement process.

Example 1   (Seed $= 1$).

Start	Step 1	Step 2
1	$s = 1, x = 1$	$s = 1, x = 2$
	$x^2 + s = 2$	$x^2 + s = 5$

Step 3   **Step 4**   ···
$s = 1, x = 5$ $s = 1, x = 26$
$x^2 + s = 26$ $x^2 + s = 677$ ···

In this example we can see that as the process continues, we get bigger and bigger numbers. We say that in this situation the Mandelbrot replacement process takes off toward infinity. (Alternatively, we can call this the *Jack-and-the-beanstalk syndrome*.) ∎

Example 2   (Seed = −1).

Start   **Step 1**   **Step 2**
−1 $s = -1, x = -1$ $s = -1, x = 0$
              $x^2 + s = 0$ $x^2 + s = -1$

              Step 3   **Step 4**   ···
              $s = -1, x = -1$ $s = -1, x = 0$
              $x^2 + s = 0$ $x^2 + s = -1$ ···

In this example the numbers at each step hop back and forth between 0 and −1. We say in this case that the Mandelbrot replacement process is *periodic* (i.e., it goes around and around in a cycle). ∎

Example 3   (Seed = −0.75).

Start   **Step 1**   **Step 2**
−0.75 $s = -0.75, x = -0.75$ $s = -0.75, x = -0.1875$
                  $x^2 + s = -0.1875$ $x^2 + s = -0.714844$

              Step 3   **Step 4**   ···
$s = -0.75, x = -0.714844$ $s = -0.75, x = -0.238998$
    $x^2 + s = -0.238998$ $x^2 + s = -0.69288$ ···

∎

As an exercise (Exercise 25), the reader may want to carry this example out, say, for another 100 or so steps and try to figure out what happens. Here is a hint as to how one can do this (as well as any other example) very efficiently with a calculator that has a memory and a square function key:

> - **Start:**   Store -0.75 ($=s$) in memory
> - **Recursive step:**   Press $\boxed{x^2}$ (squares the current value)
>   Press $\boxed{+}$
>   Press $\boxed{\text{MR}}$ (memory recall)
>   Press $\boxed{=}$ (adds the contents of the memory)

■ Complex Numbers

In Examples 1 through 3 we were careful to choose some fairly simple types of numbers (integers and decimals) for our seeds. In reality, the Mandelbrot replacement process is most interesting when applied to a more complicated category of numbers called complex numbers. You may recall having seen these numbers before (probably in intermediate algebra). These are the numbers that allow us to take square roots of negative numbers, solve quadratic equations of any kind, etc. The basic building block for complex numbers is the number $\sqrt{-1} = i$. Using i we can build all other complex numbers such as $(3+2i)$, $(\frac{5}{3} - \frac{4}{5} i)$, and the generic complex number $(a+bi)$.

Let's look at one more example of the Mandelbrot replacement process using a seed that is a complex number.

Example 4   (Seed $= i$).

Start
i

Step 1
$s = i, x = i$
$x^2 + s = i^2 + i = -1 + i$

Step 2
$s = i, x = -1 + i$
$x^2 + s = (-1 + i)^2 + i =$
$1 - 2i + i^2 + i = -i$

Step 3
$s = i, x = -i$
$x^2 + s = (-i)^2 + i = -1 + i$

Step 4
$s = i, x = -1 + i$
$x^2 + s = -i$
(see step 2)

...
...

This situation is almost identical to the one in Example 2 (the process goes around and around in a cycle), except that the cycle does not go all the way back to the seed.   ■

For our purposes, the most relevant fact about complex numbers is that each complex number can be drawn as a point in the plane. Figure

12-18 is self-explanatory. Thus, we can talk about complex numbers and points in the plane as being one and the same.

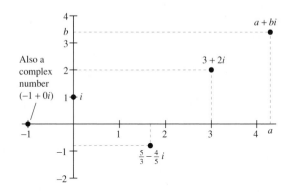

Figure 12-18 Because of this geometric interpretation of complex numbers, it is common practice to identify complex numbers with points in the plane.

Back to the Mandelbrot Set

We are ready (finally) to explain how the Mandelbrot set comes about. Examples 1 through 4 illustrate the fact that the Mandelbrot replacement process behaves in different ways depending on the choice of the seed. In particular, we want to make a distinction between seeds for which the process takes off toward infinity (the Jack-and-the-beanstalk syndrome as illustrated by Example 1) and cases where the process *does not* take off toward infinity (as in Examples 2 through 4). If we color the former seeds white and the latter black, we get the Mandelbrot set!

Let's take a deep breath now and consider what we have done. In essence, it boils down to this: We have a recursive process (the Mandelbrot replacement process) that accepts any point in the plane (i.e., any complex number) as a seed. For some seeds, the process takes off toward infinity, and we color such seeds white. All other seeds we color black. The collection of all black seeds turns out to be the Mandelbrot set. Not quite as simple as one-two-three but simple enough!

The simplicity of the Mandelbrot replacement process stands in marked contrast to the complexity of the results. The Mandelbrot set has been called the "most complex mathematical object known by man." It has also become one of the most popular and fascinating mathematical discoveries of this century.

FRACTALS

The word **fractal** (from the Latin *fractus*, "broken up, fragmented") was coined by the mathematician Benoit Mandelbrot in the mid-1970s to describe under one term objects as diverse as the Koch curve, the Sierpinski gasket, and the Mandelbrot set, as well as many shapes in nature such as those of clouds, coastlines, lightning, mountains, the vascular system in the human body, and the lungs.

All the above objects are fractals. As such, they share several characteristics. One of them is that they all have some form of symmetry of scale. (Some of the other mathematical characteristics of fractals, such as fractional dimensions, are beyond the scope of this book. Any reader who wishes to investigate other mathematical properties of fractals in depth is encouraged to look at some of the many references listed at the end of this chapter.)

A word of caution is in order: Symmetry of scale as we defined it does not by itself ensure that the object is a fractal. The picture in Fig. 12-16, for example, has symmetry of scale but is not a fractal. All fractals, however, have some form of symmetry of scale, be it exact symmetry of scale (as in the Koch curve and the Sierpinski gasket) or approximate symmetry of scale (as in the Mandelbrot set).

The discovery and study of new fractal shapes has become arguably one of the hottest mathematical topics of the last 15 years. As a field of study, fractal geometry is a mathematician's dream come true: It combines complex and interesting mathematics, beautiful graphics (see color plates 9–13) and extreme relevance to the real world.

CONCLUSION: FRACTAL GEOMETRY, A NEW FRONTIER

In his classic book, *The Fractal Geometry of Nature*, Benoit Mandelbrot wrote

Why is [standard] geometry often described as "cold" and "dry"? One reason lies in its inability to describe the shape of a cloud, a mountain, a coastline, or a tree. Clouds are not spheres, mountains are not cones, coastlines are not circles, and bark is not smooth, nor does lightning travel in a straight line.

. . . many patterns of Nature are so irregular and fragmented, that compared with [standard geometry] Nature exhibits not only a higher degree but an altogether different level of complexity. The number of distinct scales of length of natural patterns is for all practical purposes infinite.

There is a striking visual difference between the shapes of traditional geometry and the geometric shapes we discussed in this chapter. It is difficult to mistake one for the other. The shapes of traditional geometry (squares, circles, cones, etc.) and the objects we build based on them (bridges, machines, buildings, etc.) have a distinct artificial look, and no amount of artistic license can disguise this. To paraphrase Mandelbrot, one cannot make a realistic looking cloud or mountain using spheres and cones. For a long time, it was assumed that making realistic geometric models of the natural world was pretty much impossible. Consider, however, color plates 14a, 15a and, 15b. These artificial landscapes were created using the tools of a new geometry called **fractal geometry**. In this

"Clouds are not spheres, mountains are not cones, coastlines are not circles, bark is not smooth, nor does lightening travel in a straight line."—B. Mandelbrot (Myron Wood/Photo Researchers, Max and Kit Hunn/Photo Researchers)

new geometry mathematically defined fractals replace the traditional squares, circles, cones, etc., just as the computer replaces the traditional ruler, compass, protractor, etc.

Mandelbrot, who is the father of fractal geometry, was the first to realize that fantastic geometric constructions that imitate nature are possible using fractals. This realization has revolutionized many fields in which modeling nature is important. One we can all easily relate to is the field of computer animation. Viewers of the movie *Star Trek II: The Wrath of Khan* are always impressed by the Genesis planetary sequence. In this sequence an ecologically benign bomb is dropped on a barren planet and a lush tropical world emerges in front of our very eyes. This entire special effect was created in a computer using fractal geometry. Since then, fractally created landscapes have appeared in science fiction movies (*The Last Starfighter*, etc.) and in many dazzling pieces of computer art.

Because symmetry of scale is a pervasive and fundamental characteristic of many objects in nature, fractals have become an essential tool in their study—be it to predict, analyze, or imitate either the natural objects themselves or their functions. Important applications of fractal geometry have been discovered in recent years in such diverse fields as materials science, population biology, and human physiology. For details, the reader is referred to references 4, 8, 9, 11, 17, 23, and 26 in the bibliography.

Geometry as we have known it in the past was developed by the Greeks about 2000 years ago and passed on to us essentially unchanged. It was (and still is) a great triumph of the human mind, and it has allowed us to develop much of our technology, engineering, architecture, etc. As a tool and language for modeling and representing nature, Greek geometry has by and large been a failure. The discovery of fractal geometry seems to have given science the right mathematical language to remedy this failure and as such it promises to be one of the great achievements of twentieth century mathematics.

KEY CONCEPTS

chaos game
fractal
fractal geometry
Koch curve (snowflake curve)
Koch snowflake

Mandelbrot replacement process
Mandelbrot set
recursive replacement rule
Sierpinski gasket
symmetry of scale

EXERCISES

■ Walking

Exercises 1 through 4 refer to the construction of the Sierpinski gasket as described in this chapter. The following figure shows the first two steps of the construction.

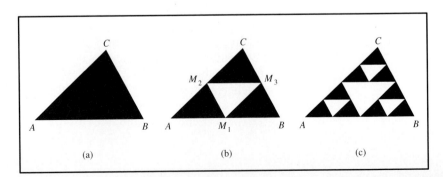

(a)   (b)   (c)

1. (a) Explain why triangle AM_1M_2 is similar to triangle ABC.
   (b) Explain why triangle $M_3M_2M_1$ is similar to triangle ABC.
   (c) Explain why triangles AM_1M_2, M_1BM_3, and M_2M_3C are congruent.
   (d) How much is the area of triangle $M_3M_2M_1$ compared to that of triangle ABC? Explain.

2. Using graph paper, start with your own arbitrary triangle ABC and draw the figures that result at steps 1, 2, and 3 in the construction of the Sierpinski gasket.

3. Suppose the area of triangle ABC is X.
   (a) Find the areas of the figures obtained in steps 1, 2, and 3 of the construction of the Sierpinski gasket.
   (b) Find the area of the figure obtained in step N in the construction of the Sierpinski gasket.
   (c) Explain why the Sierpinski gasket has zero area.

4. At each step of the construction of the Sierpinski gasket, the boundary of the figure consists of all line segments that separate a white and a black region. Suppose the original triangle ABC has perimeter P.
   (a) Find the length of the boundary at steps 1, 2, and 3.
   (b) Find the length of the boundary at step N.
   (c) Explain why the Sierpinski gasket has an infinitely long boundary.

Exercises 5 through 8 refer to the following recursive construction:

■ *Start.   Start with a solid black square.*

■ *Step 1.   Subdivide the square into nine equal subsquares and remove the central subsquare.*

■ *Steps 2, 3, etc.   Continue removing the central square of every black subsquare ad infinitum.*

*The fractal shape obtained from this construction is known as the **Sierpinski carpet**.*

5. Using graph paper, draw the figures that result at steps 1, 2, and 3 in the construction of the Sierpinski carpet.

6. Write down the recursive replacement rule for the Sierpinski carpet. (Start with a _____. Wherever you see a _____ replace it with a _____.)

7. Suppose the area of the starting square is X.
   (a) How many black squares are there in the figure obtained at step 3 in the construction?

    (b) What is the area of the figure obtained at step 3 in the construction?

    (c) How many black squares are there in the figure obtained at step N in the construction?

    (d) What is the area of the figure obtained at step N in the construction?

    (e) Explain why the area of the Sierpinski carpet is zero.

8. Suppose that the starting square has sides of length 1.

    (a) Find the length of the boundary of the figure at steps 1 and 2 in the construction.

    (b) Find the length of the boundary of the figure at step N in the construction.

    (c) Explain why the length of the boundary of the Sierpinski carpet is infinite.

Exercises 9 through 12 refer to the following recursive replacement rule:

    The fractal image obtained from this construction is known as the **quadratic Koch island**.

9. Using graph paper, draw the figures that result at steps 1 and 2 in the construction of the quadratic Koch island.

10. Suppose that the starting square encloses an area X.

    (a) Find the areas enclosed by the figures at steps 1 and 2 in the construction of the quadratic Koch island.

    (b) Find the area enclosed by the figure at step N in the construction of the quadratic Koch island.

11. The figure obtained at step N in the construction of the quadratic Koch island is a polygon. How many sides does it have?

12. (a) Find the perimeter of each of the polygons obtained at steps 1 and 2 in the construction of the quadratic Koch island.

    (b) Find the perimeter of the polygons obtained at step N in the construction of the quadratic Koch island.

Exercises 13 through 15 refer to the following recursive replacement rule:

■ *Start.   Start with a solid black square.*

■ *Step 1.   Divide each side of the square into three equal segments. Attach to the middle segment of each side a small, black square with sides one-third the sides of the starting square.*

■ *Steps 2, 3, etc.   Repeat the process (divide sides into thirds and attach appropriate-sized squares to the middle third) ad infinitum.*

13. Using graph paper, draw the figures that result at steps 1, 2, and 3 in the construction.

14. Suppose the area of the starting square is X.
   (a) What is the area of the figure obtained at step 3 in the construction?
   (b) What is the area of the figure obtained at step N in the construction?

15. Suppose that the starting square has sides of length 1.
   (a) Find the length of the boundary of the figure at steps 1 and 2 in the construction.
   (b) Find the length of the boundary of the figure at step N in the construction.

Exercises 16 and 17 refer to the chaos game as described in the chapter. Assume that we start with an arbitrary triangle ABC and that we will roll an honest die. Vertex A is assigned numbers 1 and 2; vertex B is assigned numbers 3 and 4; and vertex C is assigned numbers 5 and 6.

16. Suppose the die is rolled 16 times and the outcomes are 3, 4, 2, 3, 6, 1, 6, 5, 5, 3, 1, 4, 2, 2, 2, 3. Draw the points P_1 through P_{16} corresponding to these outcomes.

17. Suppose the die is rolled 12 times and the outcomes are 5, 5, 1, 2, 4, 1, 6, 3, 3, 6, 2, 5. Draw the points P_1 through P_{12} corresponding to these outcomes.

Exercises 18 through 20 refer to the Mandelbrot replacement process described in this chapter. (You may need a calculator to do the computations.)

18. **(a)** Apply the Mandelbrot replacement process to the seed $s = 2$ and calculate the numbers obtained in the first five steps of this process.
   (b) What happens to the numbers when we repeat the process indefinitely?

19. **(a)** Apply the Mandelbrot replacement process to the seed $s = -2$ and calculate the numbers obtained in the first five steps of this process.
   (b) What happens to the numbers when we repeat the process indefinitely?

20. **(a)** Apply the Mandelbrot replacement process to the seed $s = -0.25$ and calculate the numbers obtained in the first 20 steps of this process.
   (b) What happens to the numbers when we repeat the process indefinitely?

■ Jogging

21. This exercise refers to the construction of the Sierpinski gasket. Explain why there are $(3^N - 1)/2$ white triangles in the figure obtained at step N in the construction.

22. This exercise refers to the construction of the Sierpinski carpet discussed in Exercises 5 through 8. How many white squares does the figure obtained at step N in the construction have?

23. Explain why the construction of the Sierpinski gasket does not end up in an all-white triangle.

24. Use the formula for adding consecutive terms of a geometric sequence to show that

(a) $1 + (\frac{4}{9}) + (\frac{4}{9})^2 + \cdots + (\frac{4}{9})^{N-1} = \frac{9}{5}[1 - \frac{4^N}{9}]$

(b) $(\frac{1}{3}) A + (\frac{4}{9})(\frac{1}{3}) A + (\frac{4}{9})^2 (\frac{1}{3}) A + \ldots + (\frac{4}{9})^{N-1} (\frac{1}{3}) A = \frac{3}{5} A [1 - (\frac{4}{9})^N]$.

Exercises 25 through 30 refer to the Mandelbrot replacement process described in this chapter. (You will need a calculator.)

25. Apply the Mandelbrot replacement process to the seed $s = -0.75$. What is the long-term behavior of the numbers in this process?

26. Apply the Mandelbrot replacement process to the seed $s = 0.2$. What is the long-term behavior of the numbers in this process?

27. Apply the Mandelbrot replacement process to the seed $s = 0.25$. What is the long-term behavior of the numbers in this process?

28. Apply the Mandelbrot replacement process to the seed $s = -1.25$. What is the long-term behavior of the numbers in this process?

29. Apply the Mandelbrot replacement process to the seed $s = \sqrt{2}$. What is the long-term behavior of the numbers in this process?

30. Apply the Mandelbrot replacement process to the seed $s = -\sqrt{2}$. What is the long-term behavior of the numbers in this process?

■ **Running**

31. Suppose that we play the chaos game using triangle ABC and M_1, M_2, M_3 are the midpoints of the three sides of the triangle. Explain why it is impossible at any time during the game to land inside triangle $M_1 M_2 M_3$.

32. Consider the fractal defined by the construction given in Exercises 13 through 15. Find its area.

33. Apply the Mandelbrot replacement process to the following seeds:
(a) $s = 1 + i$. What is the long-term behavior of the numbers in this process?
(b) $s = 1 - i$. What is the long-term behavior of the numbers in this process?

34. Apply the Mandelbrot replacement process to the following seeds:
(a) $s = -0.25 + 0.25i$. What is the long-term behavior of the numbers in this process?
(b) $s = -0.25 - 0.25i$. What is the long-term behavior of the numbers in this process?

35. Show that the Mandelbrot set has a reflection symmetry. (*Hint*: See Exercises 33 and 34.)

REFERENCES AND FURTHER READINGS

1. Dewdney, A. K., "Computer Recreations: A computer microscope zooms in for a look at the most complex object in mathematics," *Scientific American*, 253 (August 1985), 16–24.

2. Dewdney, A. K., "Computer Recreations: A tour of the Mandelbrot set aboard the Mandelbus," *Scientific American*, 260 (February 1989), 108–111.

3. Dewdney, A. K., "Computer Recreations: Beauty and profundity: The Mandelbrot set and a flock of its cousins called Julia," *Scientific American*, 257 (November 1987), 140–145.

4. Dewdney, A. K., "Computer Recreations: Of fractal mountains, graftal plants and other computer graphics at Pixar," *Scientific American*, 255 (December 1986), 14–20.

5. Falconer, K. J. *Fractal Geometry: Mathematical Foundation and Applications*. New York: John Wiley & Sons, Inc., 1989.

6. Feder, Jens, *Fractals*. New York: Plenum Publishing Corp., 1988.

7. Fryde, M. M., "Waclaw Sierpinski—Mathematician," *Scripta Mathematica*, 27 (1964), 105–111.

8. Gardner, M., *Penrose Tiles to Trapdoor Ciphers*. New York: W. H. Freeman & Co., 1988, chap. 3.

9. Gleick, James, *Chaos: Making a New Science*. New York: Viking Penguin, Inc., 1987, chap. 4.

10. Gleick, James, "The Man Who Reshaped Geometry," *New York Times Magazine*, 135 (December 8, 1985), 64.

11. Goldberger, A. L., D. R. Rigney, and B. J. West, "Chaos and Fractals in Human Physiology," *Scientific American*, 262 (February 1990), 44–49.

12. Jürgens, H., H. O. Peitgen, and D. Saupe, "The Language of Fractals," *Scientific American*, 263 (August 1990), 60–67.

13. Krantz, Steven, "Fractal Geometry," *Mathematical Intelligencer*, 11 (Fall 1989), 12–16.

14. La Brecque, Mort, "Fractal Symmetry," *Mosaic*, 16 (January-February 1985), 14–23.

15. Lauwerier, Hans, *Fractals*. Princeton, N.J.: Princeton University Press, 1991.

16. Lord, E. A., and C. B. Wilson, *The Mathematical Description of Shape and Form*. New York: John Wiley & Sons, Inc., 1984, chap. 8.

17. Mandelbrot, Benoit, *The Fractal Geometry of Nature*. New York: W. H. Freeman & Co., 1983.

18. McDermott, Jeanne, "Dancing to Fractal Time," *Technology Review*, 92 (January 1989), 6–8.

19. McWorter, W. A., Jr., and J. M. Tazelaar, "Creating Fractals," *Byte*, 12 (August 1987), 123–130.

20. Peitgen, H. O., H. Jürgens, and D. Saupe, *Fractals for the Classroom.* New York: Springer Verlag Inc., 1991.

21. Peitgen, H. O., and P. H. Richter, *The Beauty of Fractals.* New York: Springer Verlag, Inc., 1986.

22. Peterson, Ivars, *The Mathematical Tourist.* New York: W. H. Freeman & Co., 1988, chap. 5.

23. Prusinkiewicz, P., and A. Lindenmayer, *The Algorithmic Beauty of Plants.* New York: Springer Verlag, Inc., 1990.

24. Schechter, Bruce, "A New Geometry of Nature," *Discover*, 3 (June 1982), 66–68.

25. Schroeder, Manfred, *Fractals, Chaos, Power Laws: Minutes from an Infinite Paradise.* New York: W. H. Freeman & Co., 1991.

26. Sorensen, Peter, "Fractals," *Byte*, 9 (September 1984), 157–172.

27. Steen, Lynn Arthur, "Fractals: A World of Nonintegral Dimensions," *Science News*, 112 (August 20, 1977), 122–123.

13

Collecting Data

> *"Data! data! data!" he cried impatiently. / "I can't make bricks without clay"*
> SHERLOCK HOLMES[*]

Censuses, Surveys, and Clinical Studies

Read up to 408 #1-8

Disclaimer: The following headlines are fictitious (but realistic).

■ More than 63% of All Americans Support New Tax Cuts

■ Average SAT Scores Up to 432 in Verbal and 491 in Math This Year

■ Consumer Prices Rose Three-tenths of One Percent in the Last Quarter

■ Nielsen Ratings Show "Wheel of Fortune" Inches Ahead of Other Game Shows With an 18.8 Rating for February

■ New Study Shows That Eating More Than One Pound of Chocolate a Day Increases the Risk of Diabetes by 40% (but what a way to go!)

What all the above headlines have in common is that they give information and that some (or all) of this information is *numerical* or *quan-*

---

[*] Arthur Conan Doyle, *Adventures of Sherlock Holmes* ("The Adventure of the Copper Beeches"). New York: Doubleday, Doran & Co. Inc., 1930, p. 297.

titative. Numerical or quantitative information is **data** and, to put it in a nutshell, statistics is the science of dealing with data.

Behind every statistical headline there is a story, and as in any story there is a beginning, a middle, an end, and a moral. This chapter is about the beginning, which in statistics typically means the process of collecting the data. Data are the raw material with which statistical information is built and if the data are flawed, then the conclusions one can draw from such data may be of little or no value.

There is a deceptive simplicity to the idea of collecting data, but in fact just the opposite is true. Collecting good data in an efficient and timely manner is often the most difficult part of the statistical story. At every turn in the process there are serious issues to deal with and potential pitfalls to avoid.

In this chapter we will discuss several systematic methods of collecting data and the issues associated with them. We will use actual case studies to illustrate the main ideas (both good and bad) in the chapter.

SIZING UP THE POPULATION

Every statistical statement refers, directly or indirectly, to some group of individuals or objects. In statistical terminology, this collection of individuals or objects is called the **population**. The population in the first headline is "all Americans." Likewise, the population for the second headline is the group of students taking the Scholastic Aptitude Test (SAT) that particular year. The population for the third headline is made up of numbers rather than people (the prices that people pay for goods and services). For the last two headlines it's hard to tell directly from the wording exactly what the populations are. Often one has to read the fine print before an exact determination of the population can be made.

■ Defining the Population

Having an exact and precise definition of who or what is the population is essential. This may seem self-evident, but it is often not done, either because of carelessness or because it is in fact impossible. What exactly is meant by the phrase "all Americans" as used in the first headline? Does it include children? How about resident aliens? How about Americans living abroad? How about people in American Samoa? Clearly, the phrase "all Americans" is very vague and subject to many possible interpretations, each of which can affect the meaning and significance of the statistical information one wants to convey.

There are occasions in which a precise definition of the population is in fact impossible. Most public opinion polls preceding an election refer to the population of "voters." But who exactly are the voters? People who are registered to vote and who may even be planning to vote may end up not voting. Is a potential voter a voter?

■ How Big Is the Population?

Along with the issue of defining the population comes the issue of size. The size of the population is often used to draw statistical conclusions. But how reliable is the information on size? When populations are small and accessible one can actually get an exact "head count" by simply counting heads the way one counts pennies in a jar. When an instructor reports to a class the average score on an exam, the population consists of all students who took the exam. The instructor can easily determine the exact size of this population—all he has to do is to check his roster.

For large populations, an accurate count is always expensive, usually difficult, and sometimes impossible. How many elephants are there in the wild? In the absence of this information, how much faith can we put in statistical statements about wild elephants? Here is another one: How many people live in the United States? Presumably this information is available from the national census which is mandated by the Constitution and conducted every 10 years. According to the 1990 U.S. Census, there are 249,632,692 people living in the United States. As we will find out later in the chapter, there is very little reason to believe that this number is accurate. In spite of this, all kinds of other important statistical information are based on this figure.

CENSUS VERSUS SURVEY

Once a population is defined, each member of the population becomes a potential source of data and there are basically two ways to go about collecting the data: (1) collect data from each and every member of the population, or (2) collect data only from a selected subgroup of the population. Method 1 is called a **census**; method 2 is called a **survey**.

A word of caution: The word "census" is associated by most people with the U.S. census which is conducted every 10 years, but any method for collecting data that uses the entire population is also a census. When an instructor reports the scores on a test to her class, the information is based on a census of the entire population (the students who took the test). To avoid confusion, we will speak of "the U.S. Census" when we refer to the national census and use "a census" for any other example.

A census is most commonly used when the population is small and accessible. Of course "small" is a relative term, but generally speaking, if the size of the population is more than a few thousand, we wouldn't want to use a census unless there is a compelling reason. At the same time, a census can be a bad idea even when the population is small. Suppose we want to collect data on the condition of a population of fish in a small pond. Conducting a census would require catching and observing each and every fish in the pond. But fish are elusive and clever creatures, and the time and effort required to do this may not be justified by the situation. Moreover, short of draining the pond (a bad idea), how would we ever know when the census is complete?

The alternative to collecting data by means of a census is to collect data using a survey. In a survey the data are collected using only a selected subgroup of the population called a **sample**. The basic idea is that questions about the entire population can be answered based on the information collected from the sample. We can think of this as using the members of the sample to act as standard-bearers for the entire population, not unlike the idea of representative government where elected officials act as spokespersons for the people.

The critical issue in surveys (just as it is in politics) is to choose a sample that is truly representative of the entire population. This is easier said than done unless the population is made up of identical individuals. For a population in which the members are not all alike, collecting data by means of a survey as opposed to a census involves a tradeoff: A survey is always subject to some error, but at the same time it requires a lot less work. The exact nature of this tradeoff depends on the situation, but if the sample is properly chosen, the margin of error can be guaranteed by statistical theory to be small and the lack of accuracy can be more than offset by the savings in cost, time, and effort. In many real-world situations, surveys are the only way to go, and their practical value has been proven time and time again.

One of the best known types of surveys are public opinion polls such as the Gallup poll and the Harris poll so often mentioned in the press. Such polls are used to report statistical information on national or regional issues ranging from voters' preferences before an election to national opinions on issues such as the environment, abortion, and the economy. Considering the fact that most modern polls are based on samples consisting of less than two thousand individuals to draw conclusions about populations of many millions, the accuracy of such polls is remarkable. (A typical Gallup poll, for example, is based on samples consisting of less than 1500 individuals.) How can such small samples give accurate information about such large populations? The secret lies in getting a good blend. George Gallup, one of the fathers of modern public opinion polls puts it this way[1]:

> *Whether you poll the United States or New York State or Baton Rouge (160,000 population), you need only the same number of interviews or samples. It's no mystery really—if a cook has two pots of soup on the stove, one far larger than the other, and thoroughly stirs them both, he doesn't have to take more spoonfuls from one than the other to sample the taste accurately.*

[1] Quoted in "The Man Who Knows How We Think," *Modern Maturity* 17, no. 2 (April–May, 1974), 11.

■ Some Basic Terminology

Before we move on to our case studies we will formalize some o basic concepts and terms. We will start with a convention: From now on we will use N to denote the size of the population, and n to denote the size of the sample. Since the sample is by definition part of the population, it must always be the case that $n < N$. The ratio n/N is called the **sampling rate**.

Let's say, for example, that we have a population of size $N = 500,000$ and we pick a sample of size $n = 1000$. In this case the sampling rate is $1000/500,000 = 1/500 = 0.2\%$. This tells us that each member of the sample represents 500 members of the population.

As we know now, using a sample is one way to collect statistical information about an entire population. Statisticians use the words "parameter" and "statistic" to distinguish between exact statistical information about a population and statistical information based on a sample. When the source of our statistical information is a sample, then any piece of data about the population obtained from that sample is called a **statistic**. A statistic is always an estimate for some fixed but unknown quantity called a **parameter**. Let's put it this way: The parameter is always the statistical information we are after—the pot of gold at the end of the statistical rainbow, so to speak. Calculating a parameter is often difficult, and sometimes impossible. The only hope for getting an exact value for it is to use a census. If we use a sample, then we can get only an estimate for the parameter, and this estimate is called a statistic.

We will use the term **sampling error** to describe the difference between a parameter and a statistic used to estimate that parameter. Sampling error can be attributed to two factors: chance error and sample bias. **Chance error** is the result of a basic fact about surveys: A statistic cannot give exact information about the population because it is by definition based on partial information (the sample). In surveys, chance error comes with the territory. While chance error is unavoidable, with careful selection of the sample and the right choice of sample size, it can be kept to a minimum.

A parallel but different cause of errors in surveys is due to sample bias. **Sample bias** is the result of having a poorly chosen sample. Even with the best intentions, getting a sample that is representative of the entire population can be very difficult, and many subtle factors can affect the "representativeness" of the sample. Sample bias is the result. As opposed to chance error, sample bias can be eliminated by using proper methods of sample selection. For the rest of this chapter, we will focus our attention on the issue of sample bias. We will touch on the concept of chance error and its applications to sampling in Chapter 16.

The essential facts about using samples to collect data about a population can be summarized as follows:

■ A parameter is an exact measure of some attribute of a population; a statistic is an estimate of the parameter obtained from a sample.

■ A parameter is a fixed quantity, whereas a statistic depends on the choice of the sample. If one chooses a different sample, one is almost certain to get a different statistic (and this is true even when the samples are chosen using exactly the same procedures). This variability among samples is called, not surprisingly, **sampling variability**.

■ Sampling error is the difference between the parameter and the statistic obtained from the sample. Sampling error is made up of two elements: chance error and sample bias.

■ Chance error is an inevitable consequence of sampling variability.

■ Sample bias is a systematic source of error caused by using improper methods for selecting the sample.

■ While chance error cannot be eliminated, it can be brought under control by choosing a suitable sample. As the size of the sample increases, chance error decreases, but surprisingly, the decrease is not in proportion to the size of the sample. After a certain point, a big increase in the size of the sample may have a very minor effect on the chance error.

■ As opposed to the situation with chance error, increasing the size of the sample does not guarantee a reduction in sample bias. In fact, with a poorly chosen sample, a large sample may actually increase sample bias.

This last proposition will be best illustrated in our first case study, one of the worst bungles in the history of public opinion polling.

CASE STUDY 1. BIASED SAMPLES: THE 1936 *LITERARY DIGEST* POLL

The presidential election of 1936 pitted Alfred Landon, the Republican governor of Kansas, against the incumbent President, Franklin D. Roosevelt. The year 1936 marked the tail end of the Great Depression, and economic issues such as unemployment and government spending were the dominant themes of the campaign. The *Literary Digest* was one of the most respected magazines of the time and had a history of accurately predicting the winners of presidential elections that dated back to 1916. For the 1936 election, the *Literary Digest* prediction was that Landon would get 57% of the vote against Roosevelt's 43% (these were the *statistics*). The actual results of the election were 62% for Roosevelt against 38% for Landon (these were the *parameters* the poll was after). The sampling error in the *Literary Digest* poll was a whopping 19%, the largest ever in a major public opinion poll. Practically all of the sampling error was the result of sample bias.

The irony of the situation is that the *Literary Digest* poll was also one of the largest and most expensive polls ever conducted, with a sample size of approximately 2.4 million people. At the same time that the *Literary Digest* poll was making its fateful mistake, George Gallup was able to predict a victory for Roosevelt using a much smaller sample of approximately 50,000 people.

These facts illustrate the proposition that bad sampling methods cannot be cured by increasing the size of the sample, which in fact just compounds the mistakes. The critical issue in sampling is not sample size but how best to reduce sample bias. There are many different ways that bias can creep into the sample selection procedure. Two of the most common ones occurred in the case of the *Literary Digest* poll.

The *Literary Digest* method for choosing its sample was as follows: Based on every telephone directory in the United States, lists of magazine subscribers, rosters of clubs and association, and other sources, a mailing list of about *10 million* names was created. Every name on this list was mailed a mock ballot and asked to return the marked ballot to the magazine.

One cannot help but be impressed by the sheer ambition of such a project. Neither is it surprising that the magazine's optimism and confidence were in direct proportion to the magnitude of the effort. In its August 22, 1936 issue (p. 3), the *Literary Digest* crowed:

> *Once again, [we are] asking more than ten million voters—one out of four, representing every county in the United States—to settle November's election in October.*
>
> *Next week, the first answers from these ten million will begin the incoming tide of marked ballots, to be triple-checked, verified, five-times cross-classified and totaled. When the last figure has been totted and checked, if past experience is a criterion, the country will know to within a fraction of 1 percent the actual popular vote of forty million [voters].*

We are now ready to analyze the sources of sample bias in the *Literary Digest* poll. There were two basic causes of the *Literary Digest's* downfall[2]: selection bias and nonresponse bias.

The first major problem with the poll was in the selection process for the names on the mailing list which, as we mentioned, were taken from telephone directories, club membership lists, lists of magazine subscribers, etc. Such a list is guaranteed to be slanted toward middle- and upper-

---

[2] This is no figure of speech—the magazine stopped publication soon after the poll.

The Literary Digest

NOVEMBER 14, 1936 · Thirty ~~few~~ CENTS

IS OUR FACE RED!

Reprinted by permission of Harper & Row, Publishers, Inc.

Owning up to a bad case of sample bias: The cover of the *Literary Digest* **issue after the 1936 election says it all.**

class voters, and by default to exclude lower-income voters. One must remember that in 1936 telephones were much more of a luxury than they are today. Furthermore, at a time when there were still 9 million people unemployed, it is obvious that the names of a very significant element of the population would not show up on lists of club memberships and magazine subscribers. At least with regard to economic status, the *Literary Digest* mailing list was far from being a representative cross section of the population. This is always a critical problem because voters are generally known to vote their pocketbooks, and it was magnified in the 1936 election when economic issues were preeminent in the minds of the voters.

When the method for choosing the sample has a built-in tendency to exclude certain segments of the population (whether intentional or not), we say that the survey suffers from **selection bias**. While it is obvious that to make good surveys selection bias must be avoided at all costs, it is not always easy to detect it ahead of time. Even the most scrupulous attempts to get a sample that is a representative cross section of the population can fall short. This fact will become quite apparent in our next case study.

The second problem with the *Literary Digest* poll was that out of the 10 million people whose names were on the original mailing list, only about 2.4 million responded to the survey. Thus, the size of the sample

(n = 2.4 million) was about one-fourth of what was originally intended. It is a well-known fact that people who respond to surveys are very different from people who don't, not only in the obvious (their attitude toward the usefulness of surveys) but also in more subtle and significant ways: They tend to be better educated and in higher economic brackets (and are in fact more likely to vote Republican).

The ratio between the number of people responding to a survey and the number of people asked to participate in the survey is called the **response rate.** For the *Literary Digest* poll, the response rate was 2,400,000/10,000,000 = 0.24, which is extremely low. When the response rate is low, a survey is said to suffer from **nonresponse bias**. Nonresponse bias is a special type of selection bias, since it means that reluctant and nonresponsive people have been automatically excluded from the sample.

Dealing with nonresponse bias presents its own set of difficulties. In a free country we cannot force people to participate in a survey, and paying them is hardly ever a solution since it can introduce other forms of bias. There are known ways, however, of minimizing nonresponse bias. The *Literary Digest* survey was conducted by mail. This approach is the most likely to magnify nonresponse bias because people often consider a mailed questionnaire just another form of junk mail. Of course, considering the size of the mailing list, the *Literary Digest* really had no other choice. Here again is another illustration of how a big sample size can be more of a liability than an asset.

Nowadays, almost all legitimate public opinion polls are conducted either by telephone or by personal interviews. Telephone polling is subject to slightly more nonresponse bias than personal interviews, but it is considerably cheaper. Even today, however, a significant segment of the population has no telephone in their homes (in fact, a significant segment of the population has no homes), so that selection bias can still be a problem in telephone surveys.[3]

The *Literary Digest* story has two morals: (1) a badly chosen big sample is much worse than a well-chosen small sample, and (2) watch out for selection bias and nonresponse bias.

Our next case study illustrates how difficult it can be to get rid of selection bias even with the very best intentions.

---

[3] The most extreme form of nonresponse bias occurs when the sample consists only of those individuals who step forward and actually "volunteer" to be in the sample. A blatant example of this is the Area Code 900 telephone polls. Here an individual not only has to step forward and volunteer to be in the sample, *he or she actually has to pay* (50 cents or more) to do so. It goes without saying that people who are willing to pay to express their opinions are hardly representative of the general public and that information collected from such polls should be considered useless. Unfortunately, the use and misuse of Area Code 900 polls is proliferating.

CASE STUDY 2. QUOTA SAMPLING: THE 1948 PRESIDENTIAL ELECTION

Soon after the 1936 fiasco, the *Literary Digest* went out of the business of polling. At the same time, the practice of using public opinion polls to measure the pulse of the American electorate was thriving. By 1948 there were several major polls competing for the big prize, that of accurately predicting the outcome of presidential elections. The best known was the Gallup poll, and the two main competitors were the Roper poll and the Crossley poll.

By this time, all major polls were using what was believed to be a much more scientific method for choosing their samples called **quota sampling**. Quota sampling had been introduced by George Gallup as early as 1935 and had been successfully used by him to predict the winner of the 1936, 1940, and 1944 elections. Quota sampling is nothing more than a systematic effort to force the sample to fit a certain national profile by using quotas: The sample should have so many women, so many men, so many blacks, so many whites, so many under 40, so many over 40, etc. The numbers in each category are taken to represent the same proportions in the sample as are in the electorate at large.

If we assume that every important characteristic of the population is taken into account when setting up the quotas, it is reasonable to expect that quota sampling will produce a good cross section of the population and therefore lead to accurate predictions. For the 1948 election between Thomas Dewey and Harry Truman, Gallup conducted a poll with a sample size of approximately 3250. Each individual in the sample was interviewed in person by a professional interviewer to minimize nonresponse bias, and each interviewer was given a very detailed set of quotas to meet. For example, an interviewer could have been given the following quotas: seven white males under 40 living in a rural area, five black males under 40 living in a rural area, six black females under 40 living in a rural area, six white females over 40 living in a rural area, etc. Other than meeting these quotas the ultimate choice of who was interviewed was left to each interviewer.

Based on the results of this poll, Gallup predicted a victory for Dewey, the Republican candidate. (The predicted breakdown of the vote was 50% for Dewey, 44% for Truman, and 6% for third-party candidates Strom Thurmond and Henry Wallace.) The actual results of the election turned out to be almost exactly reversed: 50% for Truman, 45% for Dewey, and 5% for the third-party candidates.

Truman's victory was a great surprise to the nation as a whole. So convinced was the Chicago Tribune of Dewey's victory that it went to press on its early edition for November 4, 1948 with the headline "Dewey defeats Truman"—a blunder that led to Truman's famous retort "Ain't the way I heard it." The picture of Truman holding aloft a copy of the Tribune (see photo) has become part of our national folklore. To pollsters and statisticians, the results of this election were a clear indication that

"Ain't the way I heard it," Truman gloats while holding an early edition of the Chicago Daily Tribune in which the headline erroneously claimed a Dewey victory based on the predictions of all the polls. (UPI/Bettmann)

as a method for selecting a representative sample, quota sampling can have some serious flaws.

The basic idea of quota sampling is on the surface a good one: Force the sample to be a representative cross section of the population by having each important characteristic of the population proportionally represented in the sample. Since income is an important factor in determining how people vote, the sample should have all income groups represented in the same proportion as the population at large. Ditto for sex, race, age, etc. Right away we can see one of the potential problems: Where do we stop? No matter how careful one might be, there is always the possibility that some criterion that would affect the way people vote might be missed and the sample could be deficient in this one regard.

An even more serious flaw in the method of quota sampling is the fact that ultimately the choice of who is in the sample is left to the human element. Recall that other than meeting the quotas the interviewers were free to choose whom they interviewed. Looking back over the history of quota sampling, one can see a clear tendency to overestimate the Republican vote. In 1936, using quota sampling, Gallup predicted the Republican candidate would get 44% of the vote, but the actual number was 38%. In 1940 the prediction was 48%, and the actual vote was 45%; in 1944 the prediction was 48%, and the actual vote was 46%. But in spite of the errors, Gallup was able to predict the winner correctly in 1936, 1940, and 1944. This was merely due to luck—the spread between the candidates was large enough to cover the error. In 1948 Gallup (and all the other pollsters) simply ran out of luck. It was time to ditch quota sampling.

The failure of quota sampling as a method for getting representative samples has a moral: Even with the most carefully laid plans, human intervention in choosing the sample is always subject to bias.

RANDOM SAMPLING

If human intervention in choosing the sample is always subject to bias, what are the alternatives? The answer is to let the laws of chance decide who is in the sample. There is a certain amount of irony in the fact that the best way to choose a representative cross section of the population is to, in a manner of speaking, draw the names out of a hat. In fact, people find this very hard to believe. Isn't it possible, the argument goes, to get by sheer chance a sample that is very biased (say, for example, all Republican)? The answer is yes in theory, but when the sample is large enough, the odds of it happening are so low that in practice we can pretty much rule it out. Most present-day methods of product quality control in industry, corporate audits in business, and public opinion polling are based on **random sampling** methods, that is, methods for choosing the sample in which chance intervenes in one form or another. The validity of using random sampling methods to collect reliable data is supported by both practical experience and mathematical theory. We will discuss some of the details of this theory in Chapter 16.

The most basic form of random sampling is called **simple random sampling**. Simple random sampling is based on the principle that any group of members of the population has the same chance of being in the sample as any other group as long as the two groups are of the same size. To put it another way, any member of the population has the same chance of being in the sample as any other member; any two members of the population have the same chance of being in the sample as any other two members; any three members have the same chance of being in the sample as any other three members; etc.

In principle, simple random sampling can be carried out like this: We put the name of each individual in the population in a hat, mix the names well, and then draw as many names as we need for our sample. Of course "a hat" is just a metaphor. If our population is 100 million voters and we want to choose a simple random sample of 50,000, putting all the names in a real hat and then drawing 50,000 names one by one may not be such a good idea. The modern way to do any serious simple random sampling is by computer: Make a list of members of the population, enter it in the computer, and then let the computer randomly select the names.

While simple random sampling works well in many cases, for national surveys and polls the method has some serious practical problems. To begin with, it requires us to have a list of all the members of the population. As we noted at the beginning of the chapter, the population itself may not be clearly defined and, even when it is, a complete list of its members may not be available. Can we tag all wild elephants in Africa so that we

Simple random sampling.
(The New York State
Lottery)

can make a list for a computer? Better yet, is there a list of all people living in the United States? The answer to both of these questions is no.

Another serious problem in implementing simple random sampling on a national scale is expediency and cost. Interviewing several thousand people chosen by simple random sampling means chasing people all over the country. This requires an inordinate amount of time and money. For most public opinion polls, especially those that are done on a regular basis, the time and money needed to do this are simply not available.

Our next case study illustrates the sampling method currently used for polling public opinion.

CASE STUDY 3. STRATIFIED SAMPLING: MODERN PUBLIC OPINION POLLS

Present-day methods for conducting public opinion polls need to take into account two different sets of considerations: (1) minimizing sample bias, and (2) choosing a sample that is accessible in a cost-efficient, timely manner. A random sampling method that deals in a satisfactory way with both these issues is **stratified sampling**. The basic idea of stratified sampling is to break the population into categories called **strata** and then randomly choose a sample from these strata. The chosen strata are then further divided into categories called substrata, and a random sample is taken from these substrata. The selected substrata are further subdivided, a random sample is taken from them, and so on. The process goes on for a predetermined number of layers.

In public opinion polls, the strata and substrata are usually defined by criteria that involve a combination of geographic and demographic elements. For example, at the first level, the nation is divided into "size of community" strata (big cities, medium cities, small cities, villages, rural areas, etc.). Each of these strata is then subdivided by geographical region (New England, Middle Atlantic, East Central, etc.). Within each geographical region and within each size of community stratum, some communities are selected by simple random sampling. The selected communities (called *sampling locations*) are the only places where interviews will be conducted. To further randomize things, each of the selected sampling locations is subdivided into geographical units called *wards* and within each sampling location some of its wards are once again selected by simple random sampling. The selected wards are then divided into smaller units called *precincts*, and within each ward some of its precincts are selected by simple random sampling. At the last stage, *households* are selected for interviewing by simple random sampling within each precinct. The interviewers are then given specific instructions as to which households in their assigned area they must conduct interviews in, and the order which they must follow.

The efficiency of stratified sampling compared to simple random sampling in terms of cost and time is clear. The members of the sample are clustered in well-defined and easily manageable areas, significantly reducing the cost of conducting interviews as well as the response time needed to get the data together. At the same time, every single member of the population has an equal chance of being in the sample, which guarantees a small sample bias. Note, however, that this is not simple random

The first step in a public opinion poll: collecting the data. (© 1977 The Gallup Organization, Inc., Princeton, New Jersey)

sampling because not all groups of individuals have an equal chance of being selected. (Two people living in the same precinct, for example, have a better chance of both being in the sample than two people living in different precincts).

For a large, heterogeneous nation like the United States, stratified sampling has proven to be an effective and highly reliable method for collecting national data. Most major public opinion polls today use stratified sampling methods that result in sampling errors of less than 3%. Stratified sampling methods are also used with excellent results by the Bureau of Labor Statistics to collect extremely important statistical information for business and government such as data for the monthly Consumer Price Index (CPI) and the monthly unemployment figures given by the Current Population Survey (CPS). An excellent account of how stratified sampling is used for the CPI and CPS can be found in references 7 and 3 respectively.

The Gallup Poll

Design of the Sample

The design of the sample used by the Gallup Poll for its standard surveys of public opinion is that of a replicated area probability sample down to the block level in the case of urban areas and to segments of townships in the case of rural areas.

After stratifying the nation geographically and by size of community in order to insure conformity of the sample with the 1990 Census distribution of the population, over 360 different sampling locations or areas are selected on a mathematically random basis from within cities, towns, and counties which have in turn been selected on a mathematically random basis. The interviewers have no choice whatsoever concerning the part of the city, town, or county in which they conduct their interviews.

Approximately five interviews are conducted in each randomly selected sampling point. Interviewers are given maps of the area to which they are assigned and are required to follow a specified travel pattern on contacting households. At each occupied dwelling unit, interviewers are instructed to select respondents by following a prescribed systematic method. This procedure is followed until the assigned number of interviews with male and female adults have been completed. . . .

Since this sampling procedure is designed to produce a sample which approximates the adult civilian population (18 and older) living in private households (that is, excluding those in prisons and hospitals, hotels, religious and educational institutions, and on military reservations), the survey results can be applied to this popu-

lation for the purpose of projecting percentages into numbers of people. The manner in which the sample is drawn also produces a sample which approximates the population of private households in the United States. Therefore, survey results also can be projected in terms of numbers of households.

Sampling Error
In interpreting survey results, it should be borne in mind that all sample surveys are subject to sampling error, that is, the extent to which the results may differ from those that would be obtained if the whole population surveyed had been interviewed.

Source: *The Gallup Report*. Princeton, N.J.: American Institute of Public Opinion, 1991.

Choosing the sample for the Gallup poll.

CASE STUDY 4. COUNTING THE UNCOUNTABLE: THE U.S. CENSUS.

The most ambitious (and most expensive) data collection project in the history of mankind was the 1990 U.S. Census; before that it was the 1980 U.S. Census, and before that it was the 1970 U.S. Census.

Article 1, Section 2, of the U.S. Constitution mandates that a national census be conducted every 10 years. The original intent of the census was to "count heads" for a twofold purpose: taxes and political representation. Like everything else in the Constitution, Article 1, Section 2, was a compromise of many competing interests: The count was to exclude "Indians not taxed" and to count slaves as "three-fifths of a free Person." Today, the scope of the U.S. Census has been expanded by the Fourteenth Amendment and the courts to count all persons physically present and permanently residing in the United States: citizens, legal residents, and even illegal aliens.[4]

Besides counting heads, the modern U.S. Census collects additional information about the population: sex, age, race and ethnicity, marital status, housing, income, and employment. Surprisingly, some of this information is collected by means of a survey. In the 1990 U.S. Census, two types of census forms were designed. Eighty-three percent of households received a short form consisting of 20 basic questions on sex, age, marital status, type of housing, etc.; the remaining 17% randomly selected households received longer forms asking for more detailed information on employment, family income, etc.

The importance and pervasiveness of U.S. Census data in all walks of American life cannot be overestimated. The data collected by the U.S. Census is used as the basis for

---

[4] A new law implemented with the 1990 U.S. Census requires that military personnel and other federal workers stationed overseas also be included in the census count.

■ The apportionment of Congress

■ The redrawing of state and local legislative districts

■ The distribution of billions of federal tax dollars to states, counties, cities, and municipalities

■ The collection of other official government statistics such as the Consumer Price Index and the Current Population Survey

■ The strategic planning of production and services by business and industry.

Given the fact that there is no question about the value and importance of the U.S. Census and given the tremendous resources put behind the effort (over 500,000 workers, 500 field offices, and a budget in excess of 2.5 billion dollars were used for the 1990 U.S. Census), one would expect it to be a resounding success. In fact, the exact opposite is true. The reliability of the data collected by the U.S. Census is now being openly challenged by several states, most large cities, professional statisticians, and even the public. The U.S. Census has become, in the opinion of many, a national boondoggle.

We can understand a census being expensive, that is by definition one of its drawbacks. But inaccurate and unreliable? How can one possibly go wrong when the source of the data is the entire population? Unfortunately, the notion that all individuals living in the United States can be counted like pennies in a jar is totally out of tune with the times. In 1790, when the first U.S. Census was carried out, the population was smaller and relatively homogeneous, people tended to stay in one place, and by and large felt comfortable in their dealings with the government. Under these conditions it might have been possible for census takers to accurately count heads.

Today's conditions are completely different. Getting an exact head count for a population as large and as diverse as that of the United States at present, where people are constantly on the move, where a significant number of people are trying to hide from the law, where there are hundreds of thousands of homeless, and where there is a general distrust of the government,[5] is a hopeless endeavor no matter how much money and effort one puts into it.

To make matters worse, the inability of the U.S. Census to accurately count the population and measure the nation's demographics is definitely biased against the urban poor who were undercounted by as much as 20% in some large cities. After the 1980 U.S. Census, several lawsuits were brought against the Department of Commerce and the Census Bureau by the state of New York and New York City, claiming that the undercount

---

[5] In some large cities the response rate for the 1990 U.S. Census was under 50%.

of some of their ethnic minorities was depriving them of billions of dollars in federal funds. Additional lawsuits were filed after the 1990 U.S. Census.[6] Most of these lawsuits have yet to be resolved, but regardless of the outcome the credibility of the data collected by the U.S. Census has been irreparably damaged.

Based on the 1990 U.S. Census, the official figure for the population of the United States is 249,632,692, but the spurious accuracy of such a figure is extremely misleading. Statisticians believe that this count could be off by as much as 5 million people and that the notion that the U.S. Census can produce an exact head count of the population is unrealistic. Many statisticians now suggest that the constitutional requirement to count the nation's population should be implemented by means of a statistically designed survey based on a method similar to the one used by population biologists called the "capture-recapture method" (see Exercise 25 and reference 5).

CLINICAL STUDIES

All our examples so far have dealt with situations in which the issues centered around the question, What is the *source* of the data? Should it come from every member of the population, or from some selected sample? If the latter, should the sampling rate be small or large? Should the sample be chosen by human design or by chance?

Once we got to the source of the data all our examples pretty much assumed that the data itself were available to the observer in a direct and objective manner. If the election were held today, would you vote for candidate X or candidate Y? How many teenagers live in this household? Which TV programs did you watch last night? etc.

A very different and important type of data collection involves questions for which there is no clear, immediate answer. Is smoking hazardous to your health? Will taking aspirin reduce your chances of having a heart attack? Does regular class attendance in mathematics courses help improve SAT scores? All these questions have two things in common: (1) they involve a cause and an effect, and (2) the answers require observation over an extended period of time.

The standard approach for answering questions of this sort is to set up some sort of **experiment**. An experiment is a form of controlled observation. When one wants to know if a certain cause X produces a certain effect Y, one sets up an experiment in which cause X is produced and its effects are observed. If the effect Y is observed, then it is possible that X was indeed the cause of Y. The problem, however, is the nagging pos-

---

[6] The cities of New York, Los Angeles, Chicago, Houston, the states of New York and California, the League of Cities, the U.S. Conference of Mayors, the N.A.A.C.P., and the League of United Latin American Citizens are all parties to various suits against the Census Bureau and its parent agency, the Commerce Department.

sibility that some other cause different from X sneaked in and produced the effect Y and that X had nothing to do with it.

Let's illustrate with an example. This one is fictitious but not all that far-fetched. Suppose we want to find out if too much chocolate in one's diet can increase one's chance of becoming diabetic. Here the cause X is eating too much chocolate, and the effect Y is diabetes. We set up an experiment in which 100 rats are fed a pound of chocolate a day for a period of 6 months. At the end of the 6-month period, 15 of the 100 rats have diabetes. Since in the general rat population only 3% are diabetic, we are tempted to conclude that the diabetes in the rats is indeed caused by the excessive chocolate diet. The problem is that there is no certainty that the chocolate diet was the cause. Could there be another unknown reason for the observed effect? Could it be the laboratory water the rats drank? Could it be the lack of exercise of the confined rats? Could it be that it's not chocolate itself but rather the high calorie intake that caused the diabetes?

For most cause-and-effect situations, especially those complicated by the involvement of human beings, a single effect can have many possible and actual causes. What causes heart attacks? Unfortunately, there is no single cause—diet, lifestyle, stress, and heredity are all known to be contributory causes. And the extent to which each of these causes contributes individually and the extent to which they interact with each other are extremely difficult questions that can be answered only by means of carefully designed statistical experiments.

For the remainder of this chapter we will illustrate an important type of experiment called a **clinical study**. Generally, clinical studies are concerned with determining whether a single variable or treatment (usually a vaccine, drug, therapy, etc.) can cause a certain effect (a disease, a symptom, a cure, etc.). The importance of such clinical studies is self-evident: Every new vaccine, drug, or treatment must ''prove'' itself by means of a clinical study before it is officially approved for public use. Likewise, almost everything that is bad for us (cigarettes, caffeine, cholesterol, etc.) gets its ''official'' certification of badness by means of a clinical study. (A recent newspaper headline asking ''Are Health Studies Bad for Your Health?'' was only partly meant to be a joke.)

Properly designing a clinical study can be both difficult and controversial, and as a result we are often bombarded with conflicting information produced by different studies examining the same cause-and-effect question. The basic principles guiding a clinical study, however, are pretty much established by statistical practice and are almost always followed. We will discuss them next.

The first and most important issue in any clinical study is to isolate the cause (treatment, drug, vaccine, therapy, etc.) that is under investigation from all other possible contributing causes (called **confounding variables**) that could produce the same effect. The way that this is accomplished is by performing the experiment on two different groups: a **treatment group** and a **control group**. The treatment group receives the treatment, and the control group does not. The control group is there for *comparison* purposes only: If a cause-and-effect relationship exists, then the treatment group should show the effects of the treatment and the control group should not. The comparison is most effective when the treatment and control groups are identical to each other in all other respects (except that one group is receiving the treatment and the other one isn't). If this is accomplished and the groups show differences, then the differences can be safely attributed to the treatment. In addition, both the treatment and control groups should, as much as possible, be representative of the entire population to which the experiment applies.

Any experiment in which a cause-and-effect relationship is established by comparing the results in a treatment group with the results in a control group is called a **controlled** (or **comparative**) **experiment**. Our example of a controlled experiment is a famous clinical study carried out in 1954 to determine the effectiveness of a new vaccine against polio.

CASE STUDY 5.
CONTROLLED
EXPERIMENTS:
THE 1954 SALK
POLIO VACCINE
FIELD TRIALS

Polio (infantile paralysis) has been practically eradicated in the western world. In the first half of the twentieth century, however, it was a major public health problem. Over ½ million cases of polio were reported between 1930 and 1950, and the actual number may have been considerably higher.

Because polio attacks mostly children and because its effects can be so serious (paralysis or death), eradication of the disease became a top public health priority in the United States. By the late 1940s it was known that polio is a virus and as such can best be treated by a vaccine which is itself made up of a virus. The vaccine virus can be a closely related virus that does not have the same harmful effects, or it can be the actual virus that produces the disease but which has been killed by a special treatment. The former is known as a *live-virus vaccine*, and the latter as a *killed-virus vaccine*. In response to either vaccine the body is known to produce *antibodies* which remain in the system and give the individual immunity against an attack by the real virus.

Both the live-virus and the killed-virus approaches have their advantages and disadvantages. The live-virus approach produces a stronger reaction and better immunity, but at the same time it is also more likely to cause a harmful reaction and in some cases even to produce the very disease it is supposed to prevent. The killed-virus approach is safer in terms of the likelihood of producing a harmful reaction, but it is also less effective in providing the desired level of immunity.

These facts are important because they help us understand the extraordinary amount of caution that went into the design of the experiment that tested the effectiveness of the polio vaccine. By 1953, several potential vaccines had been developed, one of the more promising of which was a killed-virus vaccine developed by Jonas Salk at the University of Pittsburgh. The killed-virus approach was chosen because there was a great potential risk in testing a live-virus vaccine in a large-scale experiment and a large-scale experiment was needed to collect enough information on polio (which in the 1950s had a rate of incidence among children of about 1 in 2000).

The testing of any new vaccine or drug creates many ethical dilemmas which have to be taken into account in the design of the experiment. With a killed-virus vaccine the risk of harmful consequences produced by the vaccine itself is small, so one possible approach could have been to distribute the vaccine widely among the population (ideally giving it to every child, but this was not possible because supplies were limited) and then follow up on whether there was a decline in the national incidence of polio in subsequent years. This is called the *vital statistics* approach and is the simplest way to test a vaccine. This is essentially the way the smallpox vaccine was determined to be effective. The problem with such an approach for polio is that polio is an epidemic type of disease, which means that there is a great variation in the incidence of the disease from one year to the next. In 1951, there were close to 60,000 reported cases of polio in the United States, but in 1952 the number of reported cases had dropped to almost half (about 35,000). Since no vaccine or treatment was used, the cause of the drop can only be attributed to the natural variability that is typical of epidemic diseases. If a totally ineffective vaccine had been tested in 1951, the observed effect—a significant drop in the incidence of polio in 1952—could have been incorrectly interpreted as a proof that the vaccine worked rather than to the real cause. The serious consequences of such a mistake are obvious.

The final decision on how best to test the effectiveness of the Salk vaccine was left to an advisory committee of doctors, public officials, and statisticians convened by the National Foundation for Infantile Paralysis and the Public Health Service.[7] In order to isolate the cause under investigation (the Salk vaccine) from other possible causes of the desired effect (a reduction in the incidence of polio), it was decided that the experiment would be a controlled experiment involving a treatment group (those receiving the actual vaccine) and a control group (those receiving a shot of a harmless salt solution). An experiment of this kind is called

---

[7] Because of disagreements as to the best way to carry out the experiment, there were actually two different but parallel experiments carried out. In this case study we will discuss only one of these, the one conducted by the Public Health Service. A complete description of both experiments and the results is given in reference 2.

a **controlled placebo experiment**. The word "placebo" refers to the fact that the members of the control group receive a fake version of the treatment (a fake vaccine, a fake pill, etc.) which is called a **placebo**. The reasons for using placebos go back to our desire that the treatment and control groups be as equal as possible in all respects, except of course that one group is receiving the vaccine and the other one isn't. It is a well-known fact that just thinking that one is getting a treatment can actually affect the way the body responds, and this effect (called the **placebo effect**) is one of the primary reasons for using controlled placebo experiments to conduct many clinical studies.

It is obvious from our preceding discussion that in a controlled placebo experiment it is always desirable that neither the members of the treatment group nor the members of the control group know to which of the two groups they belong. When this is the case the experiment is called a **blind experiment**. It is also desirable that the scientists conducting the experiment not know which individuals are given the actual treatment and which are given the placebo. The purpose of this is to make the observation, analysis, and interpretation of the results of the experiment as impartial as possible.

A controlled experiment in which neither the subjects nor the scientists conducting the experiment know which individuals are in the treatment group and which are in the control group is called a **double-blind experiment**. Making the Salk vaccine experiment double-blind was particularly important because polio is not an easy disease to diagnose—it comes in many different forms and degrees. Sometimes it can be a borderline call, and if the doctor collecting the information had had prior knowledge of whether the subject had received the real vaccine or not, the diagnosis could have been subjectively tipped one way or the other.

With all this background we can now describe the actual details of the experiment. Approximately 750,000 children were randomly selected to participate in the study. Of these, about 340,000 declined to participate and another 8500 dropped out in the middle of the experiment. The remaining children were divided into two groups—a treatment group and a control group—with approximately 200,000 children in each group. The choice of which children were selected for the treatment group and which for the control group was made by *random selection*. (An experiment in which the treatment group and control group are chosen by random selection is called a **randomized controlled experiment**.) Some of the figures and results of the experiment are shown in Table 13-1.

While Table 13-1 shows only a small part of the data collected by the Salk vaccine experiment, it can be readily seen that the difference between the treatment and control groups was significant and could rightfully be interpreted as a clear indication that the vaccine was indeed effective.

Based on the data collected by the 1954 field trials, a massive inoculation campaign was put into effect. Today, all children are routinely

	Number of Children	Number of Reported Cases of Polio	Number of Paralytic Cases of Polio	Number of Fatal Cases of Polio
Treatment group	200,745	82	33	0
Control group	201,229	162	115	4
Declined to participate in the study	338,778	182[a]	121[a]	0[a]
Dropped out in the middle	8,484	2[a]	1[a]	0[a]
Total	749,236	428	270	4

Table 13-1 Adapted from Thomas Francis, Jr., et al., "An Evaluation of the 1954 Poliomyelitis Vaccine Trials—Summary Report," *American Journal of Public Health*, 45 (1955) 25, Table 2b.

[a] These figures are not a reliable indicator of the actual number of cases—they are only self-reported cases.

inoculated against polio,[8] and polio has essentially been eradicated in the United States. A statistically designed experiment played a key role in this breakthrough.

CONCLUSION

In this chapter we discussed different methods for collecting data. In principle, the most accurate way to collect data is by means of a *census*, a method that relies on collecting data from each member of the population. In most cases, because of considerations of cost and time, a census is a completely unrealistic strategy for collecting data. When data are collected from only a subset of the population (called a *sample*), the data collection method is called a *survey*. The most important rule in designing good surveys is to eliminate or minimize *sample bias*. Today, almost all strategies for collecting data are based on surveys in which the laws of chance are used to determine how the sample is selected, and these methods for collecting data are called *random sampling* methods. Random sampling is the best way known to minimize or eliminate sample bias. Two of the most common random sampling methods are *simple random sampling* and *stratified sampling*. In some special situations other, more complicated types of random sampling can be used.

Sometimes identifying the sample is not enough. In cases in which cause-and-effect questions are involved, the data may come to the surface only after an extensive study has been carried out. In these cases isolating the cause variable under consideration from other possible causes (called *confounding variables*) is an essential prerequisite for getting reliable data. The standard strategy for doing this is a *controlled experiment* in which the sample is broken up into a *treatment group* and a *control group*.

[8] The Salk vaccine, which was used for many years, has been replaced some years ago by the Sabin vaccine, a more effective oral vaccine based on the live-virus approach.

Controlled experiments are used (and sometimes abused) in almost all clinical and scientific studies carried out today. We can thank this area of statistics for many breakthroughs in social science, medicine, and public health, as well as for the many depressing health and diet reports about all the things that are supposed to be bad for us but which we can't seem to live without.

KEY CONCEPTS

blind experiment
chance error
clinical study
confounding variable
control group
controlled experiment
controlled placebo experiment
data
double-blind experiment
nonresponse bias
parameter
population
quota sampling

randomized controlled experiment
sample
sample bias
sampling error
sample variability
sampling rate
selection bias
simple random sampling
statistic
strata
stratified sampling
treatment group

EXERCISES

■ Walking

Questions 1 through 4 refer to the following survey. In 1988 "Dear Abby" asked her readers to let her know whether they had cheated on their spouses or not. The readers' responses are summarized in the accompanying table.

	Women	Men
Faithful	127,318	44,807
Unfaithful	22,468	15,743
Total	149,786	60,550

Based on the results of this survey, Dear Abby concluded that the amount of cheating among married couples is much less than people believe (in her words, " . . . the results were astonishing. There are far more faithfully wed couples than I had surmised.")

1. (a) Describe as specifically as you can the population for this survey.
　(b) What was the size of the sample?
　(c) How was the sample chosen?

(d) 85% of the women who responded to this survey claimed to be faithful. Is the number 85% a parameter? A statistic? Neither? Explain your answer.

2. **(a)** Explain why this survey was subject to selection bias.
   (b) Explain why this survey was subject to nonresponse bias.

3. **(a)** Based on the Dear Abby data, estimate the percentage of married men who are faithful to their spouses.
   (b) Based on the Dear Abby data, estimate the percentage of married people who are faithful to their spouses.
   (c) How accurate do you think these estimates are? Explain.

4. If money were no object, could you devise a survey that might give more reliable results than the Dear Abby survey? Describe briefly what you would do.

Questions 5 through 8 refer to the following hypothetical situation. The Cleansburg Planning Department is trying to determine what percent of the people in the city want to spend public funds to revitalize the downtown mall. In order to do so they decide to conduct the following survey: Five professional interviewers (A, B, C, D, and E) are hired, and each is asked to pick a street corner of their choice within the city limits. Everyday between 4:00 and 6:00 P.M. the interviewers are to ask each passerby if he or she wishes to respond to a survey sponsored by Cleansburg City Hall and make a record of their response. If the response is yes, the person is asked to respond to the next question: Are you in favor of spending public funds to revitalize the downtown mall? Yes or no? The interviewers are asked to return to the same street corner as many days as are necessary until each one has conducted a total of 100 interviews. The data collected are as follows:

INTERVIEWER	YES[a]	NO[b]	NONRESPONDENTS[c]
A	35	65	321
B	21	79	208
C	58	42	103
D	78	22	87
E[d]	12	63	594

Table 13-2 [a] In favor of spending public funds to revitalize the downtown mall.

[b] Opposed to spending public funds to revitalize the downtown mall.

[c] Declined to be interviewed.

[d] Got frustrated and quit.

5. **(a)** Describe as specifically as you can the population for this survey.
   (b) What is the size of the sample?

6. **(a)** Calculate the response rate in this survey.
   (b) Explain why this survey was subject to nonresponse bias.

7. **(a)** Can you explain the big differences in the data from interviewer to interviewer?
   (b) One of the interviewers conducted the interviews at a street corner downtown. Which one? Explain.
   (c) Do you think the survey was subject to selection bias? Explain.
   (d) Was the sampling method used in this survey the same as quota sampling? Explain.

8. Do you think this was a good survey? If you were a consultant to the Cleansburg Planning Department, could you suggest some improvements? Be specific.

Questions 9 through 12 refer to the following survey. The dean of students at Tasmania State University wants to determine the percent of undergraduates living at home during the current semester. There are 15,000 undergraduates at TSU, so it is decided that the cost of checking with each and every one would be prohibitive. The following method is proposed to choose a representative sample of undergraduates to interview: Start with the registrar's alphabetical listing containing the names of all undergraduates. Pick randomly a number between 1 and 100 and count that far down the list, taking that name and every 100th name after it. (For example, if the random number chosen is 73, then pick the 73rd, 173rd, 273rd, etc., names on the list.) Assume the survey has a response rate of 0.95.

9. **(a)** Describe the population for this survey.
   (b) Give the exact value of N.

10. **(a)** Find the size n of the sample.
    (b) Find the sampling rate.

11. **(a)** Was this survey subject to selection bias? Explain.
    (b) Explain why the method used for choosing the sample is not simple random sampling?

12. Do you think the results of this survey will be reliable? Explain.

Questions 13 through 16 refer to the following hypothetical study. The manufacturer of a new vitamin (vitamin X) decides to sponsor a study to determine the effectiveness of vitamin X in curing the common cold. Five hundred college students in the San Diego area who are suffering from colds are paid to participate as subjects in this study. They are all given two tablets of vitamin X a day. Based on information provided by the subjects themselves, 457 out of the 500 subjects are cured of their colds within 3 days. The average number of days a cold lasts is 4.87 days. As a result of this study, the manufacturer launches an advertising campaign claiming that "vitamin X is more than 90% effective in curing the common cold."

13. **(a)** Describe as specifically as you can the population for this study.
    (b) How was the sample selected?
    (c) What was the size n of the sample?
    (d) Was this health study a controlled experiment?

14. (a) Do you think the placebo effect could have played a role in this study?
   (b) List three possible causes other than the effectiveness of vitamin *X* itself that could have confounded the results of this study.

15. List four different problems with this study that indicate poor design.

16. Make some suggestions for improving the study.

*Questions 17 through 20 refer to the following. A recent study by a team of Harvard University scientists [*Science News*, 138, no. 20 (November 17, 1990), 308] found that regular doses of beta-carotene (a nutrient common in carrots, papayas, and apricots) may help prevent the buildup of plaque producing arteriosclerosis (clogging of the arteries), which is the primary cause of heart attacks. The subjects in the study were 333 volunteer male doctors, all of whom had shown some early signs of coronary artery disease. The subjects were randomly divided into two groups. One group was given a 50-milligram beta-carotene pill every other day for 6 years, and the other group was given a similar-looking placebo pill. The study found that the men taking the beta-carotene pills suffered 50% fewer heart attacks and strokes than the men taking the placebo pills.*

17. Describe as specifically as you can the population for the study.

18. (a) Describe the sample.
   (b) What was the size *n* of the sample?
   (c) Was the sample chosen by random sampling? Explain.

19. (a) Explain why this study can be described as a controlled placebo experiment.
   (b) Describe the treatment group in this study.
   (c) Explain why this study can be described as a randomized controlled experiment.

20. (a) Mention two possible confounding variables in this study.
   (b) Carefully state what a legitimate conclusion from this study might be.

■ **Jogging**

21. Leading question bias. In many surveys, the way the questions in the survey are phrased can itself be a source of bias. When a question is worded in such a way as to predispose the respondent to provide a particular response, the results of the survey are tainted by a special type of bias called leading question bias. The following is an extreme hypothetical situation intended to drive the point home.

   The American Self-Righteous Institute is a conservative think tank. In an effort to find out how the American taxpayer feels about a tax increase, the institute conducts a "scientific" poll. The main question in the poll is phrased as follows:

   Are you in favor of paying higher taxes to bail the federal government out of its disastrous economic policies and its mismanagement of the federal budget? Yes _____ No _____.

   Ninety-five percent of the respondents answered no. The results of the survey are announced by the sponsors with the statement:

Public opinion polls show that 95% of American taxpayers oppose a tax increase.

(a) Explain why the results of this survey might be invalid.

(b) Rephrase the question in a neutral way. Pay particular attention to "highly charged" words.

(c) Make up your own (more subtle) example of leading question bias. Analyze the critical words that are the cause of bias.

22. Consider the following hypothetical survey designed to find out what percentage of people cheat on their income taxes. Fifteen hundred taxpayers are randomly selected from the Internal Revenue Service (IRS) rolls. These individuals are then interviewed in person by representatives of the IRS and read the following statement:

This survey is for information purposes only. Your answer will be held in strict confidence. Have you ever cheated on your income taxes? Yes _____ No _____.

Twelve percent of the respondents responded yes.

(a) Explain why the above figure might be unreliable.

(b) Can you think of ways in which a survey of this type might be designed so that more reliable information could be obtained? In particular, discuss who should be sponsoring the survey and how the interviews should be carried out.

23. Listing Bias. Today, most consumer marketing surveys are conducted by telephone. In selecting a sample of households that are representative of all the households in a given geographical area two basic techniques used are

1. Randomly selecting telephone numbers to call from the local telephone directory or directories.

2. Using a computer to randomly generate 7-digit numbers to try that are compatible with the local phone numbers.

(a) Briefly discuss the advantages and disadvantages of each of the two techniques. In your opinion, which of the two techniques will produce the more reliable data. Explain.

(b) Suppose that you are trying to market burglar alarms in New York City. Which of the two techniques for selecting the sample would you use? Explain your reasons.

24. The following two surveys were conducted in January 1991 in order to assess how the American public viewed media coverage of the Persian Gulf war.

Survey 1 was an Area Code 900 telephone poll survey conducted by "ABC News." Viewers were asked to call a certain 900 number if they felt the media was doing a good job of covering the war, and a different 900 number if they felt the media was not doing a good job in their coverage of the war. Each call cost 50 cents. Of the 60,000 respondents, 83% felt the media was not doing a good job.

Survey 2 was a telephone poll of 1500 randomly selected households across the United States conducted by the Times-Mirror survey organization. In this poll 80% of the respondents indicated that they approved of the press coverage of the war.

(a) Briefly discuss survey 1, indicating any possible types of bias.

(b) Briefly discuss survey 2, indicating any possible types of bias.

(c) Can you explain the discrepancy between the results of the two surveys?

(d) In your opinion, which of the two surveys gives the more reliable data?

25. **The capture-recapture method.** For many wildlife populations (including in some cases humans), it is difficult, if not impossible, to get an exact count of the size of the population. To estimate the size of wildlife populations, population biologists often use a technique called the capture-recapture method. The basic idea of this method is to first capture a sample of size n_1, tag the animals without harming them, and let them go. After an appropriate amount of time a new sample of size n_2 is captured, and the number of tagged animals is counted (we'll call this number k). Under the right conditions, the size of the entire population can be estimated from the values of n_1, n_2, and k. The purpose of this exercise is to illustrate how this method works.

(a) You have a small pond stocked with fish and want to estimate how many fish there are in the pond. Let's suppose that you capture $n_1 = 500$ fish, tag them, and throw them back in the pond. After a couple of days you go back to the pond and capture $n_2 = 120$ fish, of which $k = 30$ are tagged. Using these values of n_1, n_2, and k, estimate the fish population N in the pond.

(b) Give a formula that allows us to estimate N based on the numbers n_1, n_2, and k.

(c) For the capture-recapture method to give a reasonable estimate of N, list all the assumptions you think should be made about the two samples.

(d) Give reasons why in many situations the assumptions in part (c) may not hold true.

(e) The following real example is based on data given in D. G. Chapman and A. M. Johnson, "Estimation of Fur Seal Pup Populations by Randomized Sampling," *Transactions of the American Fisheries Society*, 97 (July 1968), 264–270. To estimate the population of fur seal pups in a rookery, 4965 fur seal pups were captured and tagged in early August. In late August, 900 fur seal pups were captured. Of these, 218 had been tagged. Based on these figures estimate the population of fur seal pups in the rookery to the nearest hundred.

26. Consider the following hypothetical situation. A potentially effective new drug for treating AIDS patients must be tested by means of a clinical study. Based on experiments conducted with laboratory animals, the drug appears to be extremely effective in treating the more serious effects of AIDS, but it also appears to have caused many side effects, including serious kidney disorders in about 20% of the laboratory animals tested.

(a) Discuss the ethical and moral issues you think should be considered in designing a clinical study to test this drug.

(b) Taking into account the ethical and moral issues discussed in part (a), describe how you would design a clinical study for this new drug. (In particular, how would you choose the participants in the study, the treatment and the control groups, etc.)

REFERENCES AND FURTHER READINGS

1. Anderson, Margo, "According to their Respective Numbers . . . for the Twenty-First Time," *Chance*, 3 (Winter 1990), 12–18.

2. Francis, Thomas, Jr., et al., "An Evaluation of the 1954 Poliomyelitis Vaccine Trials—Summary Report," *American Journal of Public Health*, 45 (1955), 1–63.

3. Freedman, D., R. Pisani, R. Purves, and A. Adhikari, *Statistics (2nd ed.)*. New York: W.W. Norton, Inc., 1991, chaps. 19 and 20.

4. Gallup, George, *The Sophisticated Poll Watchers Guide*. Princeton, N.J.: Princeton Public Opinion Press, 1972.

5. Glieck, James, "The Census: Why We Can't Count," *New York Times Magazine* (July 15, 1990), 22–26, 54.

6. Hansen, Morris, and Barbara Bailar, "How to Count Better: Using Statistics to Improve the Census," in *Statistics: A Guide to the Unknown (3rd ed.)* ed. Judith M. Tanur, et al. Belmont, Calif.: Wadsworth, Inc., 1989, 208–217.

7. McCarthy, Philip J., "The Consumer Price Index," in *Statistics: A Guide to the Unknown (3rd ed.)* ed. Judith M. Tanur, et al. Belmont, Calif.: Wadsworth, Inc., 1989, 198–207.

8. Meier, Paul, "The Biggest Public Health Experiment Ever: The 1954 Field Trial of the Salk Poliomyelitis Vaccine," in *Statistics: A Guide to the Unknown (3rd ed.)* ed. Judith M. Tanur, et al. Belmont, Calif.: Wadsworth, Inc., 1989, 3–14.

9. Mosteller, F., et al., *The Pre-election Polls of 1948*. New York: Social Science Research Council, 1949.

10. Parten, Mildred, *Surveys, Polls and Samples*. New York: Harper and Row, 1950.

11. Paul, John, *A History of Poliomyelitis*. New Haven, Conn.: Yale University Press, 1971.

12. Scheaffer, R. L., W. Mendenhall, and L. Ott, *Elementary Survey Sampling*. Boston: PWS-Kent, 1990.

13. Stephan, F. F. and P. J. McCarthy, *Sampling Opinions: An Analysis of Survey Procedure*. New York: John Wiley & Sons, Inc., 1958.

14. Warwick, D. P. and C. A. Lininger, *The Sample Survey: Theory and Practice*. New York: McGraw Hill Book Co., 1975.

15. Yates, Frank, *Sampling Methods for Censuses and Surveys*. New York: Macmillan Publishing Co., Inc., 1981.

14

Descriptive Statistics

It is a proof of high culture to say the greatest matters in the simplest way
RALPH WALDO EMERSON

Graphing and Summarizing Data

Read to 431 #3 1,3,5,19,21,23

In Chapter 13 we discussed the ins and outs of collecting statistical data. And once the data have been collected, what should we do with them?

One of the primary purposes of collecting data is to use the data to communicate some sort of a message, a message whose purpose is usually to inform and occasionally to persuade. It is a bit ironic, therefore, that after all the care and attention to detail invested in collecting data, the typical situation one is confronted with is that of having *too much* data. Having too much data presents a peculiar dilemma. There is a point past which the human mind cannot absorb or comprehend the information presented because the amount of data is much too large. There are two ways to deal with this dilemma: One is to present the data in the form of graphs (we can usually understand pictures better than numbers); the other is to use numerical summaries that serve as "snapshots" of the data.

In this chapter we will discuss ways of summarizing large amounts of data by means of graphs as well as numbers. This area of statistics is called **descriptive statistics.**

419

GRAPHICAL REPRESENTATION OF DATA

The old saying "A picture is worth a thousand words" is particularly appropriate in descriptive statistics, where a picture can be worth thousands of pieces of data. We will start our discussion of *graphical descriptions* of data with a hypothetical situation not unlike one that can routinely be found on every college campus across the land.

Example 1 (Blackbeard's Stat 101 Test Scores). Professor Groucho Blackbeard has a well-deserved reputation at Tasmania State University for tough (some say brutal) exams. The results of the most recent midterm exam in his Stat 101 course are shown in Table 14-1.

ID	Score	ID	Score	ID	Score	ID	Score	ID	Score
1257	12	2651	10	4355	8	6336	11	8007	13
1297	16	2658	11	4396	7	6510	13	8041	9
1348	11	2794	9	4445	11	6622	11	8129	11
1379	24	2795	13	4787	11	6754	8	8366	13
1450	9	2833	10	4855	14	6798	9	8493	8
1506	10	2905	10	4944	6	6873	9	8522	8
1731	14	3269	13	5298	11	6931	12	8664	10
1753	8	3284	15	5434	13	7041	13	8767	7
1818	12	3310	11	5604	10	7196	13	9128	10
2030	12	3596	9	5644	9	7292	12	9380	9
2058	11	3906	14	5689	11	7362	10	9424	10
2462	10	4042	10	5736	10	7503	10	9541	8
2489	11	4124	12	5852	9	7616	14	9928	15
2542	10	4204	12	5877	9	7629	14	9953	11
2619	1	4224	10	5906	12	7961	12	9973	10

Table 14-1 Stat 101 midterm exam scores (25 points possible). *N* = 75.

In this example we have a population of size $N = 75$ (the students taking the exam). Each of the 75 students is identified by a student ID number (rights of privacy prohibit the use of names), and the data values are midterm scores (whole numbers between 0 and 25) shown to the right of each student's ID number. We will use the data in Example 1 several times in this and later chapters. ∎

Like students everywhere, students in Professor Blackbeard's Stat 101 class have two questions foremost on their minds regarding the exam: (1) How did I do? and (2) How did the class as a whole do? The answer to question 1 can be found directly in Table 14-1, but the answer to question 2 requires a little extra effort. How can all the information given by Table 14-1 be packaged into a single intelligible whole? Let us count the ways.

■ Bar Graphs and Variations Thereof

Our first approach is to put the scores into a **frequency table** as shown in Table 14-2.

Exam scores	1	6	7	8	9	10	11	12	13	14	15	16	24
Frequency	1	1	2	6	10	16	13	9	8	5	2	1	1

Table 14-2   Frequency Table for Stat 101 Midterm Exam

The number below each score represents the **frequency** of that score, that is, the number of students getting that score. In this example there are 16 students with a score of 10, 2 students with a score of 15, etc. Note that when there are no students getting a particular score (i.e., the frequency is 0), we can omit that score from the table.

While Table 14-2 is a considerable improvement over Table 14-1 visually, it leaves something to be desired. Figure 14-1 shows the same information in a much more visually striking way called a **bar graph.**

In a bar graph the frequencies are displayed by the *heights* of the columns. This has the decided advantage that one can see in one fell swoop the overall picture. Consider, for example, the problem of detecting **outliers.** An outlier is a data value that stands out from the crowd, that is to say, a value that is noticeably larger or smaller than the rest of the data. One of the advantages of a graph such as the one in Fig. 14-1 is that outliers can be easily picked up by the naked eye. In this case there are two students, one with an abnormally low score (1) and one with an abnormally high score (24), and they are both outliers. Notice, however, that the bar graph does not identify who the outliers are (to find this out Professor Blackbeard would have to go back to Table 14-1).

Sometimes it is more convenient to show frequencies as percentages of the total population rather than as exact counts. This is particularly

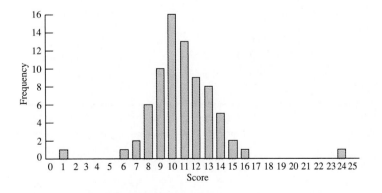

Figure 14-1 Bar graph for Stat 101 exam.

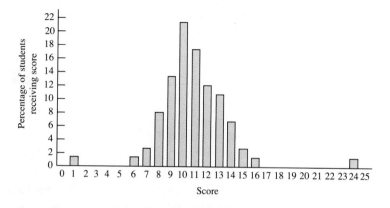

Figure 14-2 Bar graph for Stat 101 midterm exam using percentages of the population. *N* = 75.

true when we are dealing with very large data sets. Figure 14-2 shows a bar graph for Professor Blackbeard's Stat 101 exam in which the column height for each score represents the percentage of the class getting that score. The notation on the vertical axis clearly indicates that we are dealing with percentages rather than total counts. The change from total counts to percentages does not change the shape of the graph—it is basically a change in scale.

While the term "bar graph" is most commonly used for graphs like the ones in Figs. 14-1 and 14-2, devices other than bars can be used to add a little extra flair or to subtly influence the content of the information

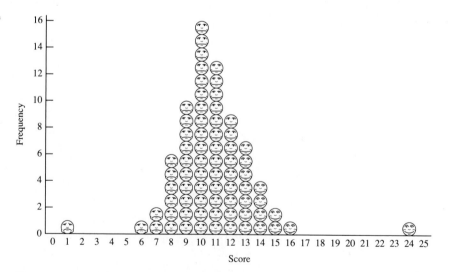

Figure 14-3 Frequency chart for Stat 101 midterm exam.

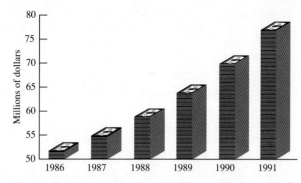

Figure 14-4 Frequency chart. XYZ Corporation annual sales (in millions of dollars).

given by the raw data. Professor Blackbeard, for example, may have chosen to display the test data using a graph like the one shown in Fig. 14-3, which gives the observer all the information given by the more staid version (Fig. 14-1) and at the same time sends a subtle little message about what he thought of the students' performance. ■

The general point here is that a bar graph is often used not only to inform but also to impress and persuade, and in such cases a clever design for the frequency columns can be more effective than just a bar. It is for this reason that we prefer to use the term **frequency chart** for graphs such as the ones in Figs. 14-3 and 14-4.

Example 2. Figure 14-4 is a frequency chart showing the "impressive" growth in the yearly sales of the XYZ Corporation over the period from 1986 to 1991. Figure 14-5 is a more sobering bar graph displaying the same information in a more realistic way. The noticeable difference between the two graphs can be attributed to several factors: (1) the choice of the

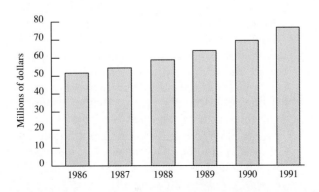

Figure 14-5 Bar graph. XYZ Corporation annual sales (in millions of dollars).

starting value on the vertical axis, (2) the choice of scale on the vertical axis, and (3) the use of graphical devices (the stacks of bills).

■

The point of Example 2 is that the graphic display of data is as much an art as it is a science and the impact that a graph can have on its intended audience is not only a function of the data values themselves but also of the way they are presented. This is what makes descriptive statistics so challenging and at the same time so dangerous. The line separating objectivity and propaganda is particularly fine in this area of statistics.

VARIABLES: QUANTITATIVE AND QUALITATIVE, CONTINUOUS AND DISCRETE

Before we continue with our discussion of graphs, we need to briefly analyze the concept of a **variable**. In statistical usage, a variable is any characteristic that varies with the members of a population. The students in Professor Blackbeard's Stat 101 course (the population) do not all perform equally on the exam. Thus, the *test score* is a variable which in this particular example is a whole number between 0 and 25. In some instances, such as when the instructor gives partial credit or when there is subjective grading, a test score may be a variable that takes on a fractional value such as $18\frac{1}{2}$ or even $18\frac{1}{4}$. Even in these cases, however, the possible increments for the values of the variable are given by some minimum amount: a quarter-point, a half-point, whatever. In contrast to this situation, consider a different variable, the *length of time* it takes a student to complete the exam. In this case the variable can take on values that differ by arbitrary small increments—a second, a tenth of a second, a hundredth of a second, etc.

When a variable represents a measurable quantity, it is called a **quantitative variable.** When the difference between the values of a quantitative variable can be arbitrarily small, we call the variable a **continuous variable,** whereas when possible values of the quantitative variable change by minimum increments, the variable is called a **discrete variable.** Examples of discrete variables are IQ, pulse, shoe size, family size, number of automobiles owned, and points scored in a basketball game. Examples of continuous variables are height, weight, foot size (as opposed to shoe size), and the time it takes one to run a mile.

Sometimes the distinction between continuous and discrete variables is blurred in the real world. Height, weight, and age are all continuous variables in theory, but in practice they are frequently rounded off to the nearest inch, ounce, year (or month in the case of babies) respectively, at which point they become discrete variables. On the other hand, money, which is in theory a discrete variable (the difference between two values of this variable cannot be less than a penny) is almost always thought of as continuous because in most real life situations, a penny can be thought of as an infinitesimally small amount of money.

Variables can also describe characteristics that cannot be measured numerically: nationality, sex, hair color, brand of automobile owned, etc. Variables of this type are called **qualitative.**

In some ways, qualitative variables must be treated differently from quantitative variables: They cannot, for example, be added, multiplied, or averaged. In other ways, qualitative variables can be treated very much like discrete quantitative variables, particularly when it comes to graphical descriptions such as bar graphs and frequency charts.

Example 3.   The bar graph in Fig. 14-6 shows undergraduate enrollments at Tasmania State University by school. In this example the variable being described is the school in which the student is enrolled.

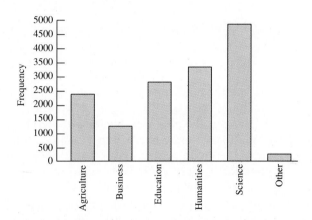

Figure 14-6 Undergraduate enrollments at Tasmania State University by school.
N = 15,000.

The bar graph in Fig. 14-7 gives exactly the same information in percentages of the total undergraduate student population. Conceptually, there is hardly any difference between these two graphs and the graphs in Figs. 14-1 and 14-2 other than the fact that the **categories** (sometimes also called **classes**) that describe the "values" of the data are not numbers.

The category labeled "Other" is one that frequently appears on bar graphs for qualitative data and deserves special attention. "Other" is used as a convenient place to put members of the population that do not fit neatly into one of the main categories (in Example 3 it is students with undeclared majors) or a place to lump together several categories that

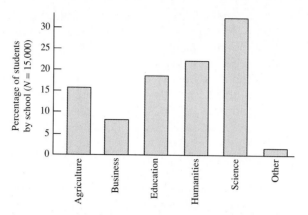

Figure 14-7 Percentage of undergraduate population at Tasmania State University by school.

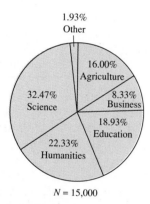

Figure 14-8 Percentage of undergraduate population at Tasmania State University by school.

contain fractions of the population that are too small to be listed on their own. ∎

When the number of categories is small, as it is in Example 3, another commonly used way to describe relative frequencies of a population by categories is the **pie chart.** Figure 14-8 shows a pie chart with exactly the same information as the one in Fig. 14-7. Note that the value of N (the size of the entire population) is listed along with the pie chart itself. This allows the reader to know the total from which the percentages in the pie chart are taken and if necessary convert these percentages to actual counts. When the size of the population N is available, it is good statistical practice to provide this information along with the pie chart.

Bar graphs and pie charts are an excellent way to graphically display qualitative data but, as always, we should be wary of jumping to hasty conclusions based on what we see on a graph. Our next example illustrates this point.

Example 4 (Television Audiences). The pie chart in Fig. 14-9 shows the percentage of prime-time (7:00 P.M. to 11:00 P.M.) viewers of network programming for each of the indicated categories.

When looking at this pie chart it is very tempting to conclude that adults watch way too much television (probably true) as opposed to children and teenagers who seem to watch a lot less (not true). All this pie chart shows is that a lot more adults watch television than children and teenagers—but then, there are a lot more adults than children and teenagers in the general population. When measured as a fraction of their respective

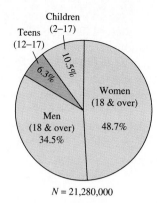

$N = 21,280,000$

Figure 14-9 Audience composition for prime-time viewership of network programs (February 1989). (*Source: The World Almanac and Book of Facts, 1990, p. 368*)

populations, children and teenagers actually watch a lot more television than adults.

Another subtle point somewhat buried in all the information is that this pie chart refers to prime-time audiences. Children do a significant amount of TV watching after school and before 7:00 P.M., and this part of the story is not shown on the pie chart.

The moral of this example is that there is more to a graph than meets the eye. Looking at the information with a healthy dose of skepticism is always a good idea. ■

■ **Class Intervals**   While the distinction between qualitative and quantitative data is important in many aspects of statistics, when it comes to deciding how best to display graphically the frequencies of a population, a critical issue is the number of categories into which the data can fall. When the number of categories is too big (say, for example, in the hundreds), then a bar graph or frequency chart can become too muddled and thereby be rendered ineffective. With qualitative data this is generally not a problem, but with quantitative data this problem does come up quite frequently: Both continuous and discrete variables can take on infinitely many values, and even when they don't, the number of values can be too large for any reasonable graph.

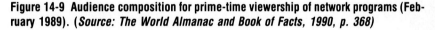

Example 5.   Suppose that, as part of a special research project, we want to look at the cumulative SAT test scores for the population of students discussed in Example 1 (those in Professor Blackbeard's Stat 101 course). Just as in Example 1, our data represents a discrete quantitative variable (in this case cumulative SAT scores). While in theory the

situation is no different from that in Example 1, in practice, because of the extremely large number of possible SAT scores (cumulative SAT scores are given in 10 point increments and range between 400 and 1600), we must deal with such data differently. The standard way to display frequency graphs in this situation is to break up the range of scores into intervals called **class intervals.** The decision as to how the class intervals are defined and how many there should be is a matter of personal choice. In this example, a sensible thing to do might be to break up the SAT scores into twelve class intervals. In this case our bar graph would look something like Fig. 14-10.

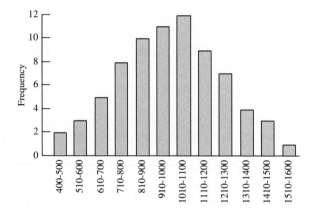

Figure 14-10 Cumulative SAT scores for N **= 75 students in Professor Blackbeard's Stat 101 course.**

∎

Note that in Example 5 we made it a point to create class intervals of the same size,[1] and this should be done as much as possible. Sometimes, however, it might make more sense to define class intervals of different lengths, as illustrated by our next example.

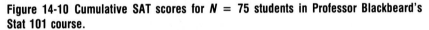

Example 6. The moment of truth for Professor Blackbeard's Stat 101 class has arrived! It is now time to turn the numerical scores on the exam into letter grades. In our terminology, this entails converting a quantitative variable (test score) into a qualitative one (letter grade) by defining class intervals associated with each grade category. In this case there is a good reason not to use class intervals of equal length. Following his own mysterious way of doing things, Professor Blackbeard defines the class in-

---

[1] A tiny exception was made for the class interval 400–500 which has one more possible test score than the others.

tervals for this particular exam according to the breakdown shown in Table 14-3.

Class Intervals	Grade
18–25	A
14–17	B
11–13	C
9–10	D
0–8	F

Table 14-3

If we combine the Stat 101 test scores in Table 14-2 with the class intervals for grades as defined in Table 14-3, we will get a new frequency table (Table 14-4) and a corresponding bar graph for the grade distribution in the exam (Fig. 14-11).

Grade	F	D	C	B	A
Frequency	10	26	30	8	1
Percentage	13.33%	34.67%	40%	10.67%	1.33%

Table 14-4

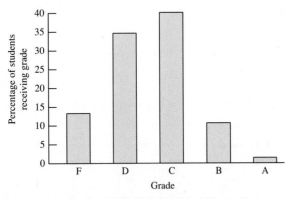

Figure 14-11 Grade distribution for Stat 101 exam.

■

■ **Histograms**   When a quantitative variable is continuous, then the possible values of the variable can vary by infinitesimally small increments. As a consequence of this, there are no gaps between the class intervals and our old way of doing things (using black columns or other types of cute stacks separated by gaps for clear viewing) will no longer work. In this case we will use a variation of a bar graph called a **histogram.** We illustrate the concept of a histogram in the next example.

Example 7.   Suppose we want to use a graph to display the distribution of starting salaries for last year's graduating class at Tasmania State University.

The starting salaries on the $N = 3258$ graduates ranged from a low of $20,350 to a high of $54,800. Based on this range and the amount of detail we want to show, we must decide on the length of the class intervals. While there are no hard-and-fast rules, any breakdown with up to 20 class intervals will produce a reasonably clean graph. Too many more than that, and the graph starts getting cluttered. Let's decide on class intervals of $5000. Table 14-5 is a frequency table for the data based on these class intervals. The third column in the table shows the data as a percentage of the population.

Salary	Number of students	Percentage
20,000 –25,000	228	7%
25,000$^+$–30,000	456	14%
30,000$^+$–35,000	1043	32%
35,000$^+$–40,000	912	28%
40,000$^+$–45,000	391	12%
45,000$^+$–50,000	163	5%
50,000$^+$–55,000	65	2%
Total	3258	100%

Table 14-5

The histogram for these data is shown in Fig. 14-12.

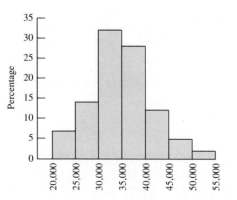

Figure 14-12 Histogram showing starting salaries for first-year graduates of Tasmania State University. N = 3258.

As Fig. 14-12 shows, a histogram is very similar to a bar graph. There are, however, several important distinctions that must be made. To begin with, because a histogram is used with continuous variables, there can be no gaps between the class intervals, and it follows therefore that the columns of a histogram must touch each other. Among other things, this forces us to make an arbitrary decision as to what happens to a value that falls exactly on the boundary between two class intervals. Should it always belong to the class interval to the left or to the one to the right? This is called the *endpoint convention*. The + marks in Table 14-5 indicate how we chose to deal with the endpoint convention in Fig. 14-12: A starting salary of $30,000, for example, should be assigned to the second rather than to the third class interval.

As with regular bar graphs, in creating histograms we should try, as much as possible, to define class intervals of equal length. When the class intervals are of unequal lengths, the rules for creating a histogram are considerably more complicated, since we can no longer use the heights of the columns to indicate the frequencies of the class intervals. We will not discuss the details of this situation here but refer the interested reader to Exercises 34 and 35 at the end of the chapter.

NUMERICAL SUMMARIES OF DATA

As we have seen, pictures can be an excellent tool for summarizing large amounts of data. Unfortunately, circumstances do not always lend themselves equally well to the use of pictures. If Professor Blackbeard were walking across campus and a student were to stop him and in casual tones ask, "How were the scores on the Stat 101 exam?" one could hardly expect Professor Blackbeard to flash a bar graph. The fact of the matter is that *numerical summaries* of data can often be used to substitute for or improve on the visual effect of graphical presentations.

In this section we will discuss several ways in which large sets of data can be summarized by a few well-chosen numbers. Before we go into specifics we will find it useful to draw an important distinction. Numerical summaries of data fall into two categories: numbers that tell us something about where the values of the data fall, and numbers that tell us something about how spread-out the values of the data are. The former are called; **measures of location,** and the latter are called **measures of spread.** The most important measures of location are the **mean** or **average,** the **median,** and the **quartiles.** The most important measures of spread are the **range,** the **interquartile range,** and the **standard deviation.** We will discuss each of these in order.

■ The Average

The best known of all numerical summaries of data is the average, sometimes also called the mean. (As much as possible, we will stick to the word "average"—it is a good down-to-earth word.) The **average** of a set of N numbers is obtained by adding the numbers and dividing by N. When the set of numbers is small, one can often calculate the average in one's head; for larger data sets, pencil and paper or a calculator is helpful. In either case the idea is very straightforward.

Example 8. In 10 games a basketball player scores 8, 5, 11, 7, 15, 0, 7, 4, 11, and 14 points, respectively. The total is 82 points in 10 games. The average is 8.2 points per game. Note that it is actually impossible for a player to score 8.2 points. As it often happens, the average taken by itself may not be a sensible possibility for the data.   ■

Example 9. We will now calculate the average score for Professor Blackbeard's Stat 101 midterm. For convenience we show once again the frequency table (Table 14-6).

Score	1	6	7	8	9	10	11	12	13	14	15	16	24
Frequency	1	1	2	6	10	16	13	9	8	5	2	1	1

Table 14-6   Frequency Table for Stat 101 Midterm Exam

The 75 data values can be totaled by taking each score and multiplying it by its corresponding frequency and adding. In this case we get

$$\text{Total} = 1\times1 + 6\times1 + 7\times2 + 8\times6 + 9\times10 + 10\times16 + 11\times13$$
$$+ 12\times9 + 13\times8 + 14\times5 + 15\times2 + 16\times1 + 24\times1 = 814.$$

The average score on the midterm exam (rounded off to two decimal places) is

$$814 \div 75 \approx 10.85 \text{ points.}$$

Intuitively, we think of this average as representing a "typical" student's score. If all test scores had been about the same, then, given the same total, each score would have been "about" 10.85 points.   ■

Table 14-7 shows a generic frequency table. To find the average of the data we do the following:

Data Value	s_1	s_2	$\cdots$	s_k
Frequency	f_1	f_2	$\cdots$	f_k

Table 14-7

■ **Step 1.** Calculate the total of the data.

$$\text{Total} = (s_1 \times f_1) + (s_2 \times f_2) + \cdots + (s_k \times f_k).$$

■ **Step 2.** Calculate N.

$$N = f_1 + f_2 + \cdots + f_k.$$

■ **Step 3.** Calculate the average.

$$\text{Average} = \text{Total} \div N.$$

As the reader can see, this is the kind of algorithm for which computers are particularly well suited, and even some of the better calculators have a built-in key for calculating averages.

So far, all our examples have involved numbers that are positive, but negative data values are also possible, and when both negative and positive data values are averaged, the results can be a little misleading.

Example 10. The monthly savings (monthly income minus monthly spending) of a family over a 1-year period is shown in Table 14-8. A negative amount indicates that, rather than saving money, the family spent more that month than their monthly income.

Month	Jan.	Feb.	Mar.	Apr.	May	Jun.	Jul.	Aug.	Sept.	Oct.	Nov.	Dec.
Savings (in $)	−732	−158	−71	−238	1839	−103	−148	−162	−85	−147	−183	500
	Christmas bills				$2000 lottery winnings							Christmas bonus from work

Table 14-8

The average monthly savings of this family over the year are

$$\frac{-732-158-71-238+1839-103-148-162-85-147-183+500}{12} = 26.$$

The $26 monthly savings paints a deceptively rosy picture of this family's finances. The truth of the matter is that this is a family living beyond its means which was bailed out by a one-shot lottery prize. ■

Our next example is a setup for an important idea that we will discuss later in the chapter.

Example 11. We're back to Professor Blackbeard's Stat 101 midterm exam. What we would now like to know is how far the test scores deviated from the average test score of 10.85 points (in other words, the difference between each test score and 10.85). Table 14-9 summarizes these deviations.

Deviation: test score—10.85	−9.85	−4.85	−3.85	−2.85	−1.85	−0.85	0.15	1.15	2.15	3.15	4.15	5.15	13.15
Frequency	1	1	2	6	10	16	13	9	8	5	2	1	1

Table 14-9

If we average these deviations, we get

Average deviation from 10.85

$$= \frac{-9.85 - 4.85 - 3.85 \times 2 - 2.85 \times 6 + \cdots + 13.15}{75} = \frac{0.25}{75} \approx 0.0033.$$

■

The fact that the average deviation from the average score in Example 11 is so small could lead someone who hasn't seen the details to conclude that the test scores were very close to each other and to the average score, which is not the case all. What really happened here is that the negative deviations and the positive deviations from the average cancelled each other out. In fact, the only reason this average deviation did not come out to be exactly 0 is because we rounded-off the average test score (10.85) to two decimal places (see Exercise 36). Once again, the problem lies in the fact that averaging numbers that are positive and negative can present a misleading picture of what the relative sizes of these numbers are. We call this the *cancellation effect*.

There are two ways out of this dilemma, and both are based on the idea that we need to get rid of the negative signs. One way to do this is to take absolute values (i.e., drop the minus signs); the other is to square the numbers (squared real numbers cannot be negative). For practical reasons the second approach is preferable and is the standard strategy used by statisticians.

The preceding comments can be summarized as follows:

1. When dealing with data values that are both positive and negative, averages can be misleading because of the possible cancellation between the positive and negative values.

2. To avoid the cancellation effect a common strategy is to square the data values first and then take the average of the squares.

Example 12. Applying the above-mentioned strategy to the deviations from the average test score on Professor Blackbeard's midterm exam, we get

Squared deviations: (test score—10.85)^2	97.02	23.52	14.82	8.12	3.42	0.72	0.02	1.32	4.62	9.92	17.22	26.52	172.92
Frequency	1	1	2	6	10	16	13	9	8	5	2	1	1

Table 14-10

When we average the data in Table 14-10 we get

Average of squared deviations

$$= \frac{97.02 + 23.52 + 14.82 \times 2 + 8.12 \times 6 + \cdots + 172.92}{75} = \frac{577.20}{75} = 7.70.$$

∎

While the squaring strategy makes all quantities positive and therefore eliminates the problem created by the cancellation effect, it creates a problem of its own: Squaring changes the order of magnitude of numbers. Whatever the original units were, we are now dealing with square units. In our last example the units were points on the exam, but the last average we calculated was square points (whatever they are). This problem can be eliminated by taking the square root (of the average of the squared data). Thus, in Example 8, if we take $\sqrt{7.70}$, we have a value that is of the same order of magnitude as the original values of the data.

■ **The Root Mean Square**

The process we have just described may appear at first sight to be both tortuous and confusing. In reality it gives an important numerical summary for a set of data called the **root mean square** (commonly denoted by rms). The rms is a useful substitute for the average only in cases where the cancellation effect must be avoided. In most typical situations the rms should not be used.

The procedure for calculating the rms consists of three steps.

■ **Step 1.** Square each of the data values. (This gets rid of the negatives.)

■ **Step 2.** Take the average (mean) of the values obtained in step 1. (This produces an average, but it is distorted because of the squaring.)

■ **Step 3.** Take the square root of the value obtained in step 2. (This brings the average back to the right order of magnitude.)

The most important use of the rms is when it is applied not to the data values themselves but to the deviations between the data values and their average as we did for the Stat 101 midterm exam scores in Example 12. The rms of the Stat 101 midterm exam deviations from the average score is $\sqrt{7.70} \approx 2.77$. This value is called the **standard deviation,** and we will come back to it toward the end of the chapter.

■ The Median

The median is another important, commonly used numerical summary of a set of data. To find the median of a set of numbers we must first sort the numbers by size. That is to say, we must rewrite the numbers in increasing order from left to right (or right to left—it makes no difference). We must then find the number in the "middle" of the sorted list. That number is the **median.**

Example 13. To find the median of the numbers 4.8, -2, 3.1, -6.5, 1.6, 0.5, and 4.25, we first sort the numbers by size and get

$$-6.5, \ -2, \ 0.5, \ 1.6, \ 3.1, \ 4.25, \ 4.8.$$

Of these seven numbers, the number in the middle is the one in the fourth position in the list, so the median is 1.6. ■

Example 14. How do we find the median of the numbers 2.2, -5.1, -2.7, 4.1, 6.2, -3, 1.2, 3.8? When we sort these numbers by size we get

$$-5.1, \ -3, \ -2.7, \ 1.2, \ 2.2, \ 3.8, \ 4.1, \ 6.2.$$

Which is the "middle" number in this list of eight numbers? Actually one could say that there are two numbers—the one in the fourth position (1.2) and the one in the fifth position (2.2)—with an equal claim to being in the "middle." In the spirit of compromise we settle the dispute by taking the number halfway between these two numbers and calling it the median. Thus, in this example the median is 1.7. ■

We can generalize what we discovered in Examples 13 and 14 in the following algorithm.

Algorithm 1.   Finding the Median of N Numbers

■   **Step 1.**   Sort the numbers by size.

■   **Step 2.**   Find the number(s) located in the "middle" position(s) in the sorted list. There are two cases:
(a) N odd. The "middle" position is position $(N + 1)/2$. The number in position $(N + 1)/2$ is the median.
(b) N even. There are two "middle" positions. They are $N/2$ and $(N/2) + 1$. The number halfway between the numbers in positions $N/2$ and $(N/2) + 1$ is the median.

Example 15.   We will now find the median score for Professor Blackbeard's Stat 101 midterm exam. For the reader's convenience we display the frequency table (Table 14-11).

Score	1	6	7	8	9	10	11	12	13	14	15	16	24
Frequency	1	1	2	6	10	16	13	9	8	5	2	1	1

Table 14-11

Having the frequency table available eliminates the need for sorting the scores—the frequency table has in fact done this for us. The total number of scores is $N = 75$, which means that the list has a single middle position. It is position $(75 + 1)/2 = 38$. We now know that the median is in the 38th position in the sorted list of the data. (It is important to understand that the median is not 38, but rather the 38th score in the sorted list.) To find the 38th score we start counting frequencies in the frequency row of Table 14-11. Counting from left to right, $1 + 1 = 2$, $1 + 1 + 2 = 4$, $1 + 1 + 2 + 6 = 10$, $1 + 1 + 2 + 6 + 10 = 20$, $1 + 1 + 2 + 6 + 10 + 16 = 36$. At this point we know that the 36th test score on the list is a 10 (the last of the 10's) and the next 13 scores are all 11's. We can conclude that the 38th test score (the median test score) is 11. In essence this says that half of the class had a score of 11 or less and half of the class had a score of 11 or more.   ■

Finding a median is not nearly as complicated as it seems, and with a little practice the reader will find this to be the case. (Start by trying Exercises 5 through 7.) As in finding the average, when large amounts of data are involved, letting a computer do the work is the most practical way to go.

A fairly common mistake is to confuse the median and the mean. The two words are quite similar, and they both define related concepts. This is one more good reason why (as much as possible) the term "average" should be used instead of "mean." Even among those who can keep the two concepts straight, a frequent misconception is to assume that the median and the average must be close to each other in value. While this is indeed the case in many types of real-life data, it is not true in general. Take, for example, the numbers, 1, 1, 1, and 97. The median of these numbers is 1, while the average is 25, a much larger number. On the other hand, if we take the numbers 1, 1, 100, 101, and 102, then the median (100) is much larger than the average (61). We can see that it is a mistake to assume, as many people do, that the median and the average are "about the same."

■ The Quartiles

The median has the effect of separating the sorted data into two halves: a **lower half** and an **upper half.** Figure 14-13 illustrates exactly what we mean.

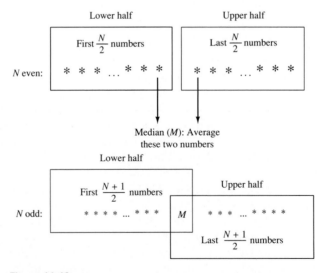

Figure 14-13

Example 16. The heights in inches of 10 children are 38, 41, 36, 49, 28, 44, 35, 43, 39, and 41. To find the lower and upper halves of these data values, we first sort them by size. This results in the list 28, 35, 36,

38, 39, 41, 41, 43, 44, 49. The lower half is then 28, 35, 36, 38, 39, while the upper half is 41, 41, 43, 44, 49. The median height of this group is 40 inches.   ■

Example 17.   Suppose that one more child of height 32 inches joins the group in Example 12. The new sorted data values are 28, 32, 35, 36, 38, 39, 41, 41, 43, 44, 49. The lower half now is 28, 32, 35, 36, 38, 39, while the upper half is 39, 41, 41, 43, 44, 49. Note that in this example the median (39) is in both halves.   ■

Now that we understand the formal meaning of the "lower half" and the "upper half" of a set of numbers, we can define the quartiles in a very straightforward manner: The **first quartile** (Q_1) is the median of the lower half; the **third quartile** (Q_3) is the median of the upper half.

Example 18.   Consider the same set of 10 children's heights discussed in Example 16. The lower half of these numbers is 28, 35, 36, 38, 39. The median of these five numbers is 36. It follows that the first quartile is 36 ($Q_1 = 36$). The upper half of the data is 41, 41, 43, 44, 49, and the median of these five numbers is 43. It follows that the third quartile is 43 ($Q_3 = 43$).   ■

Example 19.   For the 11 heights discussed in Example 17, we found that the lower half of the data consists of the numbers 28, 32, 35, 36, 38, 39. The first quartile is the median of these six numbers, which is 35.5. The upper half of the data consists of the numbers 39, 41, 41, 43, 44, 49. The third quartile is the median of these six numbers, which is 42.   ■

Just as the median serves to break up the data into halves, the purpose of quartiles is to break up the data into quarters. Saying, for example, that the first quartile of a set of numbers is 35.5 means that 25% of the numbers are smaller than or equal to 35.5. Likewise, saying that the third quartile of a set of numbers is 42 means that 75% of the numbers are smaller than or equal to 42. The median itself is the second quartile, which means that 50% of the numbers are smaller than or equal to the median. It is not hard to see why the first quartile, the median, and the third quartile are sometimes referred to as the 25th percentile, the 50th percentile, and the 75th percentile, respectively.

We will now describe the general procedure for finding the quartiles of a set of numbers.

Algorithm 2.   Finding the Quartiles of N Numbers

■   **Step 1.**   Sort the numbers by size.

■   **Step 2.**   Find the median M (see Algorithm 1).

■   **Step 3.**   Find the lower half and upper half of the data set (see Fig. 14-13).

■   **Step 4.**   Find the median of the lower half. This is the first quartile Q_1.

■   **Step 5.**   Find the median of the upper half. This is the third quartile Q_3.

■ **The Five-Number Summary**

A reasonably good summary for a large set of data can be provided numerically by giving the lowest value of the data (Min), the first quartile (Q_1), the median (M), the third quartile (Q_3), and the largest value of the data (Max). These five numbers constitute the **five-number summary** of the data.

Example 20.   Let's go back to Professor Blackbeard's Stat 101 midterm exam and see if we can come up with the five-number summary for the set of test scores. We already know that the lowest score is Min = 1, the median score is $M = 11$, and the highest score is Max = 24. We need to find the first and third quartiles. Since the total number of scores is 75 (odd), we know that the first half of the data consists of the first 38 numbers. We will use Algorithm 2. Since 38 is even, the median of the first 38 numbers is halfway between the 19th and 20th scores. To find the 19th score we go to the frequency table and start counting frequencies beginning on the left. We leave it to the reader to verify that the 19th and 20th scores are both 9. It follows that $Q_1 = 9$. To find the third quartile we need to find the median of the upper half of the scores. These are the 38th through the 75th scores. The median of these 38 numbers is the number halfway between the two middle numbers. To find it we can count in the frequency table from left to right (starting at the 38th number) or, what is much easier, count from the top down until we find the 19th and 20th numbers from the end. Using this last approach, we find that the 19th and 20th scores from the top are both 12, so we have $Q_3 = 12$. The five-number summary for the scores on the Stat 101 midterm exam is

Min = 1, $Q_1 = 9$, $M = 11$, $Q_3 = 12$, Max = 24.

Note that without Q_1 and Q_3 we would have a very distorted picture of the exam data, since both Min = 1 and Max = 24 are outliers. The test scores were not evenly spread out in the range between 1 and 24, in fact just the opposite was true. With the quartiles we can get a much better idea of what happened: The middle 50% of the test scores fell in a very narrow range of between 9 and 12 points.　∎

MEASURES OF SPREAD

An important element in summarizing data with numbers is to give an idea of how spread-out the data values are. Consider the following two sets of numbers: Set 1 = {45, 46, 47, 48, 49, 51, 52, 53, 54, 55}, and Set 2 = {1, 11, 21, 31, 41, 59, 69, 79, 89, 99). We leave it to the reader to verify that for both sets of numbers the average is 50 and the median is 50. At the same time, it is obvious that these two sets of data are very different. The numbers in Set 2 are much more spread-out than those in Set 1. What can we do to measure the amount of spread in a set of data? The most obvious approach is to take the difference between the highest and lowest values of the data (Max − Min). This difference is called the **range.** For Set 1 = {45, 46, . . . , 55) the range is 10, and for Set 2 = {1, 11, . . . , 89, 99} the range is 98.

As a measure of spread, the range is useful only if there are no outliers, since outliers can completely distort the range. On professor Blackbeard's Stat 101 exam the range of the scores is 24 − 1 = 23 points, but without the two outliers the range would be 16 − 6 = 10.

To eliminate the possible distortion caused by outliers, a common practice is to use the **interquartile range** (IQR). The interquartile range is the difference between the third quartile and the first quartile (IQR = $Q_3 - Q_1$), and its significance is that it tells us how spread-out the middle 50% of the data values are. For many types of real-world data the interquartile range is a useful measure of spread. When the five-number summary is used, both the range and the interquartile range come essentially free in the bargain.

Example 21. The five-number summary for a set of 200 test scores is Min = 4, Q_1 = 23, median = M = 27, Q_3 = 32, Max = 61. Here the range is Max − Min = 57, and the interquartile range is IQR = $Q_3 - Q_1$ = 9. What does the five-number summary tell us about this set of test scores? We know that the lowest 50 scores (the bottom 25%) are between 4 and 23; the next 50 scores are tightly bunched up between 23 and 27; likewise, the next 50 are between 27 and 32; and the last 50 scores are between 32 and 61. Could these be test scores from another one of Professor Blackbeard's courses?　∎

∎ Box Plots

A **box plot** gives a convenient way to display in a single picture all the information contained in the five-number summary. Figure 14-14 shows a generic box plot for a set of data. The box plot consists of a central

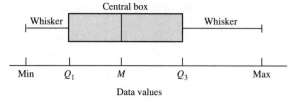

Figure 14-14 A typical box plot for a set of data.

rectangular box that goes from the first quartile Q_1 to the third quartile Q_3. A line crosses the central box indicating the position of the median M. On both sides of the central box are whiskers extending to the smallest value Min and largest value Max of the data.

Figure 14-15 shows a box plot for the data given in Example 21.

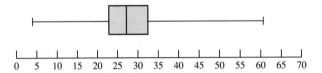

Figure 14-15 Box plot for the data in Example 21.

Box plots are particularly useful when comparing data for two or more populations. This fact is illustrated in the next example.

Example 22.   Figure 14-16 shows box plots for the starting salaries of two different populations: first-year agriculture and engineering graduates of Tasmania State University. Superimposing the two box plots on the same scale allows us to make comparisons in a particularly efficient way. It is clear, for instance, that engineering graduates are doing better overall than agriculture graduates, even though at the very top levels agriculture

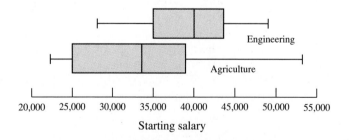

Figure 14-16 Comparison of starting salaries for Tasmania State University first-year graduates in agriculture and engineering using box plots.

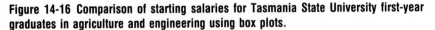

graduates are better paid. Another interesting point is that the median salary of agriculture graduates is less than the first quartile of the salaries of engineering graduates. The very short whisker on the left side of the agriculture box plot tells us that the bottom 25% of agriculture salaries are concentrated in a very narrow salary range. We can also see that agriculture salaries are much more spread-out than engineering salaries, even though most of the spread occurs at the higher end of the salary scale.

It is clear that we can get a lot of mileage out of a simple picture like the one in Fig. 14-16. (See Exercises 27 and 28.) ■

■ The Standard Deviation

Perhaps the most commonly used measure of spread for a set of data is the **standard deviation.** The reader may recall that in Examples 11 and 12 we carried out a series of calculations for the Stat 101 midterm exam scores that involved computing the rms of the deviations of the test scores from the average score. We called this value the standard deviation of the test scores.

For any set of numbers we can calculate the standard deviation using the following algorithm.

Algorithm 3.   Finding the Standard Deviation of N Numbers

■ **Step 1.**   Find the average of the N numbers. Call it A.

■ **Step 2.**   For each number x in the data set, compute $x - A$. We call this x's *deviation from the average*.

■ **Step 3.**   Using the deviations from the averages as our new data set, we compute their rms (remember, this means square the data, average the squares,[2] and then take the square root). This value is called the standard deviation (SD).

Example 23.   Let's find the standard deviation of the numbers 45, 46, 47, 48, 49, 51, 52, 53, 54, 55.

Step 1. $A = 50$.

Step 2.   Taking the deviations from the average gives the data set $-5, -4, -3, -2, -1, 1, 2, 3, 4, 5$.

---

[2] In many statistics books and statistical computer programs, this step in the algorithm is changed so that when calculating the average of the squared deviations instead of dividing by N, the division is by $N - 1$. Everything else is exactly the same. There are good reasons why this definition makes good sense, especially when the data values are obtained from a sample, but to explain them would go beyond the purpose and scope of this chapter.

Step 3.
   (a) Square: 25, 16, 9, 4, 1, 1, 4, 9, 16, 25.
   (b) Mean: [total from step 3(a)]/N = 110/10 = 11.
   (c) Root: $\sqrt{11}$.

The answer is SD = $\sqrt{11} \approx 3.317$.   ■

    It is important for the reader to understand the difference between the rms of a set of data and the standard deviation for the same data. They are not the same thing! The rms is a substitute for the average that is used primarily when the positive values of the data and the negative values of the data tend to cancel each other out. The standard deviation is a special use of the rms. Here we are not finding the rms of the data themselves but rather of a new set of data consisting of the deviations of the original data values from their average. For this data set the cancellation effect is at its peak: The negative deviations and the positive deviations always cancel out completely (Exercise 36).

    Although the standard deviation is a measure of spread that can be used with any set of numbers, it is particularly important in many situations involving real-life data. We will discuss this point in greater detail in Chapter 16.

CONCLUSION

    The basic theme of this chapter is "putting things in a nutshell." More specifically, when we have lots of data (and here we are using the word "lots" loosely), there is a need to summarize the information provided by the data in a meaningful way. In this chapter we discussed ways to summarize data sets where each data value consists of a single number (or a single category in the case of categorical variables).

    Graphical summaries of data can be produced by bar graphs, frequency charts, pie charts, histograms, etc. (There are many other types of graphical descriptions that we did not discuss in the chapter.) Which kind of graph is the most appropriate for which situation depends on many factors, and there is no substitute for experience in making these kinds of choices. One of the challenging things about descriptive statistics is that there is plenty of room for creativity.

    Numerical summaries of data fall into two categories: (1) *measures of location* such as the average, the root mean square, the median, and the quartiles, and (2) *measures of spread* such as the range, the interquartile range, and the standard deviation.

    The famous science fiction writer H. G. Wells once said,

Statistical thinking will one day be as necessary for efficient citizenship as the ability to read and write.

The concepts introduced in this chapter are to statistics what the alphabet is to reading and writing.

KEY CONCEPTS

average (mean)
bar graph
box plot
cancellation effect
class interval
continuous variable
discrete variable
five-number summary
frequency chart
frequency table
histogram
interquartile range
lower half

measures of location
measures of spread
median
outlier
pie chart
qualitative variable
quantitative variable
quartiles
range
root mean square (rms)
standard deviation
upper half

EXERCISES

■ Walking

The following sets of data Tables 14-12, 14-13, and 14-14 are used in several of the walking exercises.

Student ID	Score	Student ID	Score	Student ID	Score	Student ID	Score
1362	50	2877	80	4315	70	6921	50
1486	70	2964	60	4719	70	8317	70
1721	80	3217	70	4951	60	8854	100
1932	60	3588	80	5321	60	8964	80
2489	70	3780	80	5872	100	9158	60
2766	10	3921	60	6433	50	9347	60

Table 14-12   Chem 103: Final Exam Scores

Student ID	Score	Student ID	Score	Student ID	Score	Student ID	Score	Student ID	Score
1075	74	1998	75	3491	57	4713	83	6234	77
1367	83	2103	59	3711	70	4822	55	6573	55
1587	70	2169	92	3827	52	5102	78	7109	51
1877	55	2381	56	4355	74	5381	13	7986	70
1946	76	2741	50	4531	77	5717	74	8436	57

Table 14-13   History 3B: First Midterm Exam Scores

Student ID	Distance to School (miles)	Student ID	Distance to School (miles)	Student ID	Distance to School (miles)	Student ID	Distance to School (miles)
1362	1.5	2877	1.0	4355	1.0	6573	0.5
1486	2.0	2964	0.5	4454	1.5	8436	3.0
1587	1.0	3491	0.0	4531	1.5	8592	0.0
1877	0.0	3588	0.5	5482	2.5	8854	0.0
1932	1.5	3711	1.5	5533	1.0	8964	2.0
1946	0.0	3780	2.0	5717	8.5		
2103	2.5	3921	5.0	6307	1.5		

Table 14-14   Distance From Home to School for Students in Speech 21

1. The final exam in Chem 103 consisted of 10 questions worth 10 points each with no partial credit given. The scores on the exam are given in Table 14-12.
 (a) Make a frequency table for the exam scores.
 (b) Make a bar graph showing the actual frequencies of the scores on the exam.
 (c) Make a bar graph showing the relative frequencies of the scores on the exam (i.e., the percentage of students receiving each score).

2. The distance from home to school (measured to the closest $\frac{1}{2}$ mile) for each student in Speech 21 is given in Table 14-14.
 (a) Make a frequency table for the distances from home to school.
 (b) Make a bar graph showing the actual frequencies of the distances from home to school.
 (c) Make a bar graph showing the relative frequencies of the distances from home to school (i.e., the percentage of students at each distance).

3. The first midterm exam in History 3B consisted of 100 multiple-choice questions worth 1 point each. The scores on the exam are given in Table 14-13. Suppose the class intervals for letter grades are

90–100 A

80–89 B

70–79 C

60–69 D

Below 60 F

 (a) Make a frequency table for the letter grades on the exam.
 (b) Make a bar graph showing the actual frequencies of the letter grades on the exam.
 (c) Make a bar graph showing the relative frequencies of the letter grades on the exam (i.e., the percentage of students receiving each letter grade).
 (d) Draw a pie chart for the percentage of students receiving each letter grade.

4. Suppose that class intervals for the distances from home to school for the students in Speech 21 (Table 14-14) are defined by

0.0 miles	Very close
0.5–1.0 miles	Close
1.5–2.0 miles	Nearby
2.5–4.5 miles	Not too far
5.0–10.0 miles	Far

(a) Make a frequency table for the class intervals.
(b) Make a bar graph showing the actual frequencies of the class intervals.
(c) Make a bar graph showing the relative frequencies of the class intervals (i.e., the percentage of students in each class interval).
(d) Draw a pie chart for the percentage of students in each category.

5. The following bar graph shows the frequencies of various scores received by students in Math A on a 10-point pop quiz.

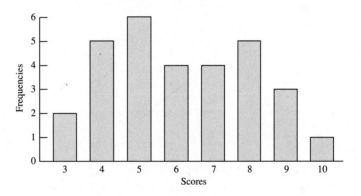

(a) How many students took the quiz?
(b) Make a frequency table for the scores on the pop quiz.
(c) What is the average score?
(d) What is the median score?

6. For the Chem 103 final exam scores in Table 14-12, find
(a) the range
(b) the median
(c) the first and third quartiles
(d) the five-number summary
(e) the interquartile range.

7. For the History 3B first midterm exam scores in Table 14-13, find
(a) the range
(b) the median
(c) the first and third quartiles
(d) the five-number summary
(e) the interquartile range.

8. For the Chem 103 final exam scores in Table 14-12, find
   (a) the average
   (b) the standard deviation.

9. For the History 3B first midterm exam scores in Table 14-13, find
   (a) the average
   (b) the standard deviation.

10. Find the average and standard deviation for each of the following three sets of numbers: $\{5, 5, 5, 5\}$, $\{0, 5, 5, 10\}$, $\{-5, 0, 0, 25\}$. Note the similarities and differences in your answers and explain.

11. Find the average and standard deviation for each of the following three sets of numbers: $\{10, 10, 10, 10\}$, $\{1, 6, 13, 20\}$, $\{1, 1, 18, 20\}$. Note the similarities and differences in your answers and explain.

12. For the data set $\{0, 1, 2, 3, 4, 5, 6, 7, 8, 9\}$ find
   (a) the average
   (b) the median
   (c) the standard deviation.

13. For the data set $\{1, 2, 3, 4, 5, 6, 7, 8, 9, 10\}$ find
   (a) the average
   (b) the median
   (c) the standard deviation.

14. For the data set $\{0, 1, 2, 3, 4, 5, 6, 7, 8, 9\}$ find
   (a) the first quartile
   (b) the third quartile
   (c) the interquartile range.

15. For the data set $\{1, 2, 3, 4, 5, 6, 7, 8, 9, 10\}$ find
   (a) the first quartile
   (b) the third quartile
   (c) the interquartile range.

16. For the data set $\{1, 2, 3, \ldots, 99, 100\}$ find
   (a) the average
   (b) the median.

17. For the data set $\{1, 2, 3, \ldots, 99, 100\}$ find
   (a) the first quartile
   (b) the third quartile
   (c) the interquartile range.

18. The pie chart in the margin shows the percentage of the undergraduate student body at Tasmania State University for each ethnic group.
   (a) Give a frequency table showing the actual frequencies for each category.
   (b) Draw the bar graph corresponding to the frequency table in part (a).

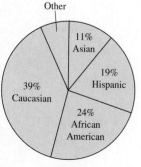

7%
Other

11%
Asian

19%
Hispanic

39%
Caucasian

24%
African
American

$N = 15,000$

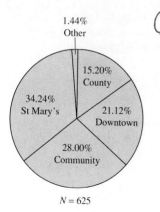

1.44%
Other

15.20%
County

34.24%
St Mary's

21.12%
Downtown

28.00%
Community

$N = 625$

19. The pie chart in the margin shows the percentage of babies born at each of the four hospitals in the city of Cleansburg in the last year.
  (a) How many babies were born at Downtown Hospital?
  (b) How many babies were born outside one of the four hospitals (at home, on the way to the hospital, etc.)?
  (c) Draw a bar graph showing the frequency for each category.

Exercises 20 through 23 refer to the data in the following table, which shows the weights (in ounces) of the 625 babies born in the city of Cleansburg in the last year.

Weight in Ounces		
More Than	Less Than or Equal to	Frequencies
48	60	15
60	72	24
72	84	41
84	96	67
96	108	119
108	120	184
120	132	142
132	144	26
144	156	5
156	168	2

20. **(a)** Give the length (in ounces) of each class interval.
  (b) Suppose a baby weighs exactly 5 pounds 4 ounces. What class interval does she belong to? Describe the endpoint convention.

21. Write a new table for these data values using class intervals of length equal to $1\frac{1}{2}$ pounds.

22. Draw the histogram corresponding to these data values using the class intervals as shown in the table.

23. Draw the histogram corresponding to the same data when class intervals of 24 ounces are used.

24. Draw a box plot for the Chem 103 final exam scores in Table 14-12.

25. Draw a box plot for the History 3B midterm exam scores in Table 14-13.

26. **(a)** Find the five-number summary, the average, and the standard deviation for the data given in the following frequency table.

Value	9 10 11 12 13 14 15 16 17 18 19 20 21 22 23 24 25
Frequency	3 5 7 4 12 10 13 11 15 13 11 4 3 0 0 0 1

  (b) Draw a box plot for this data.

Exercises 27 and 28 refer to the following figure comparing the starting salaries for Tasmania State University first-year graduates in Agriculture and Engineering using box plots. (These are the two box plots discussed in Example 22.)

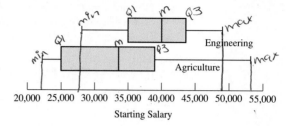

Starting Salary

27. (a) Approximately how much is the median salary for agriculture majors?
    (b) Approximately how much is the median salary for engineering majors?
    (c) Explain how we can tell that the median salary for engineering majors is more than the third quartile of the salaries for agriculture majors.

28. (a) If the number of engineering majors was 612, how many of them had a starting salary of $35,000 or more?
    (b) If the number of agriculture majors was 960, approximately how many of them made less than $25,000?

■ **Jogging**

29. The average starting salary for the 125 speech majors that graduated this year from Tasmania State University is $50,752. This is an impressive figure, but before we all rush out to change majors, consider the fact that this average is seriously distorted by the presence of an outlier—basketball star "Hoops" Tallman who just signed a $2.5 million professional contract. If we were to disregard this one outlier, what would the average starting salary be for the remaining speech majors?

30. (a) Give an example of 10 numbers such that their average is less than their median.
    (b) Give an example of 10 numbers such that their median is less than their average.
    (c) Give an example of 10 numbers such that their average is less than the first quartile.
    (d) Give an example of 10 numbers such that their average is more than the third quartile.

31. Suppose that the average of 10 numbers is 7.5 and that the smallest of them is Min = 3.
    (a) What is the smallest possible value of Max?
    (b) What is the largest possible value of Max?

32. What happens to the five-number summary of Professor Blackbeard's exam scores (Example 1) if
    (a) 2 points are added to each score
    (b) 10% is added to each score.

33. Under what conditions would the standard deviation of a set of numbers be 0?

Exercises 34 and 35 refer to histograms with unequal class intervals. When draw-ing histograms with class intervals of unequal lengths, the columns must be drawn so that the frequencies or percentages are proportional to the area of the column.

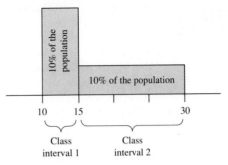

The figure illustrates what we mean. If the column over class interval 1 represents 10% of the population, then the column over class interval 2, also representing 10% of the population, must be one-third as high because the class interval is three times as large.

34. If the height of the column over the class interval 20–30 is 1 unit and the column represents 25% of the population, then
   (a) How high should the column over the interval 30–35 be if 50% of the population falls into this class interval?
   (b) How high should the column over the interval 35–45 be if 10% of the population falls into this class interval?
   (c) How high should the column over the interval 45–60 be if 15% of the population falls into this class interval?

35. Two-hundred senior citizens are tested for fitness and rated on their times on a 1-mile walk. These ratings and associated frequencies are given in the following table.

Time	Rating	Frequency
6^+–10 minutes	Fast	10
10^+–16 minutes	Fit	90
16^+–24 minutes	Average	80
24^+–40 minutes	Slow	20

Draw a histogram for these data based on the categories given by the ratings in the table.

■ **Running**

36. Given the numbers $x_1, x_2, x_3, \ldots, x_N$ with average m, show that

$$(x_1 - m) + (x_2 - m) + (x_3 - m) + \cdots + (x_N - m) = 0.$$

37. **(a)** Find two numbers whose average is m and standard deviation is s.
   (b) Find three equally spaced numbers whose average is m and standard deviation is s.

(c) Generalize the preceding by finding N equally spaced numbers whose average is m and standard deviation is s. (*Hint:* Consider N even and N odd separately.)

38. Show that the median and the average of the numbers $1, 2, 3, \ldots, N$ are always the same.

39. Suppose that the average of the numbers $x_1, x_2, x_3, \ldots, x_N$ is m and that the standard deviation of these same numbers is s. Suppose also that the average of the numbers $x_1^2, x_2^2, x_3^2, \ldots, x_N^2$ is M. Show that $s^2 = M - m^2$. (In other words, for any data set, if we take the average of the squared data values and subtract the square of the average of the data values we get the square of the standard deviation.)

40. Given that the numbers $x_1, x_2, x_3, \ldots, x_N$ have standard deviation s, explain why the numbers $x_1 + c, x_2 + c, x_3 + c, \ldots, x_N + c$ also have standard deviation s.

41. (a) Using the formula $1^2 + 2^2 + 3^2 + \cdots + N^2 = \dfrac{N(N + 1)(2N + 1)}{6}$, find the standard deviation of the data set $\{1, 2, 3, \ldots, 98, 99\}$. (*Hint:* Use Exercise 39.)

 (b) Find the standard deviation of the data set $\{315, 316, \ldots, 412, 413\}$. (*Hint:* Use Exercise 40.)

REFERENCES AND FURTHER READINGS

1. Berry, D. A., and B. W. Lindgren, *Statistics: Theory and Methods*. Pacific Grove, Calif.: Brooks/Cole Publishing Co., 1990, chap. 7.

2. Cleveland, W. S., and M. E. McGill (eds.), *Dynamic Graphics for Statistics*. Pacific Grove, Calif.: Brooks/Cole, 1988.

3. Cleveland, W. S., *The Elements of Graphing Data*. Pacific Grove, Calif.: Brooks/Cole Publishing Co., 1985.

4. Freedman, D., R. Pisani, R. Purves, and A. Adhikari, *Statistics* (2nd. ed.). New York: W. W. Norton, Inc., 1991, chaps. 3 and 4.

5. Groeneveld, Richard A., *Introductory Statistical Methods*. Boston: PWS-Kent Publishing Co., 1988, chap. 2.

6. Mosteller, F., W. Kruskal, et al., *Statistics by Example: Exploring Data*, Reading, Mass.: Addison-Wesley, 1973.

7. Sincich, Terry, *Statistics by Example*. San Francisco, Calif.: Dellen Publishing Co., 1990, chaps. 2 and 3.

8. Tanner, Martin, *Investigations for a Course in Statistics*. New York: Macmillan Publishing Co., Inc., 1990.

9. Tufte, Ed, *Envisioning Information*. Cheshire, Conn.: Graphics Press, 1990.

10. Tufte, Ed, *The Visual Display of Quantitative Information*. Cheshire, Conn., Graphics Press, 1983.

11. Tukey, J. W., *Exploratory Data Analysis*. Reading, Mass.: Addison-Wesley Publishing Co., Inc., 1977.

15

Probability

Chances are . . .

The most important questions of life are, for the most part, really only problems of probability.

PIERRE SIMON DE LAPLACE*

- ■ "There is a 90% chance of rain tomorrow." (Weather report.)

- ■ "Each year the average driver has one chance in three of being in an automobile accident." (Insurance company report.)

- ■ "The odds that the Cincinnati Reds will win the World Series are 2 to 5." (Nevada oddsmaker.)

- ■ "The probability of tossing a coin heads three times in a row is .125." (Gambler's manual.)

"Chances," "odds," "probabilities"—these words are as much a part of our everyday vocabulary as "mother," "baseball," and "apple pie." While we all probably (there it is again!) have an intuitive idea of what each of the four statements above means, giving a precise definition of terms such as "chances," "odds," and "probability" is surprisingly difficult.

* Pierre Simon, Marquis de Laplace (1749–1827) was a renowned French mathematician and astronomer. Among his many works, Laplace wrote *Théorie Analytique des Probabilités* (*The Analytical Theory of Probabilities*) one of the earliest treatises in the mathematical theory of probability,

**Photos courtesy of (From left to right) Myron Wood/Photo Researchers, Doreen E. Pugh/
Photo Researchers, © Estate of Harold Edgerton, 1965, AP/Wide World Photos**

In this chapter we will discuss probabilities, chances, and odds (they are just different ways of looking at the same thing) using both an informal and a more formal mathematical approach. Before we start our discussion in full, let's clear the air a little regarding the terminology.

The word "probability" is commonly used in two different ways. On the one hand probability is a *discipline*—in fact an important branch of mathematics—and it is in this context that we have used it in the title of this chapter. On the other hand, we will talk about the probability of something happening, and in this context "the probability" is a number such as the one in the statement "The probability of tossing a coin heads three times in a row is .125." When used in the latter context, probability can also be expressed by using the word "chance" as in "the chances of tossing a coin heads three times in a row are 12.5%."[1] The term "odds" is used to express the same concept as probability in a slightly different form. Later in the chapter we will explain how to convert odds to probabilities, and probabilities to odds.

Our discussion in this chapter is broken up into two parts. In the first part we lay down the basic concepts needed for a meaningful discussion of probability, and it isn't until the second part that we actually define and calculate probabilities using a formal approach.

WHAT IS PROBABILITY (THE DISCIPLINE)?

Probability is the study of certain kinds of random phenomena that mathematicians like to call **random experiments**. We will say that an experiment is random when it has more than one possible outcome and when we are unable to predict with certainty ahead of time which of the possible outcomes will actually occur. Typical examples of random experiments are predicting the weather, tossing a coin, rolling a pair of dice, shooting a free throw, and playing a World Series. Notice that the word "experiment" is used here in a very liberal way, and for carrying out most of these random experiments a white coat or a well-equipped lab is hardly a requirement.

Our daily lives are affected by the occurrence of random experiments in countless ways, some of which are obvious (the weather, the outcome of the World Series, etc.) and some of which are more subtle (the insurance premiums we pay, the reliability of the products we buy, etc.). Out of necessity, we will use simple examples to illustrate the basic ideas, but the reader is well advised to keep in mind that, as a subject, probability goes far beyond the tossing of coins, the rolling of dice, and the shooting of free throws.

SAMPLE SPACES

Associated with every random experiment is the set of all its possible outcomes, called the **sample space**. We illustrate this concept (and some unexpected subtleties) by means of several examples. We will use the letter S to denote the sample space, and N to denote the number of outcomes in S.

---

[1] It is customary to express probabilities in decimals (or fractions) and chances in percentages, and we will follow that custom in this chapter.

Example 1. Random experiment: toss a coin.

Sample space: $S = \{H, T\}$ (From now on we will abbreviate heads as H and tails as T.) The curly brackets are there merely to indicate that the sample space is a set.

Size of sample space: $N = 2$. ∎

Example 2. Random experiment: roll a single die.[2]

Sample space: $S = \{ \boxdot , \boxdot , \boxdot , \boxdot , \boxdot , \boxdot \}$.

Size of sample space: $N = 6$. ∎

Example 3. Random experiment: roll a pair of dice.

Sample space: $S = \{ \ \ldots \ \}$.

Size of sample space: $N = 36$.

Notice that in this example we are treating the dice as distinguishable objects (as if one were white and the other red), so that ⊡⊡ and ⊡⊡ are considered different outcomes. ∎

Example 4. Random experiment: toss a coin twice.

Sample space: $S = \{HH, HT, TH, TT\}$. (Here HT means the first toss came up H and the second toss came up T.)

Size of sample space: $N = 4$. ∎

Example 5. Random experiment: shoot a pair of free throws.

Sample space: $S = \{ss, sf, fs, ff\}$. (Here s means success and f means failure.)

Size of sample space: $N = 4$. ∎

---

[2] Singular, die; plural, dice.

Notice the similarity between Examples 4 and 5. In fact, if we were to identify H with success and T with failure, the sample spaces would be exactly the same. Examples 4 and 5 illustrate the fact that seemingly different random experiments (tossing a pair of coins, shooting a pair of free throws) can turn out to have essentially the same sample space (the symbols may be different, but the essence is the same).

Example 6.   Random experiment: Shoot as many free throws as needed until you make one, and then stop.

Sample space: $S = \{s, fs, ffs, fffs, \ldots\}$. The " $\ldots$ " here is significant. It is an indication that we are dealing with an infinite sample space. There is, after all, no limit to the number of consecutive free throws that some clown could miss.   ■

While infinite sample spaces are important in many applications of probability theory, they are beyond the level of our presentation in this chapter. The last example was given simply to illustrate the fact that infinite sample spaces are also possible. We will now return to finite sample spaces.

Example 7.   In many games of chance involving the roll of a pair of dice (such as Monopoly, craps, etc.), the outcome of the game depends on the total sum rolled. The roll ⚂⚄ corresponds to the total 8, and so on. In this situation we are not interested in the individual rolls of the die but rather in their combination. What is the sample space in this case? Since the possible outcomes are the possible totals 2, 3, . . ., 12, we have

Random experiment: Roll a pair of dice and check for the total sum rolled.

Sample space: $S = \{2, 3, 4, 5, 6, 7, 8, 9, 10, 11, 12\}$.

Size of sample space: $N = 11$.   ■

There is no contradiction between Examples 3 and 7. While the random experiment in both cases can be said to be the same (roll a pair of dice), the *observation* we make is not. In Example 3 we just roll, in Example 7 we roll and add, and so the sample spaces reflect this difference. This is our second lesson: The same basic random experiment can result in different sample spaces depending on exactly what aspect of the experiment one is observing. It is important to notice that the only unpredictable part in rolling the dice and adding the total is in the rolling of the dice (let's agree that adding is not a random experiment), and yet the combination of the two (roll, then add) results in a new sample space.

Our next example reiterates the previous point.

Example 8.   Two friends, Thelma and Mabel, go to the racetrack. There are five horses entered in race 1—let's call them A, B, C, D, and E. Thelma makes a straight bet on a winner (if her horse wins, she wins the bet). For Thelma the sample space is $S_T = \{A, B, C, D, E\}$. Mabel buys a trifecta ticket, so she must correctly get the first-, second-, and third-place finishers and in exactly that order to win her bet. For Mabel, the sample space is $S_M = \{ABC$ (ABC means horse A comes in first, horse B comes in second, and horse C comes in third), ACB, BAC, BCA, CAB, CBA, ABD, ADB, ... $\}$. Obviously, S_M is a pretty huge sample space, and for the time being we won't worry too much about writing it all out. The main point once again is that the same random experiment (the horse race) can produce more than one sample space depending on the nature of the observation that interests us.   ■

Example 8 illustrates another important point. Here we had a sample space (S_M) that was too big to write down in its entirety. This situation is not unusual. Many finite sample spaces are incredibly large, and it is inconceivable that in such cases we would even try to write down a complete list of all the outcomes. Fortunately, the critical question is the *size* of the sample space, and we can answer this question even when we don't write the sample space down. We discuss how to do this next.

■ Sizing the Sample Space: The Multiplication Principle

Example 9.   Random experiment: Toss a coin three times.
Sample space: $S = \{HHH, HHT, HTH, HTT, THH, THT, TTH, TTT\}$.
Size of sample space: $N = 8$.   ■

Example 10.   Random experiment: Toss a coin eight times.

In this example the sample space S is too big to write down. One may ask what a typical outcome might be like. Taking our cue from Example 9, we can say that a typical outcome can be described as a string of eight consecutive letters where the letters can be only H's or T's. For example, the string $THHTHTHH$ represents the following outcome: First toss came up T, second toss came up H, third toss came up H, etc. It is obvious that there are many other possible strings and that writing them all down and then counting them one by one like we count sheep would be an exhausting and very unmathematical thing to do. There is, however, another way:

■ Number of possibilities for the first toss = 2 (*H* or *T*).

■ Number of possibilities for the second toss = 2 (ditto).

.
.
.

■ Number of possibilities for the eighth toss = 2.

Total number of outcomes = 2 × 2 × 2 × 2 × 2 × 2 × 2 × 2 = 256.
We have now found in a relatively painless way that the size of our
sample space is $N = 256$.   ■

The basic principle that we used in Example 10 is called (for obvious
reasons) the **multiplication principle**. Informally stated it says that when
something takes place in several stages, to find the total number of ways
it can occur we find the number of ways each individual stage can occur
and then multiply them.

We can justify *why* the multiplication principle works with a simple
example.

Example 11.   You want a single scoop of ice cream. There are two
types of cones available (sugar and regular) and three flavors to choose
from (vanilla, chocolate chip, and chocolate). Figure 15-1 shows all the
possible combinations.

Cones   Flavors   Combinations

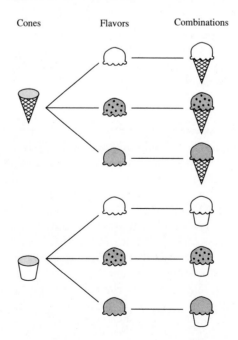

**Figure 15-1 The
multiplication principle at
work.**

■

Of course, the more complicated the scenario, the more helpful the multiplication principle is.

Example 12. Dolores goes on a business trip. She packs three different pairs of shoes, four different skirts, six different blouses, and two different jackets. If all the items are color-coordinated, how many different outfits are possible?

To answer this question we must first define what we mean by an outfit. Let's assume that an "outfit" consists of a pair of shoes, a skirt, a blouse, and a jacket. We can now use the multiplication principle to calculate the total number of possible outfits. It is $3 \times 4 \times 6 \times 2 = 144$. Color coordination obviously pays: Dolores can go for over 4 months and never have to wear the same outfit twice!

A few subtleties here and there can make the use of the multiplication principle a little more of a challenge.

Example 13. Once again, Dolores goes on a business trip. This time she packs three pairs of shoes, four skirts, three pairs of slacks, six blouses, three turtlenecks, and two jackets. As usual, everything goes with everything else, and we want to know how many different outfits are possible. This time let's define an "outfit" as consisting of a pair of shoes, some "lower wear" (either a skirt or a pair of slacks), some "upper wear" (either a blouse or a turtleneck or both), and she may or may not choose to wear a jacket (it's springtime). Let's count the possibilities for each stage that makes up the outfit separately:

■ Number of possibilities for the shoes = 3.

■ Number of possibilities for lower wear = $\underbrace{4}_{\text{Skirt}}$ + $\underbrace{3}_{\text{Slacks}}$ = 7.

■ Number of possibilities for upper wear

$$= \underbrace{6}_{\substack{\text{Blouse} \\ \text{alone}}} + \underbrace{3}_{\substack{\text{Turtleneck} \\ \text{alone}}} + \underbrace{18}_{\substack{\text{Blouse/turtleneck} \\ \text{combination (per} \\ \text{multiplication principle)}}} = 27.$$

■ Number of possibilities for jacket = $\underbrace{2}_{\text{Jacket}}$ + $\underbrace{1}_{\text{No jacket}}$ = 3.

According to the multiplication principle, the total number of different outfits is $3 \times 7 \times 27 \times 3 = 1701$. ■

Example 14.   In Example 8 we discussed a sample space S_M corresponding to all possible ways in which five horses (A, B, C, D, and E) could finish 1-2-3 in a horse race (remember Mabel's trifecta bet?). In Example 8 we started writing out the sample space:

$$S_M = \{ABC, ACB, BAC, BCA, CAB, CBA, ABD, ADB, \ldots\}$$

but we gave up because we realized it was too big. Using the multiplication principle, we can now determine exactly how big S_M is. We have

■ Number of possibilities for the 1st place horse = 5.

■ Number of possibilities for the 2nd place horse = 4 (Any other horse except the one in first place).

■ Number of possibilities for the 3rd place horse = 3 (Any horse except the ones in first and second place).

The total number of possible outcomes = 5 × 4 × 3 = 60.   ■

The multiplication principle is a simple but powerful tool that allows us to make probability calculations (coming up shortly) in a direct, efficient way. At the same time, as Examples 13 and 14 (and many of the exercises at the end of the chapter) illustrate, proper use of the multiplication principle often requires some clear thinking.

RANDOM VARIABLES

In Chapter 14 we discussed the concept of a variable associated with a given population. Variables can also be associated with a random experiment in a natural way since, after all, in a random experiment there must be, by definition, variability in the outcomes. When the value of a *quantitative* variable is determined by a random experiment, we call such a variable a **random variable**. To put it in a slightly different way, if the population we are studying is a sample space (why not?), then any quantitative variable associated with that population is called a random variable.

Let's illustrate the idea of a random variable with a few simple examples. We will follow traditional usage and describe random variables by capital letters such as X, Y, etc.

Example 15.   Random experiment: Roll a pair of dice.
Sample space: S = { [dice], [dice], [dice], [dice], [dice], [dice],

[dice], [dice], [dice], [dice], [dice], [dice],

[dice], [dice], [dice], [dice], [dice], [dice],

[dice], [dice], [dice], [dice], [dice], [dice],

[dice], [dice], [dice], [dice], [dice], [dice],

[dice], [dice], [dice], [dice], [dice], [dice] }.

Random variable: X = sum of the faces showing.
Possible values of X: 2, 3, . . ., 12.   ∎

Example 16.   Random experiment: Toss a coin eight times.
Sample space: too big to write down. N = 256 (see Example 10).
Random variable: X = number of heads tossed.
Possible values of X: 0, 1, 2, . . ., 8.   ∎

Example 17.   Random experiment: Toss a coin eight times.
Random variable: W = total winnings if you win \$1 for every head tossed and lose \$1 for every tail tossed.
Possible values of W: -8, -6, -4, -2, 0, 2, 4, 6, 8. (See Exercise 31.)   ∎

While the use of random variables appears at first glance to be a somewhat formal way to look at things that are in general quite natural, the idea of a random variable has a tremendous advantage: It allows all the tools of descriptive statistics (graphs, frequency distributions, numerical summaries, etc.) to be used in the study of random experiments.

PROBABILITIES (THE NUMBERS)

So far, we have talked about random experiments, sample spaces, the multiplication principle, and random variables, but have not said a word about probabilities. It is now time to do so.

Suppose that we toss a coin in the air. What is the probability that it will land heads? This is not a deep mathematical question, and almost everybody has an opinion on the matter. Typical answers are 50%, 1 out of 2, about one-half, etc. When asked the reason for such answers, the responses given usually fall into two categories.

One type of argument is that since the coin toss can result in two possible outcomes and heads is one of the two possibilities, the probability of heads must be 1 out of 2. We will call this type of argument the **objective approach** to defining probabilities.

The other type of argument essentially goes like this: If we toss the coin over and over again, many, many times, in the long run about half of the tosses will turn out to be heads and about half will turn out to be tails, so the probability of heads is about one-half. We call this type of argument the **frequency approach** to defining probabilities.

While both of these are legitimate arguments (assuming the coin is an honest coin), there are difficulties generalizing either of them to cover all situations that involve probabilities.

Suppose that a person who has never played basketball is asked to shoot a free throw. What is the probability that he or she will succeed? It is clear that the objective approach cannot be used in this case. Once again there are only two outcomes (success and failure), but quite obviously the probability of success cannot be forced to be 1 out of 2. We cannot split the probability pie into equal shares because success and failure are not equally likely outcomes! In this situation, any probability assigned to a successful free throw would be nothing more than the subjective opinion of an observer expressed in numerical form.

The frequency approach to defining probabilities (the one that uses expressions like "if we do it many, many times" or "in the long run") has a different kind of drawback: There are many random experiments that are not repeatable. When the weatherperson reports a "90% chance of rain tomorrow" (i.e., the probability of rain tomorrow is .9), such a statement cannot be based on a frequency-type argument (if we had many, many, tomorrows like tomorrow, in the long run it would rain on about 90% of them?). Once again, the statement "90% chance of rain tomorrow" is a subjective expression of someone's (the weatherperson's) opinion.

The debate over what is the correct way to define probabilities has been around for many years. There are *objectivists*, those who believe that probability is an intrinsic characteristic of the random experiment, and there are *subjectivists*, those who believe that probabilities are basically opinions expressed as numbers and are therefore a characteristic of the observer and not of the random experiment itself.

The mathematical way out of this dilemma is what we will discuss next. We will present only the basic ideas behind the formal mathematical treatment of probabilities because the subject is deep and covers a great deal of ground. The basic ideas originated in the early 1930s and are attributed to the Russian mathematician A. N. Kolmogorov.

■ Probability Assignments

Let's go back to the free throw example discussed earlier.

Example 18. An individual shoots a free throw. We know nothing about his or her abilities (for all we know, the person could be Michael Jordan or it could be Joe Schmoe). What is the probability that he or she will make the free throw?

We know (we feel it in our bones) that the answer need not be $\frac{1}{2}$. In fact, we know that the probability could be just about any number. Any number? Well, let's back off a little bit. It could not be a negative number (that's obvious), and it could not be a number bigger than 1 (100% chance), so we can cut our original boast down to "The probability could be any number between 0 and 1." How about 0 and 1? A probability of 0 would mean no chance of success at all; a probability of 1 would mean guaranteed success. While both of these situations are unlikely, they are nevertheless technically possible and we will allow them.

We can summarize what we have so far as follows: The probability of a successful free throw is a number between 0 and 1 inclusive. We don't know the exact value because we don't know anything about the person shooting the free throw but that's no problem—we just make it a variable (say p).

One final comment about this example. The probability of an unsuccessful free throw is also a number between 0 and 1 inclusive. In fact, the two probabilities (successful and unsuccessful) are related by the fact that they must add up to 1 (the free throw is either successful or unsuccessful—there are no other alternatives!).

Table 15-1 is a summary of all the preceding comments.

Outcomes	Probabilities
Success (s)	Probability of success $= p$
Failure (f)	Probability of failure $= 1 - p$

Table 15-1   Random Experiment: Shooting a Free Throw

Table 15-2 is the same thing in mathematical shorthand (the new notation is self-explanatory).

Outcomes		Probabilities
Sample space	$\begin{cases} s \\ f \end{cases}$	$\mathrm{Pr}(s) = p$ $\mathrm{Pr}(f) = 1 - p$

Table 15-2

■

It is worth mentioning that our new description is incredibly flexible. It works just as well when the free throw shooter is Michael Jordan (make $p = .9$) or when it is Joe Schmoe (make $p = .15$) or when it is any other Tom, Dick, or Harry in between. Each one of the choices results in a different **probability assignment** for the sample space.

Michael Jordan and Joe Schmoe shoot a freethrow. Same sample space different probabilities. (Jordan from Barry Gossage/National Basketball Association)

Example 19.   Five players, Boris, Martina, Andre, Gabriela, and Monica, enter a tennis tournament. We are interested in who is going to be the winner of the tournament. The sample space is $S = \{$Boris, Martina, Andre, Gabriela, Monica$\}$.

According to one expert, the probability assignment for this sample space is Pr(Boris) $= .25$, Pr(Martina) $= .22$, Pr(Andre) $= .14$, Pr(Gabriela) $= .18$. The value of Pr(Monica) is not given, but we can determine that Pr(Monica) $= .21$ because the total sum of the probabilities must be 1.

A second tennis expert completely disagrees with this assignment, claiming that Martina, Gabriela, and Monica all have an equal chance of winning, Andre has a 10% chance of winning, and Boris has a 27% chance of winning. These opinions translate into a second probability assignment for the sample space. What is it? Since Boris and Andre together account for a 37% chance of winning, Martina, Gabriela, and Monica must account for the remaining 63%. Since they all have an equal chance of winning, the chance for each of them is 21%. The precise formulation of the second expert's opinion is the following probability assignment for the sample space: Pr(Andre) $= .1$, Pr(Boris) $= .27$, Pr(Martina) $= .21$, Pr(Gabriela) $= .21$, and Pr(Monica) $= .21$. ∎

Examples 18 and 19 illustrate the meaning of a probability assignment for a sample space, which we now formalize. A **probability assignment** for a sample space S is a set of numbers that are assigned to the outcomes in the sample space and which satisfy the following two conditions:

1. Each number is between 0 and 1 (inclusive).

2. The numbers add up to 1.

Any set of numbers that satisfies conditions 1 and 2 is a legal probability assignment.

EVENTS

So far we have talked about the probabilities of the individual outcomes in a sample space—these are given by a probability assignment for the sample space. We will now take things one step further and talk about combinations of outcomes and their probabilities.

Example 20.   Suppose we want to determine the probability that a female player will win the tennis tournament in Example 19. Intuitively, it is clear that all we have to do is to look up the probability assigned to each female player in the sample space and add up these numbers. A slightly more mathematical description is obtained by first defining the appropriate set

Female = {Martina, Gabriela, Monica}

and then computing its probability

Pr(Female) = Pr(Martina) + Pr(Gabriela) + Pr(Monica).

Of course the exact answer depends on the probability assignment. According to the first expert, Pr(Female) = 0.22 + 0.18 + 0.21 = 0.61, but according to the second expert, Pr(Female) = 0.21 + 0.21 + 0.21 = 0.63. ∎

Most probability questions are concerned with combinations of outcomes called events. We will define an **event** as any set of outcomes taken from the sample space.

Example 21.   In Example 9 we considered the random experiment of tossing a coin three times and saw that the sample space was $S = \{HHH,$

HHT, HTH, HTT, THH, THT, TTH, TTT}. There are many possible events for this sample space. Table 15-3 shows just a few of them.

	Event	Set of Outcomes	Size of Event
1	Toss 2 or more heads	{*HHT, HTH, THH, HHH*}	4
2	Toss more than 2 heads	{*HHH*}	1
3	Toss 2 heads or less	{*TTT, TTH, THT, HTT, THH, HTH, HHT*}	7
4	Toss no tails	{*HHH*}	1
5	Toss exactly one tail	{*HHT, HTH, THH*}	3
6	Toss exactly one head	{*HTT, THT, TTH*}	3
7	First toss is heads	{*HHH, HHT, HTH, HTT*}	4
8	Toss same number of heads as tails	{ }	0
9	Toss at most three heads	*S*	8
10	First toss is heads and at least two tails are tossed	{*HTT*}	1

Table 15-3   Some of the Many Possible Events in a Sample Space

As Example 21 illustrates, there are many ways in which outcomes in a sample space can be combined to make an event, and the same event can be described (in English) in more than one way (e.g., events 2 and 4 in Table 15-3). The actual number of outcomes in an event can be as low as 0 and as high as *N* (the size of the sample space). In the case in which the number of outcomes is 0 [as in Table 15-3 (8)] the event is called the **impossible event**; in the case in which the number of outcomes is *N* [as in Table 15-3 (9)] the event is the whole sample space *S* and it is called the **certain event**.

Once a probability assignment is made for the sample space, every event is automatically assigned a probability. It is obtained by adding the probabilities of the individual outcomes that make up that event. No matter what the specific probability assignment is, the probability of the impossible event is always 0 [Pr({ }) = 0] and the probability of the certain event is always 1 [Pr(*S*) = 1].

We are now ready to describe what we will call a **formal probability model**. This is where the various concepts we have discussed so far finally come together.

FORMAL PROBABILITY MODELS

A probability model starts with a sample space, which represents the set of all possible outcomes of an observed random experiment. Next comes a probability assignment for the sample space. It bestows on each individual outcome a number between 0 and 1 inclusive, which is the probability of that outcome. In principle the mathematical model is not concerned with where these particular probabilities come from. They could come from subjective opinions of an observer, from the results of long-term frequency calculations, or from applying some complicated mathematical formula. In the eyes of the formal probability model, they are all perfectly legal as long as the rules of the game (each of the numbers is between 0 and 1 and they add up to 1) are obeyed. Once a probability assignment for the sample space is made, not only individual outcomes but also combinations of outcomes called *events* have probabilities assigned to them. The probability of any event is obtained by adding the probabilities of the individual outcomes that make up that event. In particular, the impossible event { } has probability 0 and the certain event S has probability 1. All the above facts are summarized as follows:

Formal Probability Model

■ **Sample space.** Set of all possible outcomes of a random experiment $S = \{o_1, o_2, \ldots, o_N\}$.

■ **Probability assignment for S.** To each individual outcome o_i we assign a number $Pr(o_i)$. The rules are

1. $0 \le Pr(o_i) \le 1$, and
2. $Pr(o_1) + Pr(o_2) + \cdots + Pr(o_N) = 1$.

■ **Events.** Any subset of S is an event. In particular, { } (impossible event) and S itself (certain event) are events.

■ **Probability of events.** The probability of an event is obtained by adding the probabilities of the individual outcomes that make up that event. In particular, $Pr(\{ \}) = 0$ and $Pr(S) = 1$.

■ **When Every Outcome Is Equally Likely**

The best throw of the dice is to throw them away.

ENGLISH PROVERB

One of the most common uses of probability involves situations associated with gambling. While we do not condone gambling, it does provide a rich source of examples and gives rise to many interesting mathematical questions. Many gambling situations involve the use of a physical device, a coin or coins, a die or dice, a deck of cards, a roulette wheel, etc. In these situations it is a given that the device is honest, that is, the coin or the dice are fair, the cards haven't been marked, and the roulette wheel is perfectly balanced. Mathematically speaking, this assumption translates into the fact that each individual outcome in the sample space is equally likely to occur; that is, all the outcomes have equal probabilities. Knowing that the probability of each outcome is the same and that the probabilities add up to 1 (one of the rules in any probability model) tells us (1) that the probability assignment for each outcome must be $1/N$ (where N is the size of the sample space), and (2) the probability of an event is obtained by dividing the number of outcomes in the event by N. We can describe this formal probability model mathematically as follows:

Probability Model When All Outcomes Are Equally Likely

■   Size of sample space $= N$.

■ $\Pr(\text{any outcome}) = 1/N$.

■   If E is an event, $\Pr(E) = \dfrac{\text{number of outcomes in } E}{N}$.

In this model, probabilities can be computed exactly by mathematical calculations that more often than not require use of the multiplication principle.

Example 22.   A card is drawn from an honest deck of 52 cards. What is the probability of drawing an ace?

Here $N = 52$. The event

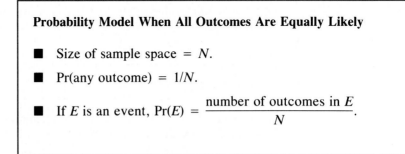

Ace = { ... }

is made up of four outcomes, each of which has probability $\frac{1}{52}$. It follows that

$$\Pr(\text{Ace}) = \frac{4}{52} = \frac{1}{13} \approx 0.077. \qquad ■$$

Example 23.   Two cards are drawn in order from an honest deck of 52 cards. What is the probability of drawing a pair of aces?

We first need to calculate the size of the sample space. Given that the cards are taken in order (first one, and then the other one), we can argue as follows:

■ The number of ways in which the first card can be drawn = 52.

■ The number of ways in which the second card can be drawn = 51.

■ N = total number of ways of drawing the two cards = 52 × 51.

Next, we calculate the size of the event E: drawing two aces.

■ The number of possibilities for the first ace = 4.

■ The number of possibilities for the second ace = 3.

■ The size of the event E = 4 × 3.

$$\Pr(E) = \frac{4 \times 3}{52 \times 51} = \frac{1}{13 \times 17} = \frac{1}{221} \approx .0045.$$ ■

Example 24. Suppose that we roll a pair of honest dice. What is the probability of rolling a total of 11? What is the probability of rolling a total of 7? What is the probability of rolling a 7 or an 11?

There are 36 ways to roll a pair of dice (Example 3), and since the dice are honest, each of these outcomes has probability $\frac{1}{36}$. There are two ways of rolling a total of 11 (\{ ▦ ▦ \}).

It follows that

$$\Pr(R11) = \frac{2}{36} = \frac{1}{18} \approx .056.$$

(We will use R11 to denote the event "rolling a total of 11," R7 to denote the event "rolling a total of 7," etc.)

The event R7 has six possible outcomes:

R7 = \{ ▦ ▦ ▦ ▦ ▦ ▦ \}

so

$$\Pr(R7) = \frac{6}{36} = \frac{1}{6} \approx .167.$$

The event R7 or R11 has eight possible outcomes (the six in R7 and the two in R11), so

$$\Pr(R7 \text{ or } R11) = \frac{8}{36} = \frac{2}{9} \approx .22.$$ ■

Example 25. Suppose that we roll a pair of honest dice. What is the probability that we will roll at least one ⊡?

We already know (Example 24) that each individual outcome has probability $\frac{1}{36}$. We will show three different ways to solve this problem.

- **Solution 1 (The Brute Force Approach).** If we just write down the event E, which is "we will roll at least one ⊡," we have

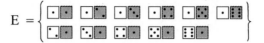

It follows that $\Pr(E) = \frac{11}{36}$.

- **Solution 2 (The Roundabout Approach).** Let's say for the sake of argument that we will win if at least one of the two dice comes up a ⊡ and we will lose otherwise. This means that we will lose if both dice come up with a number other than ⊡. Let's calculate the probability that we will lose first (this is called the roundabout way of doing things). Using the multiplication principle, we can calculate the number of individual outcomes in the event "we lose":

- Number of ways first die can come up (not a ⊡) = 5.

- Number of ways the second die can come up (not a ⊡) = 5.

- Total number of ways both dice can come up (neither a ⊡) = $5 \times 5 = 25$.
  Probability that we will lose: $\Pr(\text{lose}) = \frac{25}{36}$.
  Probability that we will win: $\Pr(\text{win}) = 1 - \frac{25}{36} = \frac{11}{36}$.

- **Solution 3 (Independent Events).** In this solution we consider each die separately. In fact, we will find it slightly more convenient to think in this case of rolling a single honest die twice (mathematically it is exactly the same thing as rolling a pair of honest dice once).

  Let's start with the first roll. The probability that we won't roll a ⊡ is $\frac{5}{6}$ (there are six possible outcomes, five of which are not a ⊡). For the same reason, the probability that the second roll will not be a ⊡ is also $\frac{5}{6}$.

  Now comes a critical observation: The probability that neither of the first two rolls will be a ⊡ is $\frac{5}{6} \times \frac{5}{6} = \frac{25}{36}$. We were able to multiply the probabilities of the two events ("first roll is not a ⊡" and "second roll is not a ⊡") because these two events are **independent**: The outcome of the first roll does not in any way affect the outcome of the second roll.

We finish the problem exactly as in solution 2: $\Pr(\text{lose}) = \frac{25}{36}$ and therefore $\Pr(\text{win}) = 1 - \frac{25}{36} = \frac{11}{36}$.

■

Of the three solutions to Example 25, solution 3 appears on the surface to be the most complicated, but in fact it is the one that shows us the most useful approach. It is based on what we will call the probability multiplication principle.

The Probability Multiplication Principle. When a complex event (the whole) can be broken down into a combination of simpler events (the parts) that are independent of each other (i.e., the outcome of any one part does not affect the outcome of any of the others), then we can calculate the probability of the whole by calculating the probabilities for each of the parts and multiplying.

The probability multiplication principle is an important and useful rule, but it works only when the parts are independent. The next two examples illustrate the usefulness of the probability multiplication principle.

Example 26. Suppose that we roll an honest die four times. What is the probability that at least once we will roll a ⊡?

Let's try to use the same approach we used in Example 25, solution 3. (If we try a brute force approach similar to the one we used in Example 25, solution 1, we will soon realize why we ought to be thankful for our new-found wisdom.) Let's say once again that we will win if we roll a ⊡ at least once, and we will lose if none of the four rolls come up ⊡. We know that

- ■ $\Pr(\text{first roll not a ⊡}) = \frac{5}{6}$.

- ■ $\Pr(\text{second roll not a ⊡}) = \frac{5}{6}$.

- ■ $\Pr(\text{third roll not a ⊡}) = \frac{5}{6}$.

- ■ $\Pr(\text{fourth roll not a ⊡}) = \frac{5}{6}$.

Because each roll is independent of the preceding ones, we can use the probability multiplication principle.

$$\Pr(\text{lose}) = \Pr(\text{not rolling any ⊡'s in four rolls}) = \left(\frac{5}{6}\right)^4 \approx .482.$$

It follows that

Pr(win) = Pr(rolling at least one ⊡ in four rolls) ≈ .518.

The practical consequence of these calculations is that if we played the above game regularly and always chose to bet that we can roll a ⊡ at least once in four rolls, we would win about 51.8% of the time. Assuming even odds (bet $1 to win $1), this is a good game to play. ■

Example 27.   Suppose that we play a game similar to the one in the preceding example but instead of rolling a single die we roll a pair of honest dice 24 times. If we roll "snake-eyes" (⊡⊡), we will win, otherwise we will lose. Should we play this game? What is the probability of winning? Once again, it would be extremely hard to solve this problem using any method other than the probability multiplication principle.

First note that we will lose if we don't roll snake-eyes on any of the 24 rolls. The probability of not rolling snake-eyes on the first roll is $\frac{35}{36}$ (see Exercise 16). It is exactly the same for the 2nd, 3rd, . . ., 24th rolls. By the probability multiplication principle (each roll is independent of the others),

$$Pr(\text{lose}) = Pr(\text{not rolling snake-eyes on any of the 24 rolls})$$
$$= \left(\frac{35}{36}\right)^{24} \approx .509.$$

It follows that

Pr(win) = 1 − Pr(lose) ≈ 1 − .509 = .491.

If we play this game regularly, we will win 49.1% of the time. Assuming even odds, we may want to think twice about playing this game on a regular basis.[3] ■

---

[3] There is an interesting historical footnote to Examples 26 and 27. In a sense the beginnings of the mathematical study of probability can be traced back to these problems. Both of the games described in Examples 26 and 27 were popular among the seventeenth-century French nobility. One of the high rollers in that crowd was a certain Antoine Gombauld, Chevalier de Méré, who happened to be acquainted with the philosopher-mathematician Blaise Pascal. De Méré asked Pascal to do a mathematical analysis of these games for him. In solving these problems a theoretical framework for the study of games of chance had to be developed. Pascal, together with two other mathematicians of the time (Pierre Fermat and Abraham de Moivre), began looking at gambling questions as mathematical problems, and it was thus that the mathematical theory of probability was born.

ODDS AND PROBABILITIES

Dealing with probabilities as numbers that are always between 0 and 1 is the mathematician's way of having a consistent terminology. To the everyday user consistency is not that much of a concern, and we know that people talk about *chances* (probabilities expressed as percentages) and *odds*, which are most frequently used to describe probabilities associated with gambling situations. Consider the statement "The odds that the Cincinnati Reds will win the World Series are 2 to 5" (or equivalently, "the odds that the Cincinnati Reds will lose the World Series are 5 to 2"). Essentially this means that if we bet $2 on the Reds, and the Reds win, we will win $5. We will also get our $2 back, so our total return will be $7. The probability of winning is the ratio between the risk ($2) and the total return if we win ($7), so in this example we can say that Pr(Reds Win) = $\frac{2}{7}$.

The general rule for converting odds to probabilities is

If the *odds in favor* of an event E are m to n, then

$Pr(E) = \frac{m}{m+n}$.

The general rule for converting probabilities into odds is

If $Pr(E) = \frac{a}{b}$ then the *odds in favor* of E are a to $b - a$ and the *odds against* E are $b - a$ to a.

Example 28.   Consider the tennis tournament first discussed in Example 19. Recall that according to the first expert the probability assignment for winning the tournament was Pr(Boris) = .25, Pr(Martina) = .22, Pr(Andre) = .14, Pr(Gabriela) = .18, and Pr(Monica) = .21. Let's convert all these probabilities into odds.

From Pr(Boris) = .25 = $\frac{1}{4}$ we find that the odds in favor of Boris winning are 1 (numerator) to 3 (denominator minus numerator). The odds against Boris winning are 3 to 1.

Likewise, Pr(Martina) = .22 = $\frac{22}{100}$ = $\frac{11}{50}$, so the odds in favor of Martina winning are 11 to 39. Note that we converted the decimal .22 to a reduced fraction before we computed the odds.

Pr(Andre) = .14 = $\frac{14}{100}$ = $\frac{7}{50}$. The odds in favor of Andre winning are 7 to 43.

Pr(Gabriela) = .18 = $\frac{18}{100}$ = $\frac{9}{50}$. The odds in favor of Gabriela winning are 9 to 41.

Pr(Monica) = .21 = $\frac{21}{100}$. The odds in favor of Monica winning are 21 to 79.

As an exercise, the reader is encouraged to calculate the odds for each player based on the probability assignment given by the second expert (refer to Example 19).

CONCLUSION

While the average citizen thinks of probabilities, chances, and odds as vague, informal concepts that are useful primarily when discussing the weather or playing the lottery, scientists, mathematicians, and statisticians think of probability as a formal framework within which the laws that govern chance events can be understood. The basic elements of this framework are a *sample space* (which represents a precise mathematical description of all the possible outcomes of a *random experiment*) and a *probability assignment* (which associates a numerical value with each of these outcomes). The numerical values represent a measure of the likelihood that a particular outcome will occur. How these numerical values come about has both mathematical and philosophical implications, but once a probability assignment is made (be it by hook or by crook), the rules of the game are strictly mathematical.

Of the many ways in which probabilities can be assigned to outcomes, a particularly important case is the one in which all outcomes have the same probability. When this happens, the critical steps in calculating probabilities revolve around two basic (but not necessarily easy) questions: (1) given a sample space, what is its size? And (2) given an event, what is its size?

To say that in this chapter we have only scratched the surface of what the mathematical theory of probability is about is an understatement. In spite of being a relatively modern branch of mathematics, probability theory is a deep body of knowledge with important applications in almost every walk of life. When one stops to think how much of life is ruled by fate and chance, the pervasive role of probability is not entirely surprising. As the famous Roman statesman Cicero once said, ''Probability is the very guide of life.''

KEY CONCEPTS

certain event	probability model
event	probability multiplication
impossible event	principle
independent events	random experiment
multiplication principle	random variable
odds	sample space
probability assignment	

EXERCISES

■ Walking

1. Consider the random experiment of tossing a coin four times.
   (a) Write out the sample space for this random experiment.
   (b) What is the size of this sample space?
   (c) Write out the event that exactly two of the coin tosses come out heads.
   (d) Assuming that the coin is honest, what is the probability of the event described in part (c)?

2. Consider the random experiment of drawing 1 card out of an ordinary deck of 52 cards.
   (a) Describe the sample space for this random experiment.
   (b) What is the size of this sample space?
   (c) Describe the event that the card drawn is a spade.
   (d) Assuming the deck is honest and the cards are shuffled, what is the probability of the event described in part (c)?

3. There are seven players (call them $P_1, P_2, \ldots, P_7$) entered in a tennis tournament. According to one expert, P_1 is twice as likely to win as any of the other players, and $P_2, P_3, \ldots, P_7$ all have an equal chance of winning.
   (a) Write down the sample space and find the probability assignment for the sample space based on this expert's opinion.
   (b) What are the odds that P_1 will win the tournament? How about P_2?

4. There are eight players (call them $P_1, P_2, \ldots, P_8$) entered in a chess tournament. According to an expert, P_1 has a 25% chance of winning the tournament, P_2 has a 15% chance of winning, P_3 has a 5% chance of winning, and all the other players have an equal chance of winning.
   (a) Write down the sample space and find the probability assignment for the sample space based on this expert's opinion.
   (b) What are the odds that P_1 will win the tournament? How about P_2?

5. (a) How many license plates can be made using three letters followed by three digits (0 through 9)?
   (b) How many license plates can be made using three letters followed by three digits if the first digit cannot be a zero?
   (c) How many license plates can be made using three letters followed by three digits if no license plate can have a repeated letter or digit?

6. (a) How many four-letter code words are there? (A code word is any string of letters—it doesn't have to mean anything.)
   (b) How many four-letter code words are there that start with the letter A?
   (c) How many four-letter code words are there that have no repeated letters?

7. A set of reference books consists of eight volumes numbered 1 to 8.
   (a) In how many ways can the eight books be arranged together on a shelf?
   (b) In how many ways can the eight books be arranged on a shelf so that at least one book is out of order?

8. Four men and four women line up at a checkout stand in a grocery store.
   (a) In how many ways can they line up?
   (b) In how many ways can they line up if the first person in line must be a woman?
   (c) In how many ways can they line up if they must alternate woman, man, woman, man, etc.? (A woman is first in line.)

9. A child's spinner has five outcomes: red, blue, yellow, purple, and orange. Experience shows that each primary color (red, blue, or yellow) comes up about 100 times in every 1000 spins and that each nonprimary color has an equal probability of occurring.
   (a) What is the sample space for the random experiment consisting of a single spin?
   (b) Write down the event that the outcome is a primary color.
   (c) Write down the event that the outcome is not a primary color.
   (d) Give a reasonable probability assignment for the sample space given in part (a).

10. A child's spinner has five outcomes: 1, 2, 3, 4, and 5.
   (a) What is the sample space for the random experiment consisting of a single spin?
   (b) Write down the event that the outcome of a spin is an odd number.
   (c) Write down the event that the outcome of a spin is an even number.
   (d) If the probability of the event described in part (b) is .3, find the probability of the event described in part (c).
   (e) If the probability of the event described in part (b) is .3 and $Pr(1) = Pr(3) = Pr(5)$ and $Pr(2) = Pr(4)$, find the probability assignment for the sample space.

11. Consider the sample space $S = \{A, B, C\}$. Make a list of all the possible events for this sample space. (Remember that an event is any subset of S including $\{\ \}$ and S itself.)

12. Consider the sample space $S = \{A, B, C, D\}$. Make a list of all the possible events for this sample space. (Remember that an event is any subset of S including $\{\ \}$ and S itself.)

13. Two friends (Thelma and Mabel) go to the racetrack. There are five horses entered in the first race. Let's call them A, B, C, D, and E. Thelma buys a "to show" ticket (this means that she picks a horse and if her horse comes in either first, second, or third she will win). Mabel buys an exacta ticket (this means that she picks two horses to come in 1-2 and if her two horses come in 1-2 in that order she will win).
   (a) Write down the sample space for Thelma.
   (b) Write down the sample space for Mabel.

14. Two friends (Thelma and Mabel) go to the racetrack. There are eight horses entered in the second race. Let's call them A, B, C, D, E, F, G, and H. Thelma buys a "to place" ticket (this means that she picks a horse and if her horse comes in either first or second she will win). Mabel buys a trifecta ticket (she must pick the first-, second-, and third-place finishers and in exactly that order to win her bet).

(a) Write down the sample space for Thelma.

(b) Without writing it all down, determine the size of the sample space for Mabel.

Problems 15 through 20 concern the random experiment of rolling a pair of dice. The random variable X is the sum of the faces showing. Assume the dice are honest.

15. (a) What is the probability that $X = 10$?
    (b) What are the odds in favor of rolling a 10?
    (c) What are the odds against rolling a 10?
    (d) What is the probability of *not* rolling a 10?

16. (a) What is the probability of rolling snake-eyes (i.e., both dice coming up 1)?
    (b) What are the odds in favor of rolling snake-eyes?
    (c) What is the probability of *not* rolling snake-eyes?
    (d) What are the odds against rolling snake-eyes?

17. (a) Describe the event $X \le 5$.
    (b) What is the probability of the event described in part (a)?
    (c) What are the odds in favor of the event described in part (a)?

18. (a) What is the probability that at least one of the two dice comes up 1?
    (b) What are the odds in favor of at least one of the two dice coming up 1?

19. Suppose the pair of dice is rolled twice and that a roll of 7 is considered a success and a roll of anything other than 7 is considered a failure.
    (a) Find the probability that both rolls will result in a success.
    (b) Give a probability assignment for the sample space $S = \{ss, sf, fs, ff\}$.

20. Suppose the pair of dice is rolled twice and that a roll of any number less than 7 is considered a success and a roll of any number larger than or equal to 7 is considered a failure.
    (a) Find the probability that both rolls will result in a failure.
    (b) Give a probability assignment for the sample space $S = \{ss, sf, fs, ff\}$.

■ **Jogging**

21. Consider the following game: We roll a pair of honest dice 25 times. If we roll "boxcars" (i.e., ⚅⚅) at least once, we will win; otherwise we will lose. What is the probability that we will win?

22. Consider the following game: We roll a pair of honest dice five times. If we roll a total of 7 at least once, we will win; otherwise we will lose. What is the probability that we will win?

23. A factory assembles car stereos. From random testing at the factory it is known that on the average 1 out of every 50 car stereos will be defective (which means that the probability that a car stereo randomly chosen from the assembly line will be defective is .02). After manufacture, car stereos are packaged in boxes of 12 for delivery to the stores.
    (a) What is the probability that in a box of 12 there are no defective car stereos? What assumptions are you making?

(b) What is the probability that in a box of 12 there is at most one defective car stereo?

24. In the game of craps, a pair of dice is rolled. A person betting "on the field" is betting that on the next roll the total rolled will be a 2, 3, 4, 9, 10, 11, or 12. What is the probability of winning a field bet?

25. There is a dice game played by some eccentric types called "subtract instead of add." In this game a pair of dice is rolled and bets are made on the difference between the numbers rolled (larger minus smaller). Let the random variable X represent this difference.
    (a) What are the possible values of X?
    (b) Assuming that both dice are honest, find the probability of the event $X = 1$.
    (c) Assuming that both dice are honest, find the probability of the event $X = 0$.

26. Consider the random experiment of drawing 2 cards from an ordinary deck of 52 cards.
    (a) What is the size of the sample space?
    (b) Assuming the deck is honest and the cards are shuffled, what is the probability that both cards will be jacks?
    (c) What is the probability that both cards will be hearts?
    (d) What is the probability that the first card will be a heart and the second card a jack?

27. A pizza parlor offers six toppings—pepperoni, Canadian bacon, sausage, mushroom, anchovies, and olives—that can be put on their basic cheese pizza. How many different pizzas can be made? (A pizza can have anywhere from no topping to all six toppings.)

28. **(a)** In how many different ways can 10 people form a line?
    (b) In how many different ways can 10 people hold hands and form a circle? [*Hint*: The answer to part (b) is much smaller than the answer to part (a). There are many different ways in which the same circle of 10 people can be broken up to form a line. How many?]

29. A club has 30 members. A committee of 3 members must be chosen.
    (a) In how many ways can this be done if the committee consists of a chairperson, a vice chairperson, and a secretary?
    (b) In how many different ways can this be done if the committee has no officers (i.e., all three members are equal)? [*Hint*: The difference between part (b) and part (a) is that in part (b) the order in which we choose the committee members does not matter. Alice, Bob, and Carla constitute the same committee as Carla, Alice, and Bob, etc. This is not the case in part (a). The answer to part (b) is smaller than the answer to part (a). In how many ways can the same committee in part (b) become different committees in part (a)?]

30. Two teams (call them X and Y) play in the World Series. The World Series is a best of seven series. This means that the two teams play against each other and the first team to win four games wins the series and the series is over. (Games cannot end in a tie.) We can describe an outcome for the World

Series by writing a string of letters that indicate (in order) the winner of each game. For example, the string *XYXXYX* represents the outcome: *X* wins game 1, *Y* wins game 2, *X* wins game 3, etc.

(a) Using the notation described above, write the sample space *S* for the World Series.

(b) Describe the event "*X* wins in five games."

(c) Describe the event "the series last seven games."

31. (a) Suppose that a coin is tossed eight times. Each time the coin comes up *H* you win $1.00, each time it comes up *T* you lose $1.00 (see Example 17). If *W* is the random variable that describes your total winnings, explain why the possible values of *W* are $-8, -6, -4, -2, 0, 2, 4, 6, 8$. (Explain why you can't, for example, end up with $W = 3$.)

(b) Suppose you win $1.50 for every *H* tossed and you lose $1.75 for every *T* tossed. What are the possible values of your winnings *W*?

■ **Running**

32. In the game of craps, the player's first roll of a pair of dice is very important. If the first roll is 7 or 11, the player wins. If the first roll is 2, 3, or 12, the player loses. If the first roll is any other number (4, 5, 6, 8, 9, 10), this number is called the players "point." The player then continues to roll until the point reappears, in which case the player wins, or a 7 shows up before the point, in which case the player loses. What is the probability that the player will win? (Assume the dice are honest.)

33. **The birthday problem.** There are 30 people in a room. What is the probability that at least two of these people have the same birthday, that is, have their birthdays on the same day and month?

34. In a sample space with three outcomes there are $2^3 = 8$ possible events (see Exercise 11), and in a sample space with four outcomes there are $2^4 = 16$ possible events (see Exercise 12). Show that in a sample space with *N* outcomes there are 2^N possible events.

35. **The Monty Hall problem.** You have just been chosen to be on a game show in which there are three doors, behind one of which is a new sports car. There is nothing behind the other two doors. You know in advance that the game show host will ask you to pick one of the doors, but before the door you picked is opened, the host will open one of the other two doors with nothing behind it. You will then be given the option of keeping the door you picked or switching to the other closed door. What should you do and why?

36. **The optimal choice problem.** A large box contains 100 tickets. Each ticket has a different number written on it. Other than the fact that the numbers are all different, we know nothing about them. The numbers can be positive or negative, integers or decimals, large or small—anything goes. Of all the tickets in the box there is, of course, one that has the biggest of all the numbers. That's the winning ticket. If we turn that ticket in, we will win a $1000 prize. If we turn any other ticket in, we will get nothing. The ground rules are that we can draw a ticket out of the box, look at it, and if we think it's the winning ticket we can turn it in. If we don't, we get to draw again, but before we do so we must tear the other ticket up—once we pass on a ticket we can't use

it again! We can continue drawing tickets this way until we find one we like or run out of tickets.

(a) What is the probability that the first ticket we draw will be the winning ticket?

(b) What's the probability that after we have drawn 50 tickets, the winning ticket will still be in the box?

(c) It is very surprising, but there is a strategy for playing this game that will give us a better than 25% chance of winning. Describe such a strategy.

37. Tickets for a certain lottery contain the integers from 1 to 50 as shown in the figure. Players pay $1.00 for a ticket and then mark 6 of the 50 numbers. At

Number AQJ47391									
THE MILLION DOLLAR LOTTERY									
Mark 6 Numbers									
1	2	3	4	5	6	7	8	9	10
11	12	13	14	15	16	17	18	19	20
21	22	23	24	25	26	27	28	29	30
31	32	33	34	35	36	37	38	39	40
41	42	43	44	45	46	47	48	49	50

the end of the week after all the tickets have been marked, 6 numbers are drawn at random without replacement from a large vat containing 50 balls numbered from 1 to 50. If a player's ticket matches all 6 numbers drawn, the player wins $1 million (otherwise nothing). (a) What is the probability of winning this lottery? (b) What are the odds against winning this lottery?

REFERENCES AND FURTHER READINGS

1. David, F. N., *Games, Gods and Gambling*. Buckinghamshire, England: Charles Griffin & Co., 1962.

2. di Finetti, B., *Theory of Probability*. New York: John Wiley & Sons, Inc., 1970.

3. Freedman, D., R. Pisani, R. Purves, and A. Adhikari, *Statistics* (2nd ed.). New York: W. W. Norton, Inc., 1991, chaps. 13 and 14.

4. Gnedenko, B. V. and A. Y. Khinchin, *An Elementary Introduction to the Theory of Probability*. New York: Dover Publications, Inc., 1962.

5. Levinson, Horace C., *The Science of Chance*. New York: Rinehart & Co., 1950.

6. Maistrov, L. E., *Probability Theory: A Historical Perspective*. New York: Academic Press, Inc., 1974.

7. McGervey, John D., *Probabilities in Everyday Life*. New York: Ivy Books, 1986.

8. Trefil, James, "Odds Are Against Your Breaking That Law of Averages," *Smithsonian* (September, 1984), 66–75.

9. Weaver, Warren, *Lady Luck: The Theory of Probability*. New York: Dover Publications, Inc., 1963.

16 Normal Distributions

Variety is the spice of life.

ANONYMOUS

Everything Is Back to Normal (Almost)

Variation is at the very heart of statistics. If there were no variation in data, there would be very little for statisticians to do. Fortunately, the common thing for a variable, be it a test score, a grade, a height, etc., is to take many different values throughout the population (that is why it is called a variable!). The pattern of variation of the variable over the entire population is called the variable's **distribution**.

A variable's distribution can follow a very irregular, unpredictable pattern, or it can follow a pattern in the true sense of the word, showing a behavior that is fairly regular and quite predictable. The latter happens surprisingly often, and when it does, we can visualize the distribution as fitting into a certain established "mold." In this case, the bar graphs or histograms that describe the distribution have a predictable *overall shape*—maybe a straight line, maybe a bell shape, maybe something else. A statistician thinks of this nice, regular mold into which his or her data more or less fit as a mathematical ideal for the way the data are expected to come out. This mathematical ideal is most commonly referred to as a *mathematical model* for the data.

Thinking of data in terms of mathematical models may seem to be a form of mathematical wishful thinking, but the usefulness of this idea cannot be overstated. The characteristics of a mathematical model are abstract, theoretical facts which can be applied to the specific distribution at hand where they translate into useful, concrete information about the population.

Of the many different mathematical models for data, by far the most pervasive, is the **normal** model. This is the model used to analyze distributions whose bar graphs or histograms follow a "bell-shaped" pattern, and it is the focus of our discussion for this chapter.

APPROXIMATELY NORMAL DISTRIBUTIONS

We start with a pair of examples.

Example 1. We are back to Professor Blackbeard's Stat 101 exam which we first discussed in Chapter 14 (see Table 14-1 on p. 420). Let's look at the distribution of test scores for the exam, but let's throw out the two outliers (Fig. 16-1). By definition they are exceptional values of the data, and it's OK to take this liberty.

We can see that the bar graph in Fig. 16-1 has a very distinctive shape—it is that of a bell. The bell-shaped curve superimposed on the bar graph is the mathematical idealization of the distribution of test scores. One shouldn't worry too much about the fact that the fit isn't perfect. As we will see in the next section, having such an idealization still gives us a powerful tool for describing and analyzing the data.

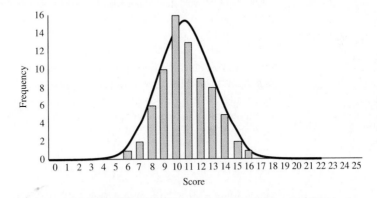

Figure 16-1 Distribution of Stat 101 test scores (without outliers).

To those readers who are experienced test takers, the fact that the test scores on the Stat 101 midterm follow a bell-shaped pattern is not all that surprising—test scores often do that. (At the same time, we should be wary of generalizations: It is wrong to assume that test scores *always* follow a bell-shaped pattern.) ∎

Example 2.   In this example we will consider the distribution of the heights (in inches) of the top 270 professional basketball players in the United States for the 1990–1991 NBA season. Our data values were obtained as follows: For each of the 27 teams in the National Basketball Association we selected the top 10 players (the starting 5 plus the first 5 substitutes off the bench). Table 16-1 on the next page shows the heights (to the nearest inch) of the top 10 players on each team.

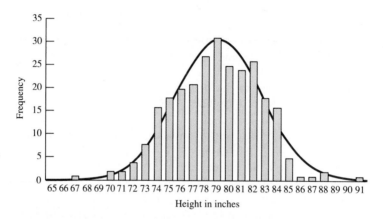

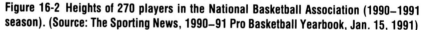

Figure 16-2 Heights of 270 players in the National Basketball Association (1990–1991 season). (Source: The Sporting News, 1990–91 Pro Basketball Yearbook, Jan. 15, 1991)

Figure 16-2 is a bar graph for the 270 heights shown in Table 16-1. Once again there is a rough bell-shaped pattern to the data. A mathematical idealization of the pattern is given by the curve superimposed on the bar graph. Once again the fit isn't perfect, but it is good enough to be useful in practice.   ■

Examples 1 and 2 illustrate two different situations (different populations and different variables) where the distributions can be described as fitting a bell-shaped mold. When this happens, we say that the distribution is **approximately normal**. A **normal distribution** is the idealized mathematical model we use to analyze any distribution that is approximately normal. A normal distribution is described by a bell-shaped curve called a **normal curve**.

ATLANTIC DIVISION

Team	Top 10 Players' Heights (inches)									
Boston	72	81	77	84	79	82	84	81	78	78
Miami	78	79	81	72	75	80	80	79	83	74
New Jersey	73	85	81	83	77	80	78	78	79	79
New York	73	84	75	82	81	77	80	80	82	78
Philadelphia	79	91	78	74	83	75	82	76	80	78
Washington	75	77	82	75	83	81	81	81	79	76

CENTRAL DIVISION

Team	Top 10 Players' Heights (inches)									
Atlanta	74	84	80	82	76	74	75	67	80	78
Charlotte	80	76	77	80	77	81	78	79	81	78
Chicago	74	85	82	74	77	78	83	74	84	79
Cleveland	79	79	84	79	82	75	75	82	72	83
Detroit	78	83	75	85	74	74	83	80	83	73
Indiana	77	79	78	80	78	82	88	82	78	79
Milwaukee	82	82	75	82	79	76	82	76	83	84

MIDWEST DIVISION

Team	Top 10 Players' Heights (inches)									
Dallas	74	78	75	86	79	76	75	75	80	80
Denver	70	78	76	79	73	78	80	76	84	81
Houston	75	78	79	74	76	84	80	82	76	77
Minnesota	87	71	79	78	78	79	82	73	84	78
Orlando	83	80	83	83	80	80	73	77	74	76
San Antonio	80	81	80	79	77	85	85	75	79	77
Utah	83	81	80	88	77	76	76	81	74	73

PACIFIC DIVISION

Team	Top 10 Players' Heights (inches)									
Golden State	72	79	84	77	79	82	75	77	77	79
LA Clippers	84	76	74	75	78	76	82	80	82	82
LA Lakers	83	83	74	81	81	81	76	82	81	77
Phoenix	78	82	76	79	73	83	78	81	80	82
Portland	77	81	82	79	84	79	77	75	80	76
Sacramento	80	81	84	78	77	83	79	74	81	84
Seattle	71	81	79	70	82	79	81	77	76	84

Table 16-1   The heights of the top 270 National Basketball Association players listed by team. (Source: *The Sporting News*, 1990–91 Pro Basketball Yearbook, Jan 15, 1991).

The long and the short of it in the NBA: Outliers Manute Bol (7′, 7″) and Spud Webb (5′, 7″) size each other up. (Manny Millan/*Sports Illustrated*, © Time, Inc.)

Because we are using the word "normal" in so many contexts here, let's summarize the three concepts we have just introduced.

■ **Approximately Normal Distribution.** This is a distribution of real-life data with a bar graph or histogram that more or less follows a bell-shaped pattern.

■ **Normal Distribution.** This is a mathematical model, an idealization of how the data would come out in a perfect world.

■ **Normal Curve.** A graphical representation of a normal distribution. It is always a bell-shaped curve.

NORMAL CURVES AND THEIR PROPERTIES

In this section, we will briefly discuss the main mathematical features of normal curves and then see how these facts can be used to obtain important information about any variable with an approximately normal distribution. All the variables that we will discuss from here on will be quantitative variables.

Normal curves (Fig. 16-3) can come in many different packages (some are skinny and tall, others are squat and short), but they all share every one of their important mathematical characteristics. In fact, whether a normal curve is skinny and tall or squat and short or somewhere in between depends on the way we scale the units on the axes. With the proper choice of scale on the axes, any two normal curves can be made to look like one another.

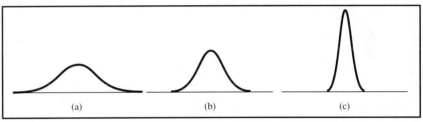

Figure 16-3 Three normal curves. (a) Short and squat. (b) In between. (c) Skinny and tall.

(a) (b) (c)

What follows is a summary of the most important facts about normal distributions and normal curves.

■ **Symmetry.** Every normal curve has exactly one reflection symmetry about a vertical axis (Fig. 16-4). This axis of symmetry splits the bell-shaped region outlined by the curve into two identical halves.

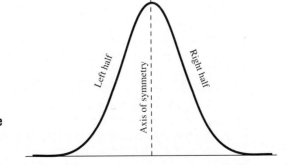

Figure 16-4 A normal curve has one vertical reflection symmetry.

Left half Axis of symmetry Right half

■ **Center.**   The point where the vertical axis of reflection cuts the horizontal axis is the center of the normal distribution. The data value corresponding to this point is both the *median* and the *average* of the distribution. To avoid showing any favoritism, we will call this value the **center**, and we will denote it by the Greek letter μ (mu) (Fig. 16-5). The fact that the median and the average are the same is a consequence of the fact that a normal curve has a vertical reflection symmetry. Any distribution (be it normal or not) that has such a symmetry will have the median equal to the average.

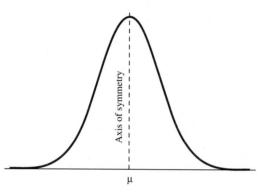

Figure 16-5 The median and the average of a normal distribution are equal.

■ **Standard Deviation.**   For any normal distribution the standard deviation is the key to measuring spread. We discussed the standard deviation in Chapter 14. Among other things, we saw that calculating the standard deviation for a large distribution of data values can be a great deal of work and that it's something best left to a good calculator or a statistical software package.

The easiest way to describe the standard deviation of a normal distribution is geometrically. Suppose we have a piece of wire we want to bend so that it is shaped exactly like the right half of a normal curve (if we can do the right half, then we also do the left half since they are symmetric.) At the very top we must bend the wire downward [as illustrated in Fig. 16-6(a)] while at the bottom we must bend the wire upward [Fig. 16-6(b)]. There is exactly one point P where the curvature of the wire changes from down to up [Fig 16-6(c)]. Locating the point P gives us a way to find the standard deviation of the normal distribution—it is the distance from P to the axis of symmetry of the distribution as shown in Fig. 16-7. We will use the Greek letter σ (sigma) to denote the standard deviation.

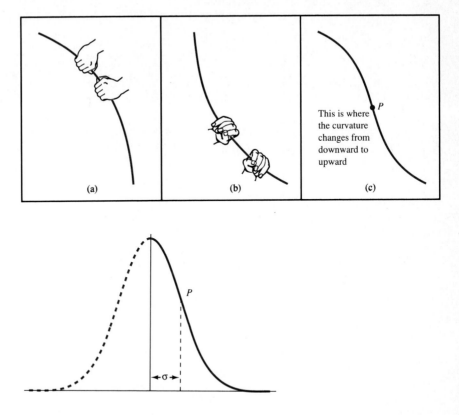

Figure 16-6

(a)　　　(b)　　　(c)

This is where the curvature changes from downward to upward

P

Figure 16-7 The standard deviation (σ) is the distance between the point *P* where the bell changes curvature and the axis of symmetry of the bell.

P

σ

In general, calculating the standard deviation of a normal distribution requires mathematical tools that go beyond the scope of this book. For the purposes of our discussion we will take it for granted that the standard deviation is known.

■ **Areas Under a Normal Curve. The 68-95-99.7 Rule.** For any normal curve the following three facts are always true.

1. The area under the curve and within one standard deviation to the left and right of the center ($\mu \pm \sigma$) is equal to 68% of the total area under the curve [Fig. 16-8(a)].

2. The area under the curve and within two standard deviations to the left and right of the center ($\mu \pm 2\sigma$) is equal to 95% of the total area under the curve [Fig. 16-8(b)].

3. The area under the curve and within three standard deviations to the left and right of the center ($\mu \pm 3\sigma$) is equal to 99.7% of the total area under the curve [Fig. 16-8(c)].

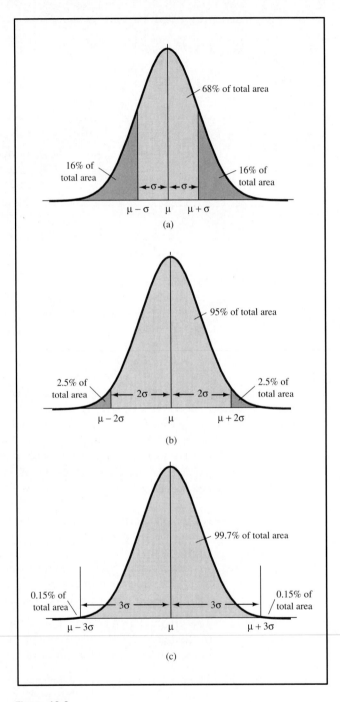

Figure 16-8

Figure 16-8 illustrates the 68-95-99.7 rule, as well as some of its consequences, in complete detail. The reader is advised to study Fig. 16-8 carefully.

One of the important consequences of the 68-95-99.7 rule is that for all practical purposes almost all the area under the normal curve (99.7%) is concentrated in a section that is plus or minus three standard deviations from the center. Outside that interval there simply isn't that much left of the distribution.

Any normal distribution is completely described by two numbers: the center (μ) and the standard deviation (σ). If someone were to give us two numbers for μ and σ (σ must be a positive number), say $\mu = 16.5$ and $\sigma = 2.3$, we could talk about *the* normal distribution with center 16.5 and standard deviation 2.3 because there is one and only one such beast.

**AREAS
AND
PERCENTILES**

We will now discuss a very critical question: Why should we care about areas under the normal curve? Or, to put it in a different way, what do areas under a normal curve tell us about the associated distribution? Remember that our starting point is some variable which when evaluated over a population produces an approximately normal distribution. The bar graph or histogram for this distribution is roughly bell-shaped, and the normal curve that we associate with it is a mathematical idealization of the bar graph. The curve itself follows (roughly) the heights of the bars, and the areas of the various segments under the curve correspond to percentages of the population associated with those segments. It all sounds terribly complicated, but it isn't.

Let's start with the fact that the total area under a normal curve represents the total population (100%) (Fig. 16-9). Next let's look at the center

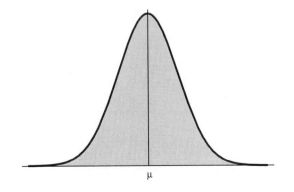

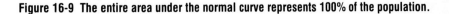

Figure 16-9 The entire area under the normal curve represents 100% of the population.

μ. We know that the normal curve is symmetric about a vertical axis passing through μ, which tells us that the shaded area in Fig. 16-10 must be one-half the total area under the curve. This tells us that for 50% of the population the variable takes values less than or equal to μ, and for 50% of the population the variable takes values bigger than or equal to μ. This explains why μ is the median of the distribution.

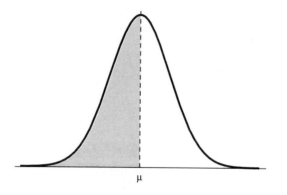

Figure 16-10 The vertical axis through μ divides the area under the curve into two pieces of equal area.

How about the quartiles? Following the same line of thinking that we used with the median, we guess that the first quartile (Q_1) must be located at the place where the area to its left and under the curve is 25% of the total area. This is indeed the case, as illustrated in Fig. 16-11(a). Likewise, the third quartile (Q_3) is located so that the area to its left and under the curve is 75% of the total area.

When we know the value of μ and σ, then we can locate the quartiles using the following two rules:

$$Q_3 \approx \mu + (0.675)\sigma$$
$$Q_1 \approx \mu - (0.675)\sigma.$$

We won't explain where the magic number 0.675 comes from other than to say that there are tables that give not only the location of the quartiles but also of any other arbitrary **percentile** of the population. (The xth percentile is the value of the variable for which x percent of the population is at or below that value.) For a very simplified version of such a table the reader is referred to Exercises 26 and 27.

We illustrate some of the ideas of this section with a couple of examples.

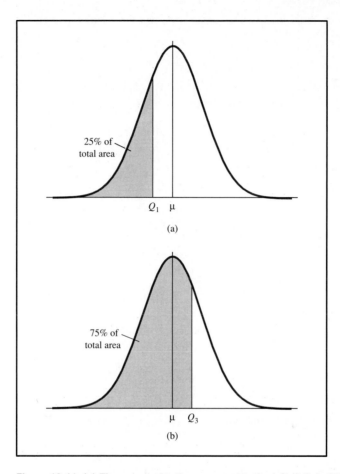

25% of
total area

Q_1 μ

(a)

75% of
total area

μ Q_3

(b)

Figure 16-11 (a) The area under the curve and to the left of Q_1 is 25% of the total area. (b) The area under the curve and to the left of Q_3 is 75% of the total area.

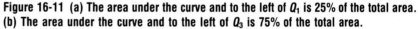

Example 3. Let's analyze the normal curve with center located at 500 and standard deviation $\sigma = 100$. The median is $\mu = 500$, the first quartile is found by taking $Q_1 \approx 500 - 0.675 \times 100 = 432.5$, and the third quartile is $Q_3 \approx 500 + 0.675 \times 100 = 567.5$. The 68-95-99.7 rule tells us that 99.7% of the area under the curve falls between the values of 200 and 800, that 95% of the area under the curve falls between the values of 300 and 700, and that 68% of the area under the curve falls between the values of 400 and 600. (See Fig. 16-12 on the next page.)

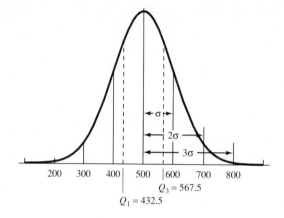

Figure 16-12 The normal curve with center μ = 500 and standard deviation σ = 100.

Example 4. Let's suppose that we are told that the distribution of scores on the mathematics part of the SAT for the entering class at Tasmania State University had an approximately normal distribution with center $\mu = 500$ points and standard deviation $\sigma = 100$ points.[1] We can think of the normal curve shown in Fig. 16-12 as a mathematical model for the distribution of the SAT scores, and we can put all the abstract things we learned in Example 3 to good use. (See Fig. 16-13.) Here are

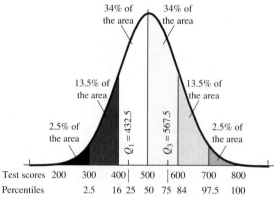

Figure 16-13 Scores on the SAT math and associated percentiles for the entering class at Tasmania State University.

---

[1] The reader is warned that these are ficticious (and conveniently chosen) values. In 1990, for example, for the entire national population of college bound high school seniors, the median math score was 476.

A typical SAT score report. The first row shows SAT-verbal score, the second row shows SAT-mathematics score. The two columns on the right show this student's percentile for two different reference populations: college-bound seniors across the nation as opposed to college-bound seniors from the same state the student is from.

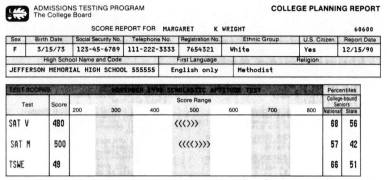

ADMISSIONS TESTING PROGRAM
The College Board

COLLEGE PLANNING REPORT

SCORE REPORT FOR MARGARET   K WRIGHT   60600

Sex	Birth Date	Social Security No.	Telephone No.	Registration No.	Ethnic Group	U.S. Citizen	Report Date
F	3/15/73	123-45-6789	111-222-3333	7654321	White	Yes	12/15/90

High School Name and Code	First Language	Religion
JEFFERSON MEMORIAL HIGH SCHOOL 555555	English only	Methodist

Test	Score	Score Range 200–800	National	State
SAT V	480	<<<>>>	68	56
SAT M	500	<<<<>>>>	57	42
TSWE	49		66	51

(© 1987 by College Entrance Examination Board and Educational Testing Service)

some of the things that we can say:

- Of all the students taking the test 50% scored 500 points or less (in the mathematics part), and 50% scored 500 points or more.

- About 25% of all the students scored below $Q_1 = 432.5$ points, and about 25% scored above $Q_3 = 567.5$ points.

- About 99.7% of all students scored somewhere between 200 and 800 points. Actually, since 200 is the minimum score (they give 200 points just for showing up) and 800 is the maximum score reported, we can go out on a limb and say that 100% of the students scored between 200 and 800.

- About 95% of all students scored between 300 and 700 points. The remaining 5% was split equally between those who scored 300 or below (2.5%) and those who scored 700 or above (2.5%).

- About 68% of all students scored between 400 and 600 points. By symmetry it follows that the remaining 32% was split equally between those who scored 400 or below (16%) and those who scored 600 or above (16%). ■

Let's put our new formal knowledge to good use by considering the case of three TSU freshpersons (Vanessa, Rudy, and Cleo) who scored 490, 570, and 710 points, respectively, on the mathematics portion of the SAT. In what percentiles of the Tasmania State University entering class would these scores roughly place them?

A score of 490 is close to 500, which would place Vanessa a little bit under the 50th percentile, a good guess would be somewhere around the 47th or 48th percentile. Rudy's score of 570 is slightly above the third quartile ($Q_3 = 567.5$). We would guess that Rudy's score places him somewhere around the 75th or 76th percentile. As for Cleo, her score of 710 is above the 97.5th percentile. So we can figure that Cleo placed somewhere around the 98th or 99th percentile. Our guesses are actually very good. The exact percentiles for any score can be calculated using special statistical tables for normal distributions (or a good piece of statistical software). It turns out that Vanessa's score of 490 places her in the 46th percentile, Rudy's score of 570 puts him in the 76th percentile, and Cleo's score of 710 puts her in the 98th percentile. (Way to go Cleo!)

We conclude with a parting comment: When we started Example 4, there was precious little that we had to work with (the scores followed an approximately normal distribution with mean 500 and standard deviation 100). From these facts and the little bit of theory that we learned, we were able to get plenty of mileage. Most remarkably, we were able to do this without looking at any actual data.

NORMAL DISTRIBUTIONS OF RANDOM EVENTS

We are now ready to take up another important aspect of normal distributions—their connection with the behavior of random events. Our starting point is the following important example.

Example 5 (A Coin Tossing Experiment).   We are going to take an honest coin and toss it 100 times. Of these 100 tosses, how many times will the coin come up heads? Fifty? Maybe, maybe not. Forty-five? Why not? Seventy-five? Possible, but it wouldn't be wise to bet on it.

For convenience, we will use X to represent the number of heads that come up when we toss the coin 100 times. What does common sense tell us about the possible values of the random variable X? First of all, we cannot predict the exact value of X—in principle it could be anything from $X = 0$ (the 100 tosses come up tails) to $X = 100$ (the 100 tosses come up heads). Of course some values of X are less likely to occur than others: We suspect that tossing 100 heads is very unlikely; that tossing 55 heads is a lot more likely; and that tossing 50 heads is even more likely.

One way that we can test our intuition about how likely the various values of X are is to repeat the experiment of tossing the coin 100 times many times and check the frequencies of the various outcomes. To do this all one needs is an honest coin, some pencil and paper for record keeping, and lots and lots of time. The South African mathematician John Kerrich was imprisoned by the Germans during World War II, and while in prison he carried out such an experiment. He tossed a coin 10,000 times and kept records of the number of heads in each successive set of 100 tosses.

Rather than reproduce Kerrich's data here, we decided to do our own high-tech version of the experiment by letting a computer do the coin tossing as well as all the record keeping. (It is very easy to have a computer toss a pretend coin, and the results are just as valid as those we would

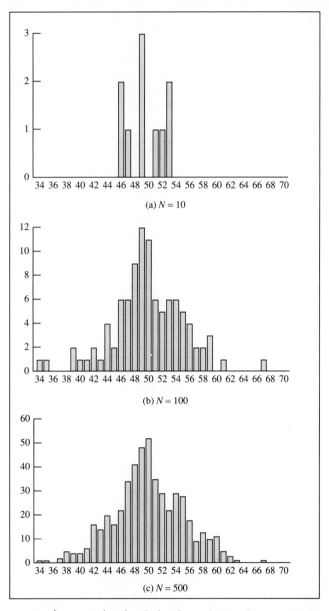

(a) $N = 10$

(b) $N = 100$

(c) $N = 500$

Figure 16-14 The 100 toss experiment: Toss an honest coin 100 times and count the number of heads (X). Repeat the experiment N times and chart the frequencies of X.

get tossing a real coin. It is also a lot easier on the thumb!).

Figure 16-14 shows the frequencies for the random variable X when we repeated the experiment (a) 10 times, (b) 100 times, (c) 500 times, (d) 1000 times, (e) 5000 times, and (f) 10,000 times.[2]

---

[2] This means 10,000 × 100 individual tosses. It would take about 5 minutes (working fast) to toss a real coin 100 times and record the data. If one were to toss coins without a break for 8 hours a day, 7 days a week, it would take about 3 months of nonstop coin tossing to collect this amount of data by hand.

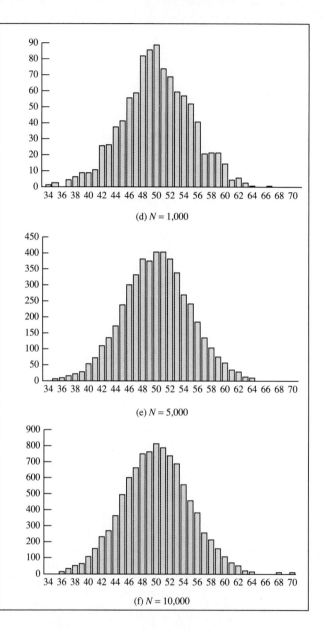

(d) $N = 1,000$

(e) $N = 5,000$

(f) $N = 10,000$

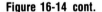

Figure 16-14 cont.

Figure 16-14 paints a pretty clear picture of what happened. When we repeated the experiment 10 times, we got $X = 46$ twice, $X = 47$ once, $X = 49$ three times, $X = 51$ once, $X = 52$ once, and $X = 53$ twice. A bar graph for these data values is shown in Fig. 16-14(a). This distribution may look a little surprising, but with such a small number of trials just about anything can happen. As we increased the number N of repetitions of the experiment, the picture started to come into sharper focus, and by the time we got to $N = 5000$ the unmistakable shape of a normal distribution showed up loud and clear. When $N = 10,000$ [Fig. 16-14(f)], we got an almost perfect normal distribution!   ■

As the reader can probably guess, what happened in this last example was not an accident. For small values of N things are pretty unpredictable, but as N gets bigger, the normal nature of the distribution is guaranteed to come out. Let's put it this way: Suppose someone else decided to do the same experiment all over, once again tossing an honest coin (be it by hand or by computer) 100 times, counting the value of the random variable X = the number of heads, and then repeating the whole thing N times. For $N = 10$ their results are likely to be quite different from the results we show in Fig. 16-14(a), but as N gets bigger, the results will begin to look more and more alike. By the time they get to $N = 10,000$ their data will be almost identical to the data we show in Fig. 16-14(f). In a sense, we can say that doing these experiments is a waste. For large values of N we can predict what the distribution of X will be like with an extremely high degree of accuracy without ever tossing a coin!

The most important conclusion that we can draw from Example 5 is that for sufficiently large values of N, the random variable X has a normal distribution. For a complete grasp of the situation we need two more pieces of information: the values of the center μ and the standard deviation σ of this distribution. Looking at Fig. 16-14(f), we can pretty much see where the center of this normal distribution is—right at 50. The fact that

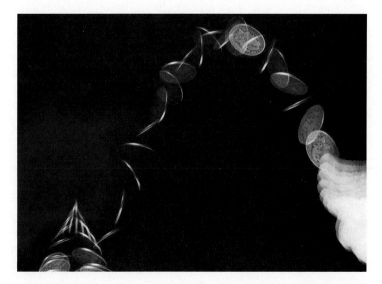

Figure 16-15 Collecting the data the hard way. (© Estate of Harold Edgerton, 1965)

$\mu = 50$ should be no surprise. Since the coin is an honest coin, we know that the likelihood of tossing fewer than 50 heads is the same as that of tossing fewer than 50 tails which in turn is the same as that of tossing more than 50 heads. Clearly the axis of symmetry of the distribution has to be at 50.

The value of the standard deviation is less obvious. For now, let's accept the fact that it is $\sigma = 5$. We will explain how we got this value shortly.

Let's summarize what we now know: An honest coin is tossed 100 times. The number of heads in the 100 tosses is a random variable which we call X. If we repeat this experiment a large number of times (N), the random variable X will have an approximately normal distribution with center $\mu = 50$ and standard deviation $\sigma = 5$, and the larger the value of N, the better this approximation will be.

The real significance of these facts is that they are valid not just because we went to the trouble (or rather had the computer take the trouble for us) of tossing a coin hundreds of thousands of times. The truth of the matter is that even if we did not toss a coin at all, all of the above would still be true: For a sufficiently large number of repetitions of the experiment of tossing an honest coin 100 times, the number of heads X is a random variable that has an approximately normal distribution with center $\mu = 50$ and standard deviation $\sigma = 5$. This is a mathematical fact.

The next step we are going to take is critical. Suppose that we have an honest coin and are going to toss it 100 times. We are going to do this just once—period. Can we make any predictions about the number of heads that are going to come up? The answer is most definitely yes. By taking advantage of the fact that the number of heads that will come up is a random variable with an approximately normal distribution we will be able to make excellent predictions about the outcome of such an experiment. For starters, we can predict that the chances that the number of heads will fall somewhere between 45 and 55 are 68% (it follows from the fact that this is one standard deviation to the left and right of the center), the chances that the number of heads will fall somewhere between 40 and 60 are 95%, and the chances that the number of heads will fall somewhere between 35 and 65 are a whopping 99.7%. All we are using here is the knowledge that we are dealing with a variable that has a normal distribution with center $\mu = 50$ and standard deviation $\sigma = 5$, together with some of the things that we learned about normal distributions (such as the 68-95-99.7 rule).

THE HONEST AND DISHONEST COIN PRINCIPLES

In Example 5 everything was based on tossing an honest coin 100 times. What if we were to toss a coin 500 times? Or 1000 times? Or n times? Not surprisingly, everything we discovered in Example 5 would still be true, except that the values of μ and σ would be different. Specifically, we can make the following general statement for which we have coined the name the "honest coin principle."

The Honest Coin Principle.

An experiment consists of tossing an honest coin n times and counting the number of heads (X). If we repeat this experiment a large number of times (N), then

1. The random variable X has an approximately normal distribution, and the larger N is, the closer the distribution is to a normal distribution.

2. The center of the distribution is $\mu = n/2$.

3. The standard deviation of the distribution is $\sigma = \sqrt{n}/2$.

(This is why for $n = 100$ we found that the center is $\mu = \frac{100}{2} = 50$ and the standard deviation is $\sigma = \sqrt{100}/2 = \frac{10}{2} = 5$.)

Example 6.   An honest coin is going to be tossed 256 times. Before this is done there is an opportunity to make some bets. Let's say that we can make a bet (with even odds) that if the number of heads falls somewhere between 120 and 136 we will win, otherwise we will lose. Should we make such a bet?

By the honest coin principle we know that the number of heads in 256 tosses of an honest coin is a random variable having a distribution that is approximately normal with center $\mu = \frac{256}{2} = 128$ and standard deviation $\sigma = \sqrt{256}/2 = \frac{16}{2} = 8$. The values 120 to 136 are exactly one standard deviation to the right and left of 128, and by the 68-95-99.7 rule there is a 68% chance that the number of heads will fall somewhere between 120 and 136. We should indeed make this bet! ■

By a similar calculation we can argue that there is a 95% chance that the number of heads will fall somewhere between 112 and 144, and the chance that the number of heads will fall somewhere between 104 and 152 is an overwhelming 99.7%.

Our next step is to see if we can discover a **dishonest coin principle**. (We can't always count on an even break!) Would the ideas that we have seen so far apply if we have a dishonest coin? The answer is yes. In fact the nature of the coin only plays a role in calculating the center and the standard deviation.

The Dishonest Coin Principle.

An experiment consists of tossing a dishonest coin n times and counting the number of heads (X). This particular coin has a probability of landing heads equal to p (and therefore the probability of landing tails is $1-p$). If we repeat this experiment a large number of times (N), then

1. The random variable X has an approximately normal distribution, and the larger N is, the closer the distribution is to a normal distribution.

2. The center of the distribution is $\mu = n \cdot p$.

3. The standard deviation of the distribution is $\sigma = \sqrt{n \cdot p \cdot (1-p)}$.

Example 7. A coin is rigged so that it comes up heads only 20% of the time (i.e., $p = .20$). The coin is tossed 100 times ($n = 100$).

According to the dishonest coin principle, the distribution of the number of heads is approximately normal with center $\mu = 100 \times .20 = 20$ and standard deviation $\sigma = \sqrt{100 \times .20 \times .80} = 4$.

Note that in this case heads and tails are no longer symmetric, but the dishonest coin principle will work just as well for tails as it does for heads. The distribution for the number of tails is approximately normal with center $\mu = 100 \times .80 = 80$ and standard deviation $\sigma = \sqrt{100 \times .80 \times .20} = 4$.

Based on these facts we can now make the following assertions:

(a) There is a 68% chance that the number of heads will fall somewhere between 16 and 24.

(b) There is a 95% chance that the number of heads will fall somewhere between 12 and 28.

(c) The number of heads is almost guaranteed (a 99.7% chance) to fall somewhere between 8 and 32. ∎

The dishonest coin principle can be applied to any coin, even one that is fair ($p = \frac{1}{2}$). In the case $p = \frac{1}{2}$ the honest and dishonest coin principles say the same thing (Exercise 25).

The dishonest coin principle is a down-to-earth version of one of the most important facts in all of statistics, known by the somewhat intimidating name of the *central limit theorem*. We will now briefly illustrate why the importance of the dishonest coin principle goes beyond the tossing of coins.

SAMPLING AND THE DISHONEST COIN PRINCIPLE

Example 8.   A large container (sometimes called an urn) is filled with 100,000 beads, 20,000 of which are red and 80,000 of which are white. The beads in the urn are thoroughly mixed. We will now draw a sample from this urn as follows: (1) We stick our hand in the urn and randomly pick a bead, (2) we check and record the color of the bead, (3) we put the bead back in the urn. We repeat this process $n = 100$ times. Let the random variable X represent the number of red beads counted in the 100 draws. What can we say about X?

A moment's reflection will show that statistically this example is identical to Example 7—each red bead drawn (probability .20) can be identified with the tossing of a head (also probability .20). It follows that the dishonest coin principle can be applied to make the following assertions.

(a)   There is a 68% chance that the number of red beads drawn will fall somewhere between 16 and 24.

(b)   There is a 95% chance that the number of red beads drawn will fall somewhere between 12 and 28.

(c)   The number of red beads drawn is almost guaranteed (a 99.7% chance) to fall somewhere between 8 and 32.

Probably the most important point of all this is that each of the above facts can be rephrased in terms of sample errors. Suppose that the purpose of drawing the 100 beads from the urn is to use the number of red beads in the sample to estimate the percentage of red beads in the entire population. Let's say, for the sake of argument, that we draw 24 red beads in the sample of 100. If this statistic (24%) were used to estimate the percent of red beads in the urn, then the sample error would be 4% (the estimate is 24% and the exact value of the parameter is 20%). By the same token, if we had drawn 16 red beads, the sample error would be -4%. Since the standard deviation of $\sigma = 4$ beads (which we computed in Example 7) is exactly 4% of the sample size ($n = 100$) we can rephrase assertions a through c in terms of sample errors as follows.

(a)   When estimating the proportion of red beads in the urn using a sample of 100 beads, the chance that the sample error will fall somewhere between -4% and 4% is 68%.

(b) When estimating the proportion of red beads in the urn using a sample of 100 beads, the chance that the sample error will fall somewhere between −8% and 8% is 95%.

(c) We can guarantee with great confidence (99.7% certainty) that, when estimating the proportion of red beads in the urn using a sample of 100 beads, the sample error will fall somewhere between −12% and 12%. ◼

Example 9.   Suppose we have the same urn as in Example 8, but this time we are going to draw a sample of size $n = 1600$. Before we even count the number of red beads in the sample, let's see how much mileage we can get out of the dishonest coin principle. Theoretically, we are doing the same thing as tossing a dishonest coin (with $p = .2$) 1600 times. The standard deviation here is $\sqrt{1600 \times .2 \times .8} = 16$, which is exactly 1% of the sample. This means that when we estimate the proportion of red beads in the urn using this sample, we can once again use the dishonest coin principle to make certain qualified assurances about the size of the sample error, namely,

(a) We can say with some confidence (68%) that the sample error will fall somewhere between −1% and 1%.

(b) We can say with a lot of confidence (95%) that the sample error will be between −2% and 2%.

(c) We can say with a tremendous amount of confidence (99.7%) that the sample error will be between −3% and 3%. ◼

The ideas discussed in this section are fundamental to an understanding of how to estimate the reliability of any sample. We end this section, however, with a disclaimer: When using a random sample to estimate a population proportion out there in the real world, the standard deviation σ given by the dishonest coin principle cannot be calculated directly (the way we did in Examples 8 and 9) because we don't know p. If we did, we wouldn't have to use a sample! It is possible, however, to get a very good estimate of σ and to use it to make the same types of qualified assurances we gave in the last set of statements a through c. (See Exercise 30.)

CONCLUSION

The wheels of science do not always spin forward. In previous chapters we learned about all the wonderful things that we can do with data. In a sense, this chapter was about all the wonderful things that we can do without data. The process of drawing conclusions with limited information is an essential aspect of statistics. The way the process works is by looking at a mathematical idealization of the data. In many real-life situations one can safely predict ahead of time what the overall pattern of the data will be like—this overall pattern is what we call a *mathematical model* for the data. A good understanding of the mathematical properties of a model is an invaluable tool for analyzing the real data.

Perhaps the most commonly used mathematical model of real-life data is the normal distribution. Contrary to what many people think, the normal distribution is not guaranteed to always be the right model for looking at real-life data. It is nevertheless true that when we look at almost any variable that measures a characteristic of a population (height, IQ, blood pressure, etc.), it is often the case that its distribution is approximately normal. When this happens, we can use the mathematical theory of normal distributions to gain a better understanding of the data.

A second context in which normal distributions came up in this chapter was in connection with random experiments. Many types of random experiments exhibit exactly the same long-run behavior as that shown by the repeated tossing of a coin. In these cases the *dishonest coin principle* (which includes the *honest coin principle* as a special case) tells us that this long-run behavior can be described by a normal distribution. The importance of the dishonest coin principle in real-life applications of statistics cannot be overstated. It is, among other things, the mathematical principle that allows us to use random sampling to draw reliable conclusions about a population, and, in a more sobering vein, it is also the principle that guarantees that in the long run gamblers will always lose and casinos will always win.

KEY CONCEPTS

68-95-99.7 rule
approximately normal
    distribution
center
dishonest coin principle
distribution

honest coin principle
normal curve
normal distribution
percentile
standard deviation

EXERCISES

■ Walking

Exercises 1 through 4 refer to the following: 250 students take a college entrance exam. The scores on the exam have an approximately normal distribution with center $\mu = 52$ points and standard deviation $\sigma = 11$ points.

1. **(a)** Estimate the average score on the exam.
   (b) Estimate what percent of the students scored 52 points or more.
   (c) Estimate what percent of the students scored between 41 and 63 points.
   (d) Estimate what percent of the students scored 63 points or more.

2. **(a)** Estimate how many students scored between 30 and 74 points.
   (b) Estimate how many students scored 74 points or more.
   (c) Estimate how many students scored 85 points or more.

3. **(a)** Estimate the first-quartile score for this exam.
   (b) Estimate the third-quartile score for this exam.
   (c) Estimate the interquartile range for this exam.

4. For each of the following scores, estimate in what percentile of the students taking the exam the score would place you.
   (a) 51
   (b) 64
   (c) 60
   (d) 85.

Exercises 5 through 8 refer to the following: As part of a research project, the blood pressures of 2000 patients in a hospital are recorded. The systolic blood pressures (given in millimeters) have an approximately normal distribution with center $\mu = 125$ and $\sigma = 13$.

5. **(a)** Estimate the number of patients whose blood pressure was between 99 and 151 millimeters.
   (b) Estimate the number of patients whose blood pressure was 99 millimeters or less.

6. **(a)** Estimate the third quartile (Q_3) for the distribution of blood pressures.
   (b) Estimate the interquartile range for the distribution of blood pressures.

7. For each of the following blood pressures, estimate the percentile of the patient population to which they correspond.
   (a) 100 millimeters
   (b) 112 millimeters
   (c) 115 millimeters
   (d) 138 millimeters
   (e) 164 millimeters.

8. **(a)** Estimate the value of the lowest (Min) and highest (Max) blood pressures. (Assume there were no outliers and use the 68-95-99.7 rule.)
   (b) Assuming there were no outliers, give an estimate of the five-number summary (Min, Q_1, μ, Q_3, Max) for the distribution of blood pressures.

Exercises 9 through 12 refer to the following: Packaged foods sold at supermarkets are not always the weight indicated on the package. Variability always crops up in the manufacturing and packaging process. Suppose that the exact weight of a "12-ounce" bag of potato chips is a random variable that has an approximately normal distribution with center μ = 12 ounces and standard deviation σ = 0.5 ounces.

9. If a "12-ounce" bag of potato chips is chosen at random, what are the chances that
    (a) it weighs somewhere between 11 and 13 ounces?
    (b) it weighs somewhere between 12 and 13 ounces?
    (c) it weighs more than 11 ounces?

10. If a "12-ounce" bag of potato chips is chosen at random, what are the chances that
    (a) it weighs somewhere between 11.5 and 12.5 ounces?
    (b) it weighs somewhere between 12 and 12.5 ounces?
    (c) it weighs more than 12.5 ounces?

11. Suppose that 500 bags of potato chips are chosen at random. Estimate the number of bags with weight
    (a) 11 ounces or less.
    (b) 11.5 ounces or less.
    (c) 12 ounces or less.
    (d) 12.5 ounces or less.
    (e) 13 ounces or less.
    (f) 13.5 ounces or less.

12. Suppose that 1500 bags of potato chips are chosen at random. Estimate the number of bags of potato chips with weight
    (a) between 11 and 11.5 ounces.
    (b) between 11.5 and 12 ounces.
    (c) between 12 and 12.5 ounces.
    (d) between 12.5 and 13 ounces.
    (e) between 13 and 13.5 ounces.

13. Find μ and σ for the normal curve shown in the figure.

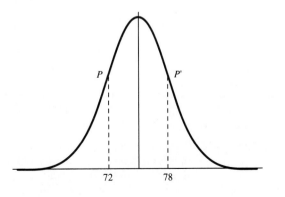

14. Find μ and σ for the normal curve shown in the figure.

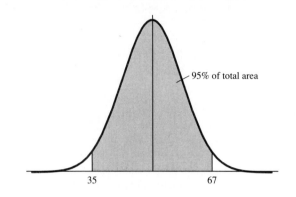

95% of total area

35   67

15. Find μ and σ for the normal curve shown in the figure.

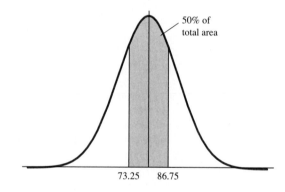

50% of total area

73.25   86.75

16. Find μ and σ for the normal curve shown in the figure.

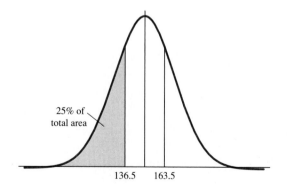

25% of total area

136.5   163.5

Exercises 17 through 20 refer to the following: The distribution of weights for children of a given age and sex is approximately normal. This fact allows a doctor or nurse to find from a child's weight the weight percentile of the population (all children of the same age and sex) to which the child belongs. Typically, this is done using special charts provided to the doctor or nurse, but these percentiles can also be computed using facts about approximately normal distributions such as the ones we learned in this chapter. (The figures in these examples are approximate values taken from tables produced by the National Center for Health Statistics, U.S. Department of Health and Human Services.)

17. The distribution of weights for 6-month-old baby boys is approximately normal with center μ = 17.25 pounds and standard deviation σ = 2 pounds.
   (a) Suppose that a 6-month-old boy weighs 15.25 pounds. Approximately what weight percentile is this child in?
   (b) Suppose that a 6-month-old boy weighs 21.25 pounds. Approximately what weight percentile is this child in?
   (c) Suppose that a 6-month-old boy is in the 75th percentile. Estimate his weight.

18. The distribution of weights for 12-month-old baby girls is approximately normal with center μ = 21 pounds and standard deviation σ = 2.2 pounds.
   (a) Suppose that a 12-month-old girl weighs 16.6 pounds. Approximately what weight percentile is she in?
   (b) Suppose that a 12-month-old girl weighs 18.8 pounds. Approximately what weight percentile is she in?
   (c) Suppose that a 12-month-old girl is in the 75th percentile. Estimate her weight.

19. The distribution of weights for 1-month-old baby girls is approximately normal with center μ = 8.75 pounds and standard deviation σ = 1.1 pounds.
   (a) Suppose that a 1-month-old girl weighs 11 pounds. Approximately what weight percentile is this child in?
   (b) Suppose that a 1-month-old girl weighs 12 pounds. Approximately what weight percentile is this child in?
   (c) Suppose that a 1-month-old girl is in the 25th percentile. Estimate her weight.

20. The distribution of weights for 12-month-old baby boys is approximately normal with center μ = 22.5 pounds and standard deviation σ = 2.2 pounds.
   (a) Suppose that a 12-month-old boy weighs 24 pounds. Approximately what weight percentile is he in?
   (b) Suppose that a 12-month-old boy weighs 21 pounds. Approximately what weight percentile is he in?
   (c) Suppose that a 12-month-old boy is in the 84th percentile. Estimate his weight.

■ **Jogging**

21. Over the years, the eggs produced by the Fibonacci Egg Company have been known to have weights with an approximately normal distribution whose center is μ = 1.5 ounces and standard deviation is σ = 0.35 ounces. Eggs are classified by weight as follows:

   ■ Extra large: more than 2.2 ounces

   ■ Large: between 1.5 and 2.2 ounces

   ■ Discount (sold to wholesalers): less than 1.5 ounces.

Out of 5000 eggs, approximately how many would you expect would fall in each of the three categories.

22. An honest coin is tossed $n = 3600$ times. Let the random variable Y denote the number of tails tossed.

(a) Find the center μ and the standard deviation σ for the distribution of the random variable Y.

(b) What are the chances that the value of Y will fall somewhere between 1770 and 1830?

(c) What are the chances that the value of Y will fall somewhere between 1800 and 1830?

(d) What are the chances that the value of Y will fall somewhere between 1830 and 1860?

23. An honest die is rolled. If the roll comes out even (2, 4, or 6), you will win \$1; if the roll comes out odd (1, 3, or 5), you will lose \$1. Suppose that in one evening you play this game $n = 2500$ times in a row.

(a) What is the probability that by the end of the evening you will not have lost any money?

(b) What is the probability that the number of even rolls will fall between 1250 and 1300?

(c) What is the probability that you will win \$100 or more?

(d) What is the probability that you will win exactly \$101?

24. A dishonest coin with probability of heads $p = .4$ is tossed $n = 600$ times. Let the random variable X represent the number of times the coin comes up heads.

(a) Find the center and standard deviation for the distribution of X.

(b) Find the first and third quartiles for the distribution X.

(c) Suppose that you could choose between one of the following bets (\$1 to win \$1):

(i) The number of heads will fall somewhere between 230 and 250.

(ii) The number of heads will be less than or equal to 230 or more than or equal to 250.

Which of the two bets would you choose? Explain your answer.

25. Explain why when the dishonest coin principle is applied with an honest coin, we get the honest coin principle.

Exercises 26 and 27 refer to the following table. It is a simplified version of a more elaborate statistical table that gives, for every value of x, the approximate location of the xth percentile for a normal distribution with center μ and standard deviation σ.

Percentile	Approximate Location	Percentile	Approximate Location
99th	$\mu + 2.33\sigma$	1st	$\mu - 2.33\sigma$
95th	$\mu + 1.65\sigma$	5th	$\mu - 1.65\sigma$
90th	$\mu + 1.28\sigma$	10th	$\mu - 1.28\sigma$
80th	$\mu + 0.84\sigma$	20th	$\mu - 0.84\sigma$
75th	$\mu + 0.675\sigma$	25th	$\mu - 0.675\sigma$
70th	$\mu + 0.52\sigma$	30th	$\mu - 0.52\sigma$
60th	$\mu + 0.25\sigma$	40th	$\mu - 0.25\sigma$
50th	μ		

26. The distribution of weights for 6-month-old baby boys is approximately normal with center $\mu = 17.25$ pounds and standard deviation $\sigma = 2$ pounds.
    (a) Suppose that a 6-month-old boy weighs in the 95th percentile of his age group. Find his weight in pounds approximated to two decimal places.
    (b) Suppose that a 6-month-old boy weighs 17.75 pounds. Determine in what percentile of the weight distribution for his age group this child is in.
    (c) Suppose that a 6-month-old boy weighs 15.2 pounds. Estimate in what percentile of the weight distribution for his age group this child is in.

27. Five-thousand students took a college entrance exam. The scores on the exam have an approximately normal distribution with center $\mu = 55$ points and standard deviation $\sigma = 12$ points.
    (a) Suppose a student's score places her in the 60th percentile. Find her score on the exam.
    (b) Suppose that a student scored 45 points on the exam. Estimate the percentile which this score places him in.
    (c) Suppose that a student scored 83 points on the exam. Estimate the percentile which this score places her in.
    (d) Approximately how many students scored 83 points or more?
    (e) Approximately how many students scored 52 points or less?

■ Running

28. On an American roulette wheel, there are 18 red numbers and 18 black numbers, plus 2 green numbers (0 and a 00). Thus, the probability of a red number coming up on a spin of the wheel is $p = \frac{18}{38} \approx .47$. Suppose that we go on a binge and we bet \$1 on red 10,000 times in a row. (A \$1 bet wins \$1 if red comes up, otherwise we lose the \$1.)

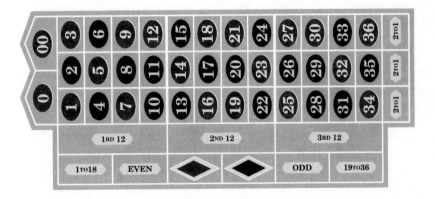

    (a) Let Y represent the number of times we lose (i.e., red does not come up). Use the dishonest coin principle to describe the distribution of the random variable Y.
    (b) Approximately what are the chances that we will lose 5300 times or more?
    (c) Approximately what are the chances that we will lose somewhere between 5150 and 5450 times?
    (d) Explain why the chances that we will break even or win in this situation are essentially zero.

29. An urn contains 10,000 beads, 20% of which are red and the rest white. Suppose that we draw a sample of size $n = 400$ (drawing a bead and each time replacing it in the urn before drawing again).
    (a) If Y is the number of white beads in the sample, find the center and standard deviation for the distribution of Y.
    (b) What are the chances that the number of white beads in the sample will fall somewhere between 304 and 332?
    (c) When estimating the proportion of white beads in the urn using a sample of 400 beads, what are the chances that the sample error will fall somewhere between -4% and 4%?

30. An urn contains a large number of beads which are either red or white. Suppose that we draw a sample of 1200 beads, of which 300 (25%) are red. We want to use the statistic $25\% = .25$ as an estimate of the number of red beads in the urn. To estimate the sample error we can use the dishonest coin principle and use $\hat{p} = .25$ as an approximation for the exact value of p.

    Using this value of $\hat{p}$ and the dishonest coin principle, explain why we can say with a lot of confidence (95%) that our estimate of 25% is within $\pm 2.5\%$ of the exact value of p.

REFERENCES AND FURTHER READINGS

1. Berry, D. A. and B. W. Lindgren, *Statistics: Theory and Methods*. Pacific Grove, Calif.: Brooks/Cole Publishing Co., 1990, chap. 8.

2. Converse, P. E. and M. W. Traugott, "Assessing the Accuracy of Polls and Surveys," *Science*, 234 (1986), 1094–1098.

3. Freedman, D., R. Pisani, R. Purves, and A. Adhikari, *Statistics (2nd ed.)*. New York: W. W. Norton, Inc., 1991, chaps. 16 and 18.

4. Groeneveld, Richard A., *Introductory Statistical Methods*. Boston: PWS-Kent, 1988, chap. 5.

5. Larsen, J. and D. F. Stroup, *Statistics in the Real World*. New York: Macmillan Publishing Co., Inc., 1976.

6. Mosteller, F., W., et al., *Statistics by Example: Detecting Patterns*. Reading, Mass.: Addison-Wesley Publishing Co., Inc., 1973.

7. Sincich, Terry, *Statistics by Example*. San Francisco, Calif.: Dellen Publishing Co., 1990, chaps. 6 and 7.

8. Tanner, Martin, *Investigations for a Course in Statistics*. New York: Macmillan Publishing Co., Inc., 1990.

Answers to Selected Problems

CHAPTER 1

■ Walking

1. (a)

Number of voters	6	2	1	3
1st choice	A	C	B	C
2nd choice	B	D	D	B
3rd choice	C	B	C	D
4th choice	D	A	A	A

    (b) no   **(c)** The Atrium (*A*)

3. (a) 51   **(b)** *A*

5. *B*

7. *C*

9. Winner: *A*. Second place: *C* and *D* (tied). Fourth place: *B*. Last place: *E*.

11. Winner: *A*. Second place: *B*, *C*, and *D* (tied). Last place: *E*.

13. Winner: *A*. Second place: *D*. Third place: *B*. Fourth place: *C*. Last place: *E*.

15. *H*

17. Winner: *R*. Second place: *H*. Third place: *C*. Fourth place: *S*. Last place: *O*.

19. Winner: *H*. Second place: *R*. Third place: *C*. Fourth place: *S*. Last place: *O*.

21. A

23. Winner: A. Second place: B. Third place: C. Fourth place: D. Last place: E.

25. Winner: A. Second place: B. Third place: E. Fourth place: D. Last place: C.

27. Winner: A. Second place: B. Third place: D. Fourth place: E. Last place: C.

29. Winner: C. Second place: B. Third place: D. Last place: A.

■ Jogging

31. **(a)** 1225

(b) 2450

33. **(a)** Since there are only *two* columns in the preference schedule, one of the columns must represent the votes of 11 or more voters and so the first choice in that column is the first choice of more than half of the voters.

(b) The majority winner is the first choice of more than half of the voters and so will win every pairwise comparison and be the winner under the method of pairwise comparisons.

(c) The argument given in (a) can be repeated provided there are an odd number of voters (3 or more) since then one of the *two* columns must represent the votes of more than half of the voters (there cannot be a tie). The argument in (b) applies to any election in which there is a majority winner.

35. Suppose that the results of the election under the Borda count method (1st place: 5 points, 2nd place: 4 points, 3rd place: 3 points, 4th place: 2 points, and 5th place: 1 point) were

Candidate	A	B	C	D	E
Points	a	b	c	d	e

Under the revised scheme (1st place: 4 points, 2nd place: 3 points, 3rd place: 2 points, 4th place: 1 point, and 5th place: 0 points), A loses a point for every voter, so does B, etc. It follows that the results of the election under the revised scheme must be

Candidate	A	B	C	D	E
Points	$a - 21$	$b - 21$	$c - 21$	$d - 21$	$e - 21$

Since these numbers have the same values relative to each other as the original numbers the outcome of the election is still the same.

37. If there is a candidate that is the first choice of a majority of the voters then using the plurality with elimination method, that candidate will be declared the winner of the election in the first round.

39. If there is a candidate that is the first choice of a majority of the voters then that candidate will win every one to one comparison with any other candidate and so will be a Condorcet winner. If a voting method violates the majority criterion, then there is an election for which a candidate is the first choice of a majority of the voters (and hence a Condorcet winner), and yet is not the winner of the election. Consequently, the voting method also violates the Condorcet criterion.

CHAPTER 2

■ **Walking**

1. (a) 3   **(b)** 10   **(c)** 10   **(d)** $\{P_1, P_2\}, \{P_1, P_3\}, \{P_1, P_2, P_3\}$
   (e) P_1 only   **(f)** $P_1: \frac{3}{5}; P_2: \frac{1}{5}; P_3: \frac{1}{5}.$

3. (a) $P_1: \frac{3}{5}; P_2: \frac{1}{5}; P_3: \frac{1}{5}$
   (b) $P_1: \frac{1}{2}; P_2: \frac{1}{2}; P_3: 0.$

5. $P_1: \frac{1}{3}; P_2: \frac{1}{3}; P_3: \frac{1}{3}; P_4: 0; P_5: 0.$

7. (a) $\langle P_1, P_2, P_3 \rangle, \langle P_1, P_3, P_2 \rangle, \langle P_2, P_1, P_3 \rangle,$
      $\langle P_2, P_3, P_1 \rangle, \langle P_3, P_1, P_2 \rangle, \langle P_3, P_2, P_1 \rangle.$
   (b) $\langle P_1, \underline{P_2}, P_3 \rangle, \langle P_1, \underline{P_3}, P_2 \rangle, \langle P_2, \underline{P_1}, P_3 \rangle,$
      $\langle P_2, \underline{P_3}, \underline{P_1} \rangle, \langle P_3, \underline{P_1}, P_2 \rangle, \langle P_3, \underline{P_2}, \underline{P_1} \rangle.$
   (c) $P_1: \frac{2}{3}; P_2: \frac{1}{6}; P_3: \frac{1}{6}.$

9. (a) $P_1: \frac{2}{3}; P_2: \frac{1}{6}; P_3: \frac{1}{6}.$
   (b) $P_1: \frac{1}{2}; P_2: \frac{1}{2}; P_3: 0.$

11. (a) P_3 is a dummy. A motion will pass if and only if both P_1 and P_2 vote for
      it.
   (b) P_2, P_3, P_4 are all dummies. P_1 is a dictator.

13. (a) 16   **(b)** 31   **(c)** 31   **(d)** 120

15. (a) $P_1: 1; P_2: 0; P_3: 0.$
   (b) $P_1: \frac{2}{3}; P_2: \frac{1}{6}; P_3: \frac{1}{6}.$
   (c) $P_1: \frac{1}{2}; P_2: \frac{1}{2}; P_3: 0.$
   (d) $P_1: \frac{1}{2}; P_2: \frac{1}{2}; P_3: 0.$
   (e) $P_1: \frac{1}{3}; P_2: \frac{1}{3}; P_3: P\frac{1}{3}$

17. (a) 40,320
   (b) 479,001,600
   (c) 6,227,020,800
   (d) $26A$
   (e) 13!

19. $A: \frac{1}{3}; B: \frac{1}{3}; C: \frac{1}{3}; D: 0.$

■ **Jogging**

21. (a) 63
   (b) Only the coalition consisting of all the players (the grand coalition), and
      every player is critical in this coalition.
   (c) The Banzhaf power index of each player is $\frac{1}{6}.$
   (d) If the quota equals the sum of all the weights ($q = w_1 + ... + w_N$) then
      the only winning coalition is the grand coalition and every player in it is
      critical. It follows that each of the N players is critical once and so the
      Banzhaf power index of each player is $\frac{1}{N}.$

23. (a) [8: 6,3,3,3]. Banzhaf power distribution: $P_1: \frac{1}{2}; P_2: \frac{1}{6}; P_3: \frac{1}{6}; P_4: \frac{1}{6}.$
   (b) [9: 6,3,2,2]. Banzhaf power distribution: $P_1: \frac{1}{2}; P_2: \frac{3}{10}; P_3: \frac{1}{10}; P_4: \frac{1}{10}.$
   (c) [12: 6,3,3,3]. Banzhaf power distribution: $P_1: \frac{2}{5}; P_2: \frac{1}{5}; P_3: \frac{1}{5}; P_4: \frac{1}{5}.$
   (d) [12: 6,3,3,2]. Banzhaf power distribution: $P_1: \frac{1}{3}; P_2: \frac{1}{3}; P_3: \frac{1}{3}; P_4: 0.$

25. (a) P_5 is a dummy. It takes (at least) three of the first four players to pass a motion. (P_5's vote doesn't make any difference.)

   (b) Both the same. $P_1: \frac{1}{4}$; $P_2: \frac{1}{4}$; $P_3: \frac{1}{4}$; $P_4: \frac{1}{4}$; $P_5: 0$.

   (c) $q = 21$, $q = 31$, $q = 41$.

27. (a) Both have Shapley-Shubik power distribution $P_1: \frac{1}{2}$; $P_2: \frac{1}{6}$; $P_3: \frac{1}{6}$; $P_4: \frac{1}{6}$.

   (b) In the weighted voting system $[q:w_1, w_2, \dots, w_N]$, P_k is pivotal in the sequential coalition $\langle P_1, P_2, \dots, P_k, \dots, P_N \rangle$ means $w_1 + w_2 + \dots + w_k \geq q$ but $w_1 + w_2 + \dots + w_{k-1} < q$. In the weighted voting system $[cq:cw_1, cw_2, \dots, cw_N]$, P_k is pivotal in the sequential coalition $\langle P_1, P_2, \dots, P_k, \dots, P_N \rangle$ means $cw_1 + cw_2 + \dots + cw_k \geq cq$ but $cw_1 + cw_2 + \dots + cw_{k-1} < cq$. These two statements are equivalent since $cw_1 + cw_2 + \dots + cw_k = c(w_1 + w_2 + \dots + w_k) \geq cq$ if and only if $w_1 + w_2 + \dots + w_k \geq q$, and $cw_1 + cw_2 + \dots + cw_{k-1} = c(w_1 + w_2 + \dots + w_{k-1}) < cq$ if and only if $w_1 + w_2 + \dots + w_{k-1} < q$. This same reasoning applies to any sequential coalition and so the pivotal players are exactly the same. (Of course, we are assuming $c > 0$.)

29. (a) P_5 is a dummy if $q = 14$.   (b) $q = 15$.

CHAPTER 3

■ Walking

1. (a) Three answers possible as shown in the following table.

Chooser 1	Chooser 2	Divider
s_2	s_1	s_3
s_3	s_1	s_2
s_3	s_2	s_1

   (b) Two answers possible as shown in the following table.

Chooser 1	Chooser 2	Divider
s_1	s_3	s_2
s_2	s_3	s_1

   (c) The only possible fair division is

Chooser 1	Chooser 2	Divider
s_1	s_3	s_2

   (d) The divider can pick between s_1 and s_3—let's say the divider picks s_1. Then s_2 and s_3 can be combined again into a *cake* that may then be divided between Chooser 1 and Chooser 2 using the divider/chooser method.

3. (a) One possible fair division of the cake is

Chooser 1	Chooser 2	Chooser 3	Divider
s_2	s_4	s_1	s_3

(b) Another possible fair division of the cake is

Chooser 1	Chooser 2	Chooser 3	Divider
s_4	s_1	s_2	s_3

(c) Since none of the choosers have chosen s_3, s_3 can only be given to the divider.

5. (a) A fair division of the cake is

Chooser 1	Chooser 2	Chooser 3	Chooser 4	Divider
s_3	s_4	s_2	s_5	s_1

(b) Another fair division of the cake is

Chooser 1	Chooser 2	Chooser 3	Chooser 4	Divider
s_4	s_3	s_2	s_5	s_1

(c) Since none of the choosers chose s_1, s_1 can only be given to the divider.

7. (a) no   **(b)** P_4 **(c)** P_1 **(d)** P_3 **(e)** 3

9. *A* gets the desk and the tapestry and must pay \$295.55; *B* gets the dresser and receives \$111.11; *C* gets the vanity and receives \$184.44.

11. (a) *A* ends up with all four items and must pay \$173,777.78.
(b) *B* ends up with \$89,888.89.
(c) *C* ends up with \$83,888.89.

13. (a) P_1 gets items 10, 11, 12, 13; P_2 gets items 1, 2, 3; P_3 gets items 5, 6, 7.
(b) items 4, 8, and 9

15. (a) P_1 gets items 1, 2; P_2 gets items 10, 11, 12; P_3 gets items 4, 5, 6, 7.
(b) items 3, 8, and 9

17. (a) The allocation of items to each player is as follows:

Player	Items received
P_1	19 and 20
P_2	15, 16, and 17
P_3	1, 2, and 3
P_4	11, 12, and 13
P_5	5, 6, 7, and 8

(b) Items 4, 9, 10, 14, and 18 are left over.

19. (a) only (iii)   **(b)** (i): I; (ii): I; (iii): I; (iv) either

■ Jogging

21. (a) After deciding who is the chooser and who are the three dividers, the three dividers divide the cake into three slices using the lone chooser method (for three players as described in the chapter). Each considers her slice worth at least 1/3 of the total. Each divider divides her piece into four subpieces that she considers to be of equal value. The chooser now picks one subpiece from each divider.

(b) After deciding who is the chooser and who are the $N - 1$ dividers, the dividers divide the cake into $N - 1$ slices using the lone chooser method (for $N - 1$ players). Each considers his slice worth at least $1/(N - 1)$ of the total. Each divider divides his piece into N subpieces that he considers to be of equal value. The chooser now picks one subpiece from each divider.

23. (a) The total area is 30,000 m² and the area of C is only 9,000 m². Since P_2 and P_3 value the land uniformly, each thinks that a fair share must have an area of at least 10,000 m².

(b) Since there are 21,000 m² left, any cut that divides the remaining property in parts of 10,500 m² will work.

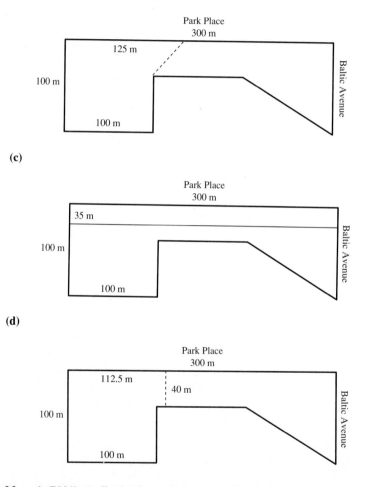

(c)

(d)

25. Move 1 (Bidding). Each player makes a sealed bid giving his or her honest assessment of the dollar value of each of the items in the estate. **Move 2 (Original Allocation).** Each item goes to the highest bidder for that item. (In case of ties, a predetermined tie-breaking procedure should be invoked.) Each player's fair share is calculated by multiplying the total of that player's bids by the percentage that player is entitled to. (Multiply the total of P_1's bids by $r_1/100$, etc.) Each player puts in or takes out from a common pot (*the estate*) the difference between his or her fair share and the total value of the items allocated to that player. **Move 3 (Dividing the surplus).** After the original allocations are completed there may be a surplus of cash in the estate. This surplus is divided among the players according to the percentage each is entitled to. (P_1 gets $r_1/100$ of the surplus, etc.)

27. **(a)**

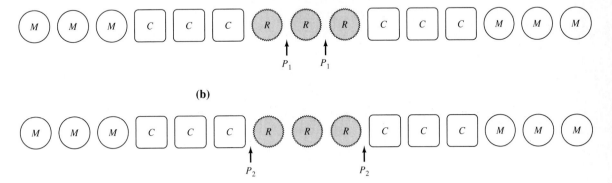

(b)

(c)

(d) P_1 gets 1 Reese's Pieces; P_2 gets 3 caramels and 3 mints; P_3 gets 1 caramel and 3 mints.

(e) 2 caramels and 2 Reese's Pieces.

29. Think of this problem as an ordinary fair division problem with ten players with A controlling seven of the players (P_1, P_2, P_3, P_4, P_5, P_6, P_7) and B controlling the three other players (P_8, P_9, P_{10}). Using the lone divider method, A will receive seven pieces each of which is worth (to A) at least $1/10$ of the cake and B will get three pieces each of which is worth (to B) at least $1/10$ of the cake.

31. **(a)** No. The chooser is not assured a fair share in her own value system.

   (b) The divider. (The divider is assured a fair share of the cake if he divides it into two pieces each of which represent a fair share.)

CHAPTER 4

■ Walking

1. **(a)** 50,000

   (b) A: 26.6; B: 53.4; C: 14.2; D: 65.8

   (c) A: 27; B: 53; C: 14; D: 66

3. (a) A: 26.923; B: 54.049; C: 14.372; D: 66.599
   (b) A: 26; B: 54; C: 14; D: 66
5. (a) The standard divisor (50,000) works.
   (b) A: 27; B: 53; C: 14; D: 66
7. A: 13; B: 38; C: 20; D: 31; E: 33; F: 15
9. A: 13; B: 38; C: 20; D: 31; E: 33; F: 15
11. (a) 9.5156 The standard divisor represents the average number of patients per nurse.
   (b) A: 86.49; B: 69.26; C: 40.14; D: 29.11
   (c) A: 87; B: 69; C: 40; D: 29
13. (a) 124
   (b) 200,000
   (c) A: 5,052,000; B: 3,664,000; C: 516,000; D: 7,432,000; E: 8,136,000
   (d) A: 25; B: 19; C: 3; D: 37; E: 40
15. A: 25; B: 18; C: 2; D: 38; E: 41
17. A: 25; B: 18; C: 3; D: 37; E: 41
19. (a) Bob: 7; Peter: 2; Ron: 1
   (b) Bob: 8; Peter: 3; Ron: 0
   (c) With 10 pieces of candy, Ron was to get one piece, but with 11 pieces, Ron ends up with none.

■ Jogging

21. Answers will vary. One such example is: Apportion 10 seats among the four states A, B, C, and D with populations given in the following table.

State	A	B	C	D
Population (in millions)	2.24	2.71	2.13	2.92

Hamilton's and Lowndes' methods both result in the apportionment A: 2, B: 3, C: 2, D: 3.

23. Same as answer to Exercise 21.
25. (a) The two fractional parts must add up to 1. So, either they are both 0.5, or one is more than 0.5 and the other is less than 0.5.
   (b) Since the two fractional parts add up to 1, the surplus to be allocated using Hamilton's method is 1, and it will go to the state with largest fractional part, that is, the state with fractional part more than 0.5. This is the same result that is obtained by just rounding off in the conventional way, which in this case happens to be the result given by Webster's method. (Anytime that rounding off the quotas the conventional way produces integers that add up to M, Webster's method reduces to conventional rounding off of the quotas.)
   (c) When there are only 2 states, Hamilton's and Webster's methods agree and Webster's method can never suffer from the Alabama or population paradox.
   (d) When there are only 2 states Webster's and Hamilton's methods agree and Hamilton's method can never violate the quota rule.
27. (a) A: 5; B: 10; C: 15; D: 21.
   (b) For $D = 100$ the modified quotas are A: 5, B: 10, C: 15, D: 20 which are

all integers and so remain unchanged when we round upward—giving a total of 50. For $D < 100$, each of the modified quotas above will increase and so rounding upward will give at *least* A: 6, B: 11, C: 16, D: 21 or a total of at *least* 54. For $D > 100$, each of the modified quotas above will decrease and so rounding upward will give at *most* A: 5, B: 10, C: 15, D: 20 or a total of at *most* 50.

    (c) From part (b) we see that there is no divisor such that after rounding the modified quotas upward, the total is 51.

29. (a) In Jefferson's method the modified quotas are larger than the standard quotas and so rounding downward will give each state at least the integer part of the standard quota for that state.

    (b) In Adams' method the modified quotas are smaller than the standard quota and so rounding upward will give each state at most one more than the integer part of the standard quota for that state.

    (c) If there are only two states, an upper quota violation for one state results in a lower quota violation for the other state (and vice-versa). Since neither Jefferson's nor Adams' method can have both upper and lower violations of the quota rule, neither can violate the quota rule when there are only two states.

CHAPTER 5

■ Walking

1. (a) Vertices: A, B, C, D; Edges: e_1, e_2, e_3, e_4
  (b) Vertices: A, B, C; Edges: none
  (c) Vertices: V, W, X, Y, Z; Edges: XX, XY, XZ, XV, XW, WY, YZ

3. (a)

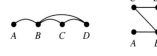

  (b)

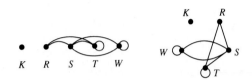

5. (a) $\deg(A) = 3$; $\deg(B) = 2$; $\deg(C) = 1$; $\deg(D) = 2$
  (b) $\deg(A) = 0$; $\deg(B) = 0$; $\deg(C) = 0$
  (c) $\deg(X) = 6$; $\deg(Y) = 3$; $\deg(Z) = 2$; $\deg(V) = 1$; $\deg(W) = 2$

7. (a) Both graphs have four vertices A, B, C, and D and (the same) edges AB, AC, AD, BD.

(b)

9. (a)   **(b)**

11. (a) C, B, A, H, F **(b)** C, B, D, A, H, F **(c)** 4   **(d)** 3   **(e)** 12

13. (a) D, C, B, A, D **(b)** 6   **(c)** HA and FE

15. (a) None of them.   **(b)**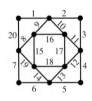

17.

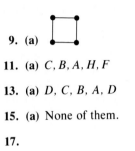

Has an Euler circuit since all vertices have even degree.

Has no Euler circuit, but has an Euler path since there are exactly two vertices of odd degree.

Has neither an Euler circuit nor an Euler path since there are four vertices of odd degree.

19.

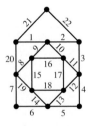

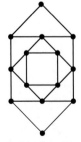

Can't be done since there are four vertices of odd degree.

21.

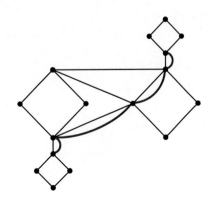

23.

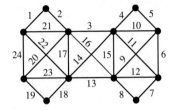

25.

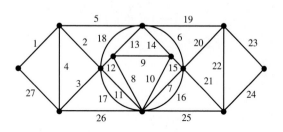

■ **Jogging**

27. (a)

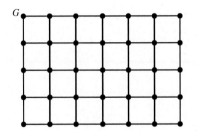

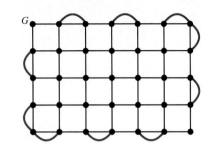

(b)

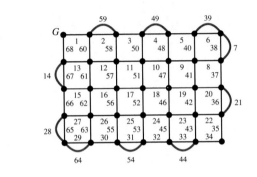

(c)

29. Eulerizing the graph shown in Fig. 5-10(b) requires the addition of two edges so the cheapest walk will cost $9.00. One possible such walk is shown in the following figure.

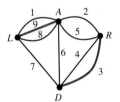

(Note: Traveling the edge added between L and A means going back over either one of the bridges between L and A.)

31. (a)

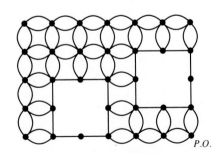

(b) The graph is already eulerized.

(c)

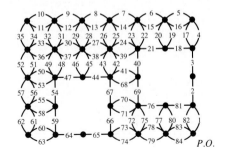

33. This problem can be represented by the graph

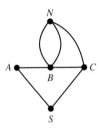

where N is the North Bank, S is the South Bank, and A, B, and C are the three islands. Since N and C are the only vertices of odd degree, this graph has an Euler path (but no Euler circuit). So it is possible to take a walk and cross each bridge exactly once as long as you start at either N or C and end at the other one. (You cannot start and end at the same place.)

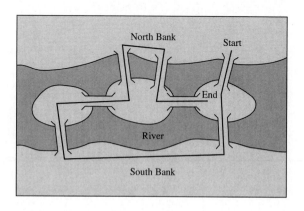

35. (a)

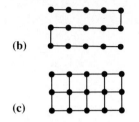

(b)

(c)

CHAPTER 6

■ **Walking**

1. *A, D, B, E, C, F, G, A* and *A, G, B, D, C, E, F, A*

3. *A, B, C, D, E, F, G, A* | *A, G, F, E, D, C, B, A*
    A, B, E, D, C, F, G, A | *A, G, F, C, D, E, B, A*
    A, F, C, D, E, B, G, A | *A, G, B, E, D, C, F, A*
    A, F, E, D, C, B, G, A | *A, G, B, C, D, E, F, A*

5. (a) 6   **(b)** 4   **(c)** *A, B, C, D, E, A*; 32   **(d)** *A, D, B, C, E, A*; 27

7. (a) $6! = 5! \times 6 = 120 \times 6 = 720$

(b) $9! = \dfrac{10!}{10} = 362{,}880$

(c) $9! = 362{,}880$

9. (a) *A, C, B, D, A* with weight 62
   (b) *A, D, C, B, A* with weight 80
   (c) *A, B, D, C, A* with weight 74
   (d) .290; .194 (rounded to 3 decimal places).

11. *D, A, C, E, B, D*

13. (a) *A, D, B, E, C, A* with weight 103
   (b) *A, B, C, D, E, A* with weight 82

15. *C, D, E, A, B, C*

17. *A, B, F, C, D, E, A*

19. (a)

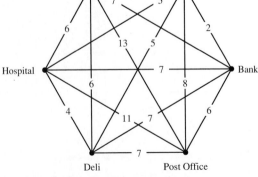

(b) Home, Bank, Post Office, Deli, Hospital, Karl's, Home (30 miles, Nearest Neighbor Algorithm).

21.

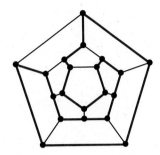

23. Any circuit passing through vertex B must contain edge AB. Any circuit passing through vertex D must contain edge AD. Any circuit passing through vertex G must contain edge AG. Consequently, any circuit passing through vertices B, D, and G must contain at least three edges meeting at A and hence would pass through vertex A more than once.

25. (a) E, D, C, B, A, F
    (b) A, B, C, D, E, J, G, I, F, H

27. (a) $2^3 = 8 > 6 = 3!$
    (b) $2^4 = 16 < 24 = 4!$
    (c) $N!$ is bigger.
$$2^5 = 2 \cdot 2^4 < 2 \cdot 4! < 5 \cdot 4! = 5!$$
$$2^6 = 2 \cdot 2^5 < 2 \cdot 5! < 6 \cdot 5! = 6!$$
$$2^7 = 2 \cdot 2^6 < 2 \cdot 6! < 7 \cdot 6! = 7!$$
$$\vdots$$
$$2^{k+1} = 2 \cdot 2^k < 2 \cdot k! < (k+1) \cdot k! = (k+1)!$$
$$\vdots$$

In other words, as k increases by 1, 2^k increases by a factor of 2, but $k!$ increases by a factor of $(k+1)$.

29. Dallas, Houston, Memphis, Louisville, Columbus, Chicago, Kansas City, Denver, Atlanta, Buffalo, Boston, Dallas.

CHAPTER   7

1. (a) tree
   (b) not a tree—has a circuit and also is not connected
   (c) not a tree—has a circuit
   (d) tree

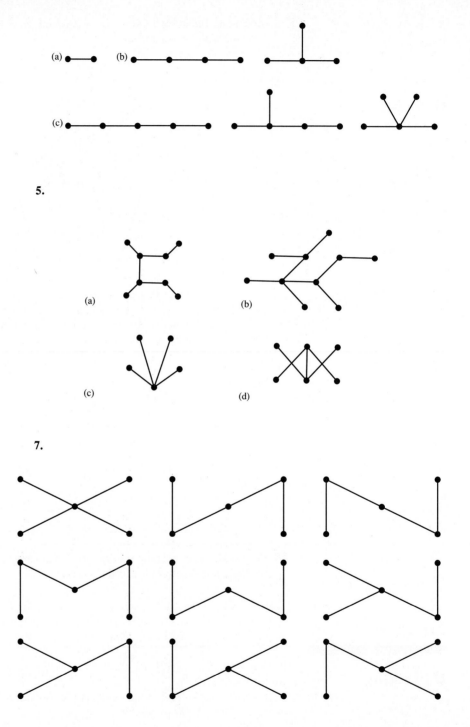

5.

7.

9.

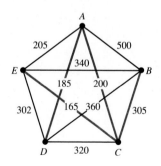

11.

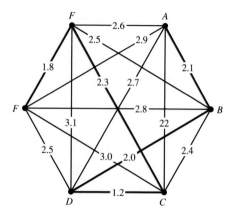

13. (a) *CE* + *ED* + *EB* is larger since *CD* + *DB* is the shortest network connecting the points *C*, *D*, and *B*.
   (b) *CD* + *DB* is the shortest network connecting the points *C*, *D*, and *B* since angle *CDB* is 120° and so the shortest network is the same as the minimum spanning tree.
   (c) *CE* + *EB* is the shortest network connecting the points *C*, *E*, and *B* since angle *CEB* is more than 120° and so the shortest network is the same as the minimum spanning tree.

15. 88.2

17. Call $\angle SAB = x$. Then, $\angle SBA = 60 - x$; $\angle SBC = 60 - (60 - x) = x$ and $\angle SCB = 60 - x$. Also, $AB = BC$. It follows that the triangles *ASB* and *BSC* are congruent. An identical argument shows that *BSC* and *CSA* are congruent.

19. (a) From Exercise 18, it follows that triangle *ABJ* is a 30°-60°-90° triangle with hypotenuse $AB = a$. It follows that $BJ = a/2$ and $JA = (\sqrt{3}/2)a \approx 0.866a$.

(b) Triangle *BSJ* is also a 30°-60°-90° triangle with longer leg *BJ* = *a*/2. It follows that the shorter leg *SJ* = $a/2\sqrt{3}$ and the hypotenuse *BS* = $a/\sqrt{3}$ Since *BS* = *SA* we have *SA* = $a/\sqrt{3}$ = $(\sqrt{3}/3)a \approx 0.577a$

■ Jogging

21. 230 miles

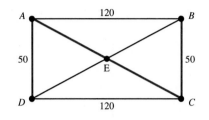

23. In a tree there is one and only one path joining any two vertices (Property 1). Consequently, the only path joining two adjacent vertices is the edge connecting them and so if that edge is removed, the graph will become disconnected.

25. (a) No. A tree with four vertices must have three edges and hence the sum of the degrees of all the vertices is 6.

(b) 6

(c) 8

(d) $2N - 2$

27.

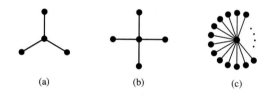

(a)   (b)   (c)

29. Length of the network is $4x + (100 - x) = 3x + 100$, where $50^2 + \left(\dfrac{x}{2}\right)^2 = x^2$. (See figure.)

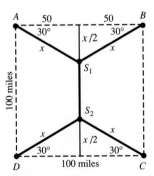

Solving the equation gives $x = 100\sqrt{3}/3$, and so the length of the network is $100\sqrt{3} + 100$.

31. (a) $\angle BFA = \angle AEC = \angle CGB = 120°$

(b) In $\triangle ABF$, $\angle BFA = 120°$ [from part (a)] and so $\angle BAF + \angle ABF = 60°$. Therefore $\angle BAF < 60°$ and $\angle ABF < 60°$. Similarly, $\angle ACE < 60°$, $\angle CAE < 60°$, $\angle BCG < 60°$, and $\angle CBG < 60°$. Consequently, $\angle A = \angle BAF + \angle CAE < 60° + 60° = 120°$. Likewise, $\angle B < 120°$ and $\angle C < 120°$.

(c) Any point X inside or on $\triangle ABF$ (except vertex F) will have $\angle AXB > 120°$. Any point X inside or on $\triangle ACE$ (except vertex E) will have $\angle AXC > 120°$. Any point X inside or on $\triangle BCG$ (except vertex G) will have $\angle BXC > 120°$. If S is the Steiner point, $\angle ASB = \angle ASC = \angle BSC = 120°$, and so S cannot be inside or on $\triangle ABF$ or $\triangle ACE$ or $\triangle BCG$. It follows that the Steiner point S must lie inside $\triangle EFG$.

CHAPTER 8

■ Walking

1. (a)

Vertex	Degree	Indegree	Outdegree	Incident to	Incident from
A	3	2	1	C	B, D
B	2	0	2	A, D	—
C	1	1	0	—	A
D	2	1	1	A	B

(b)

Vertex	Degree	Indegree	Outdegree	Incident to	Incident from
A	3	2	1	C	B, C
B	2	0	2	A, D	—
C	4	1	3	A, D, E	A
D	3	3	0	—	B, C, E
E	2	1	1	D	C

(c)

Vertex	Degree	Indegree	Outdegree	Incident to	Incident from
A	1	1	0	—	B
B	3	2	1	A	E
C	2	1	1	F	E
D	1	1	0	—	E
E	5	0	5	B, C, D, F	—
F	2	2	0	—	C, E

3.

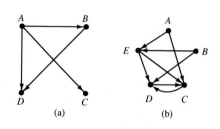

(a)   (b)

5.

Tasks	F	AF	IF	AW	AR	PL	IW	E	IR	HC	PT	EC	FM	C	FC
Time 0	7	12	17	23	31	35	42	46	51	54	58	60	63	64	70

7. According to the precedence relations, E needs to be completed before starting HC, so E cannot be placed after HC in the schedule.

9. At time 16 there is an available job for Xavier (PL) and so Xavier cannot remain idle.

11.

Time	0		9		15	17			26
Processor 1		C(9)		E(6)		G(2)		idle	
Processor 2		A(8)		B(5)		D(12)			F(1)
Time	0			8	13				25 26

13.

Time	0		15	20	25	30		45		65
Processor 1		D		H	J	idle	I		K	
Processor 2		C	B	F	G	A	E	idle		
Time	0		10	14	17	21 23	30			65

15.

Time	0	2	4		11	14	19		29		49
Processor 1	B		E	F		I			K		
Processor 2	A	C		idle	G		idle				
Processor 3		D			H	J		idle			
Time					12	15	20	25			49

17.

Time	0		8	15	20	27	32	35	38	44
X		AR		F	IF	IW	IR	HC	FM	FC
Z		AW	AF	idle		PL	idle	E	PT	EC c	idle

| Time | 0 | 6 | 11 | | 20 24 | 27 | 31 | | 35 37 38 | 44 |

19. (a)

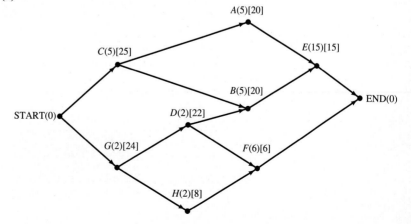

$A(4)$

$B(4)$

$C(4)$

START(0)

$D(7)$ $G(15)$

$E(7)$ $H(15)$

$F(7)$ $I(15)$

$J(15)$

END(0)

(b)

Time	0		7		22	26
Processor 1		D		G		C
Processor 2		E		H		idle
Processor 3		F		I		idle
Processor 4	A	B		J		idle

| Time | 0 | 4 | 8 | | 23 | 26 |

21. (a)

$A(5)[20]$

$C(5)[25]$

$E(15)[15]$

$B(5)[20]$

START(0)

$D(2)[22]$

END(0)

$G(2)[24]$

$F(6)[6]$

$H(2)[8]$

(b)

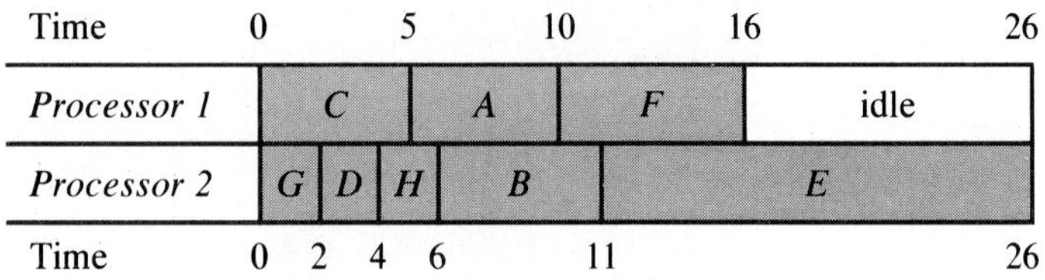

Time	0	5	10	16	26
Processor 1	C	A	F	idle	
Processor 2	G D H	B	E		

Time 0 2 4 6 11 26

23.

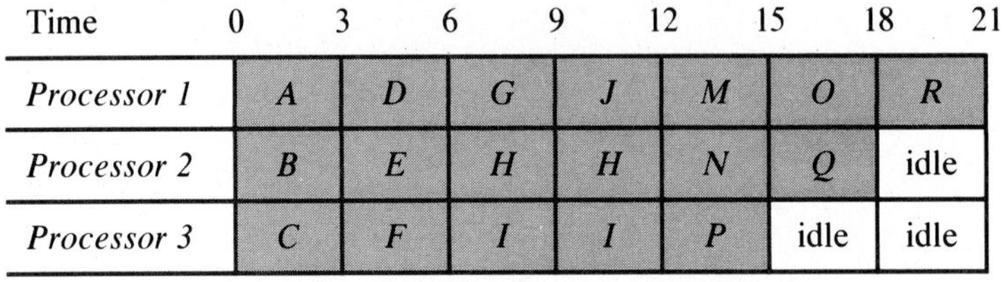

Time	0	3	6	9	12	15	18	21
Processor 1	A	D	G	J	M	O	R	
Processor 2	B	E	H	H	N	Q	idle	
Processor 3	C	F	I	I	P	idle	idle	

25.

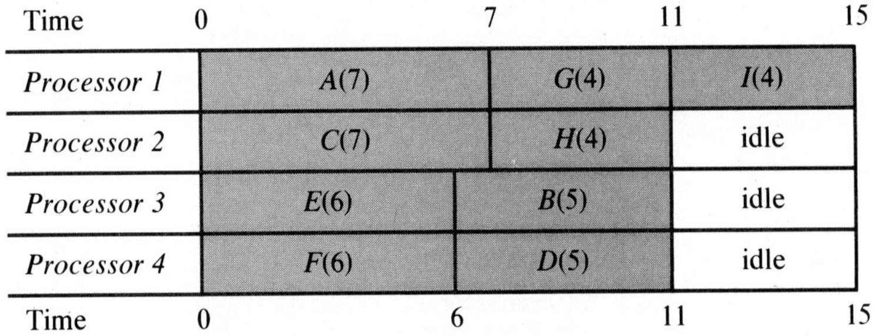

Time	0		7	11	15
Processor 1	A(7)		G(4)	I(4)	
Processor 2	C(7)		H(4)	idle	
Processor 3	E(6)	B(5)		idle	
Processor 4	F(6)	D(5)		idle	

Time 0 6 11 15

27.

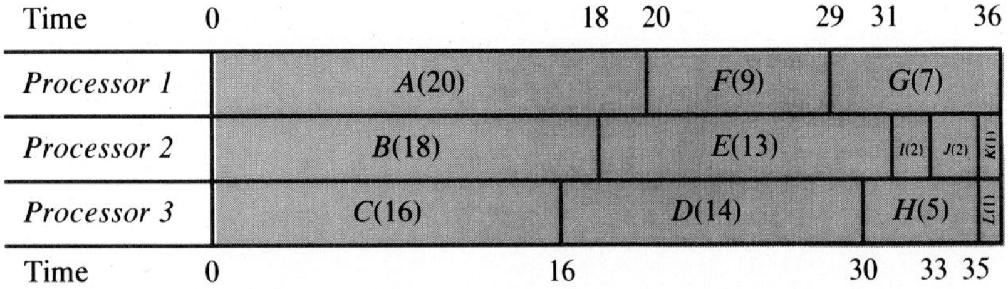

Time	0	18	20	29	31	36
Processor 1	A(20)		F(9)		G(7)	
Processor 2	B(18)		E(13)		I(2) J(2) K(1)	
Processor 3	C(16)	D(14)		H(5)	L(1)	

Time 0 16 30 33 35

This schedule is optimal because the sum of the lengths of all the tasks is
108 and so 3 processors cannot complete the job any sooner than time 108/
3 = 36.

29. (a)

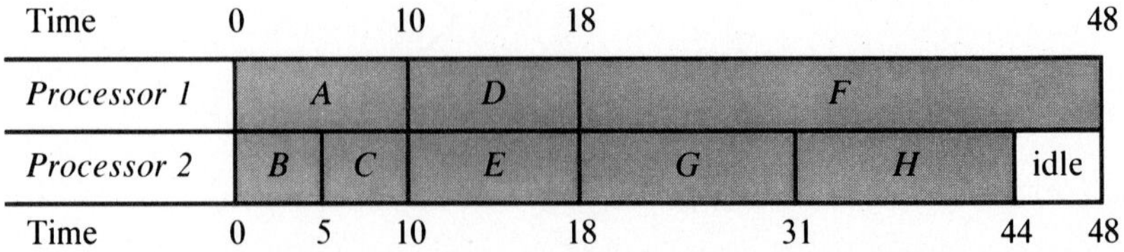

Time	0	10	18	48
Processor 1	A	D	F	
Processor 2	B	C	E	G	H	idle

Time 0 5 10 18 31 44 48

(b) The critical path for this project has length 48 and so the job cannot be completed any sooner.

(c)

Time	0		10	18	26		56
Processor 1		A		D	E	F	
Processor 2	B		G		idle		
Processor 3	C		H		idle		
Time	0	5		18			56

(d) Processors cannot remain idle if there is an available task to be done. Consequently, tasks A, B, C must be assigned first (they are the only 3 tasks available at the beginning) and then G and H must be assigned to the processors that finished B and C (since all other tasks require that A is completed first). This means that task D or E must be assigned to the processor that finishes task A and then all 3 processors will be free at time 18. But there are two tasks left, F and either D or E. F cannot be started until both E and D are completed and so the project cannot be completed before time $18 + 38 = 56$.

(e) The optimal completion time with 3 processors is more than the optimal completion time with 2 processors. This paradoxical situation is a consequence of the requirement that a processor cannot remain idle if there is an available task. Tasks D and E should be completed early so that the long task F can be started, but processor 2 and processor 3 were forced to start other jobs and so were not available to start tasks D or E when they became available.

CHAPTER 9

■ Walking

1. (a) $F_{16} = 987$
   (b) $F_{35} = 9227465$ and $F_{38} = 39088169$
3. (a) 5, 8, 11, 14, 17, 20, 23, 26, 29, 32
   (b) $A_{500} = 1502$
   (c) Is $3053 = 3N + 2$? Yes, $N = 1017$. $A_{1017} = 3053$.
5. (a) $-1, 3, -1, 3, -1$
   (b) -1
   (c) 100
7. (a) 0, 4, 12, 28, 60, 124, 252, 508, 1020, 2044
   (b) $A_{N+1} = 2A_N + 4$
   (c) No (all terms are even)
9. 1, -1, 1, -1, 1
11. 3, 0, 1, 10, 8, -9, 5, 47, 10, -69
13. (a) $\frac{1}{2}, \frac{1}{3}, \frac{1}{4}, \frac{1}{5}$
    (b) $\frac{1}{214}$

15. (a) 2, 4, 6, 8, 10, 12
(b) 200
(c) $V_N = 2N$
17. $x = 12$, $y = 10$
19. $x = 1.2$

■ Jogging

21. $x = 6$, $y = 12$, $z = 10$
23. $x = 3$, $y = 5$
25. $\Phi(\Phi - 1) = \left(\dfrac{1 + \sqrt{5}}{2}\right)\left(\dfrac{1 + \sqrt{5}}{2} - 1\right) = \left(\dfrac{1 + \sqrt{5}}{2}\right)\left(\dfrac{-1 + \sqrt{5}}{2}\right) =$
$\dfrac{-1 + 5}{4} = 1$. Therefore $\dfrac{1}{\Phi} = \Phi - 1$.
27. (a) 4, 7, 12, 20, 33, 54, 88
(b) $F_1 + F_2 + \cdots + F_N = F_{N+2} - 1$
29. Since AM and AC are the same length, triangle I is isosceles and $\angle AMC = 72°$ and so $\angle MAC = 180° - 72° - 72° = 36°$. Consequently triangle ABC is similar to triangle I, making triangle II a gnomon to triangle I.
31. (a) 1 **(b)** -1 **(c)** 3 **(d)** 4
33. (a) 13, 21, 34 **(b)** $A_N = F_{N+4}$
35. (a) $N = 1$: $F_3^2 - F_2^2 = 2^2 - 1^2 = 4 - 1 = 3 = 1 \cdot 3 = F_1 \cdot F_4$
$N = 2$: $F_4^2 - F_3^2 = 3^2 - 2^2 = 9 - 4 = 5 = 1 \cdot 5 = F_2 \cdot F_5$
$N = 3$: $F_5^2 - F_4^2 = 5^2 - 3^2 = 25 - 9 = 16 = 2 \cdot 8 = F_3 \cdot F_6$
(b) $F_{N+2}^2 - F_{N+1}^2 = (F_{N+2} - F_{N+1})(F_{N+2} + F_{N+1}) = F_N \cdot F_{N+3}$

CHAPTER 10

■ Walking

1. (a) $40.50 **(b)** 40.5% **(c)** 40.5%
3. 35%
5. $4587.64
7. (a) $9083.48 **(b)** 12.6825%
9. The Great Bulldog Bank: 6%; The First Northern Bank: $\approx 5.9\%$; The Bank of Wonderland: $\approx 5.65\%$
11. $\approx$$1133.56
13. (a) 716 **(b)** $16 + 7N$ **(c)** 3,519,500 **(d)** 3,510,141
15. (a) 213 **(b)** $137 + 2N$ **(c)** $7124 **(d)** $2652
17. (a) 3^{99} **(b)** 3^{N-1} **(c)** $(3^{100} - 1)/2$ **(d)** $(3^{100} - 3^{49})/2$
19. (a) $\frac{10}{29}$ **(b)** $\frac{19}{29}$ **(c)** stable at $\frac{19}{29}$ **(d)** $\frac{25}{29}$

■ Jogging

21. 100%
23. $10,737,418.23
25. $\approx$14,619 snails
27. 6425
29. (a) $804.63 **(b)** $771.36

CHAPTER 11

■ Walking

1.

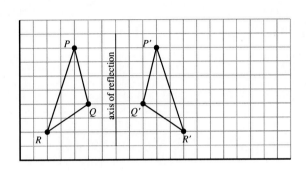

3. (a) 140°   **(b)** 279°

5.

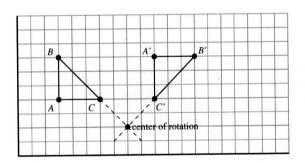

7.

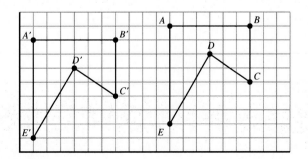

9.

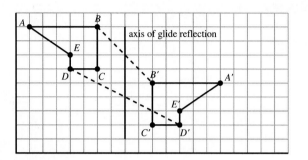

11. (a) vertical reflection
(b) horizontal reflection
(c) horizontal reflection, vertical reflection, reflection about an axis in the northwest direction, reflection about an axis in the northeast direction, rotation of 90° (180°, 270°, etc.)
(d) 180° rotation
(e) identity only

13. (a) C   **(b) V**   **(c) I**   **(d) S**   **(e) J**

15.

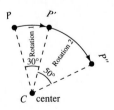

17. (a) translation, vertical reflection, horizontal reflection, 180° rotation
(b) translation, horizontal reflection
(c) translation, vertical reflection
19. (a) translation, vertical reflection
(b) translations, horizontal reflection

■ **Jogging**

21. (a) Rotation 1 moves point P to point P' (see figure).
Rotation 2 moves point P' to point P'' (see figure).
Rotation 1 followed by rotation 2 moves point P to point P'' and so is equivalent to a rotation with center C and a clockwise angle of 80°.

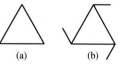

(b) Rotation 1 moves point P to point P' (see figure).

Rotation 2 moves point P' to point P'' (see figure).

Rotation 1 followed by rotation 2 moves point P to point P'' and so is equivalent to a rotation with center C and a clockwise angle of $\alpha + \beta$.

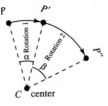

(c) Rotation 2 followed by rotation 1 is a rotation with center C and clockwise angle $\beta + \alpha$ (which is equal to $\alpha + \beta$). (See figure.)

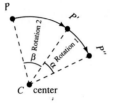

23. (a)

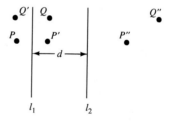

(b)

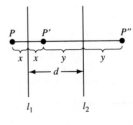

Distance from P to P'' is $x + x + y + y = 2x + 2y = 2(x + y) = 2d$. This is a translation by a vector perpendicular to l_1 and l_2 of length $2d$ and going from left to right.

(c)

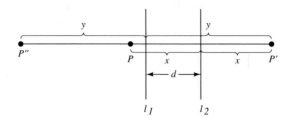

Distance from P to P'' is $y - (x + x - y) = 2y - 2x = 2(y - x) = 2d$. This is a translation by a vector perpendicular to l_1 and l_2 of length $2d$ and going from right to left.

25.

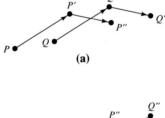

(a)

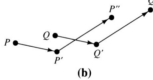

(b)

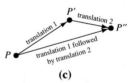

(c)

27. (a) translation, 180° rotation
   (b) translation, 180° rotation
   (c) translation, vertical reflection, 180° rotation, glide reflection
29. (a) Rotations and translations are proper rigid motions, and hence preserve clockwise-counterclockwise orientations. The given motion is an improper rigid motion (it reverses the clockwise-counterclockwise orientation).
   (b) If the rigid motion was a reflection, then PP', RR', and QQ' would all be perpendicular to the axis of reflection and hence would all be parallel.
   (c) It must be a glide reflection (the only rigid motion left).

CHAPTER 12

■ Walking

1. **(a), (b), (c)** A line segment joining the midpoints of two sides of a triangle is parallel to the third side and has length one-half the third side. This implies that all the triangles are congruent.

   (d) $\frac{1}{4}$; all 4 triangles are congruent and so have equal areas.

3. **(a)** $\frac{3}{4}X$, $\frac{9}{16}X$, $\frac{27}{64}X$

   (b) $\left(\frac{3}{4}\right)^N X$

   (c) $\left(\frac{3}{4}\right)^N$ gets closer and closer to 0 as N gets bigger and bigger.

5.

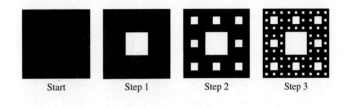

Start   Step 1   Step 2   Step 3

7. **(a)** 512

   (b) $\left(\frac{8}{9}\right)^3 X = \frac{512}{729} X$

   (c) 8^N

   (d) $\left(\frac{8}{9}\right)^N X$

   (e) $\left(\frac{8}{9}\right)^N$ gets closer and closer to 0 as N gets bigger and bigger.

9.

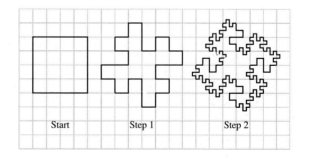

Start   Step 1   Step 2

11. 4×7^N

13.

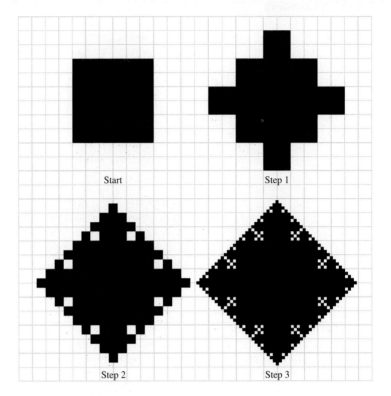

Start

Step 1

Step 2

Step 3

15. **(a)** $\frac{20}{3}$, $\frac{100}{9}$ **(b)** $4\left(\frac{5}{3}\right)^N$

17.

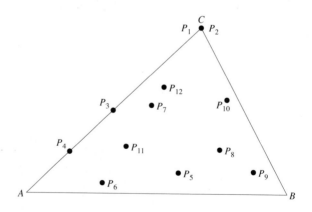

19. **(a)** 2, 2, 2, 2, 2   **(b)** stable at 2

■ **Jogging**

21. At each step, the number of black triangles is tripled and one new white triangle is introduced for each black triangle. The following table summarizes the first few steps.

	White △'s	Black △'s
Step 1	1	3
Step 2	$1 + 3$	3^2
Step 3	$1 + 3 + 3^2$	3^3
Step 4	$1 + 3 + 3^2 + 3^3$	3^4

We see that the Nth step there will be

$$1 + 3 + 3^2 + 3^3 + \cdots + 3^{N-1} = \frac{3^N - 1}{2}$$

white triangles.

23. There will be infinitely many points left. For example, the 3 vertices of the original triangle will be left as well as the vertices of every black triangle that occurs at each step of the construction.

25. Step 99: -0.4299; step 100: -0.5652. Long-term behavior, oscillates above and below -0.5 and gets closer and closer to -0.5.

27. Step 99: 0.4906; step 100: 0.4907. Long-term behavior, gets closer and closer to 0.5.

29. Step 1: $2 + \sqrt{2}$. Long-term behavior, grows without bound.

CHAPTER 13

■ **Walking**

1. (a) Answer 1: All married people. Answer 2: All married people who read Dear Abby's column. Note: While both answers are acceptable, it is clear that Dear Abby was trying to draw conclusions about married people at large so Answer 1 is better.

   (b) 210,336

   (c) self-selection

   (d) 85% is a statistic, since it is based on data taken from a sample.

3. (a) 74.0%

   (b) 81.8%

   (c) Not very accurate. The sample was far from being representative of the entire population.

5. (a) The citizens of Cleansburg.   **(b)** 475

7. (a) The choice of street corner could make a great deal of difference in the responses collected.

   (b) D. (We are making the assumption that people who live or work downtown are much more likely to answer yes than people in other parts of town.)

   (c) Yes, for two main reasons: (i) people out on the street between 4 P.M. and 6 P.M. are not representative of the population at large. For example, office and white collar workers are much more likely to be in the sample than homemakers and school teachers. (ii) The five street corners were chosen by the interviewers, and the passersby are unlikely to represent a cross section of the city.

   (d) No. No attempt was made to use quotas to get a representative cross section of the population.

9. (a) All undergraduates at Tasmania State University.

   (b) $N = 15,000$

11. (a) No. The sample was chosen by a random sampling method.

   (b) In simple random sampling, any two members of the population have as much chance of both being in the sample as any other two. But in this sample, two people with the same last name—say Len Euler and Linda Euler—have almost no chance of both being in the sample. (By the way, the sampling method described in this exercise is frequently used and goes by the technical name of **systematic sampling**.)

13. (a) Anyone who could have a cold and would consider buying vitamin X (i.e., pretty much all adults).

   (b) Presumably they volunteered. (We could infer this from the fact that they are being paid.)

   (c) $n = 500$

   (d) No. There was no control group.

15. (i) Using college students. (College students are not a representative cross section of the population in terms of age and therefore in terms of how they would respond to the treatment.)

   (ii) Using subjects only from the San Diego area. (Same as (i)).

   (iii) Offering money as an incentive to participate. (Bad idea!)

   (iv) Allowing self-reporting (the subjects themselves determine when their colds are over) is a very unreliable way to collect data and is especially bad when the subjects are paid volunteers.

17. Anyone who could potentially suffer from arteriosclerosis.

19. (a) There was a treatment group (the ones getting the beta-carotene pill) and there was a control group. The control group received a placebo pill. These two elements make it a controlled placebo experiment.

   (b) The group that received the beta carotene pills.

   (c) Both the treatment and control groups were chosen by random selection.

■ Jogging

21. (a) The question was worded in a way that made it almost impossible to answer yes.

   (b) This is harder than it seems. A reasonably neutral question might be: "Do you support some form of tax increase as a way to balance the federal budget?"

23. (a) Under method 1, people whose phone numbers are unlisted are automatically ruled out from the sample. At the same time, method 1 is cheaper and easier to implement than method 2.

   (b) For this particular situation, method 2 is likely to produce much more reliable data than method 1. The two main reasons are: (1) People with unlisted phone numbers are very likely to be the same kind of people that would seriously consider buying a burglar alarm, and (2) the listing bias is more likely to be significant in a place like New York City. (People with unlisted phone numbers make up a much higher percentage of the population in a large city such as New York than in a small town or rural area.)

25. (a) 2000

(b) $N = \dfrac{n_2}{k} \cdot n_1$

(c) Both samples should be a representative cross section of the same population. In particular, it is essential that the first sample, after being released, be allowed to disperse evenly throughout the population and that the population should not change between the time of the capture and the time of the recapture.

(d) It is possible (especially when dealing with elusive types of animals) that the very fact that the animals in the first sample allowed themselves to be captured makes such a sample biased (they could represent a slower, less cunning group). This type of bias is compounded with the animals that get captured the second time around. A second problem is the effect that the first capture can have on the captured animals. Sometimes the animal may be hurt (physically or emotionally) making it more (or less) likely to be captured the second time around. A third source of bias is the possibility that some of the tags will come off.

(e) 20,500

CHAPTER   14

■ **Walking**   **1. (a)**

Score	10	50	60	70	80	100
Frequency	1	3	7	6	5	2

(b)

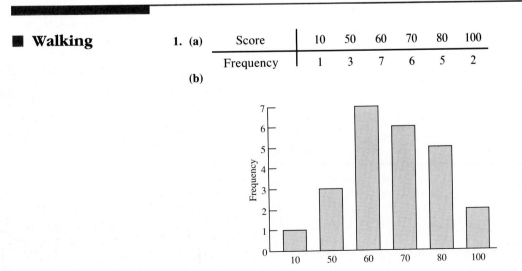

(c)

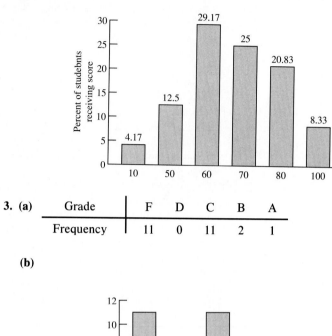

3. (a)

Grade	F	D	C	B	A
Frequency	11	0	11	2	1

(b)

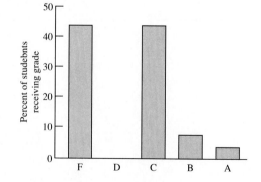

(c)

(d)

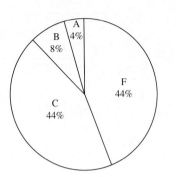

5. (a) 30

(b)

Score	3	4	5	6	7	8	9	10
Frequency	2	5	6	4	4	5	3	1

(c) 6.17

(d) 6

7. (a) 79

(b) 70

(c) $Q_1 = 55$, $Q_3 = 76$

(d) Min = 13, $Q_1 = 55$, $M = 70$, $Q_3 = 76$, Max = 92

(e) IQR = 21

9. (a) ≈65.32   **(b)** ≈15.76

11. Average = 10, SD = 0; Average = 10, SD ≈ 7.18; Average = 10, SD ≈ 9.03. Averages are all the same but the standard deviations get larger as the numbers are more spread out.

13. (a) 5.5   **(b)** 5.5   **(c)** 2.87

15. (a) 3   **(b)** 8   **(c)** 5

17. (a) 25.5   **(b)** 75.5   **(c)** 50

19. (a) 132

(b) 9

(c)

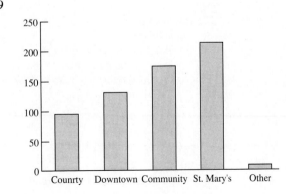

21.

Weight in Pounds

More Than	Less Than or Equal to	Frequencies
3	$4\frac{1}{2}$	39
$4\frac{1}{2}$	6	108
6	$7\frac{1}{2}$	303
$7\frac{1}{2}$	9	168
9	$10\frac{1}{2}$	7

23.

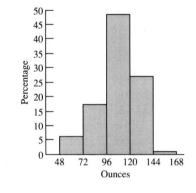

25.

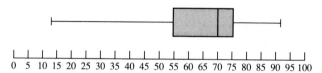

27. (a) Between $33,000 and $34,000
   (b) $40,000
   (c) The line crossing the central box (indicating the median salary) of the engineering box plot is to the right of the end of the central box (indicating the third quartile salary) of the agriculture box plot.

■ **Jogging**

29. $31,000
31. (a) 8   **(b)** 48
33. When all the numbers are equal.

35.

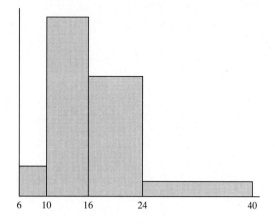

CHAPTER 15

■ **Walking**

1. (a) {*HHHH, HHHT, HHTH, HHTT, HTHH, HTHT, HTTH, HTTT, THHH, THHT, THTH, THTT, TTHH, TTHT, TTTH, TTTT*}
    (b) 16
    (c) {*HHTT, HTHT, HTTH, THHT, THTH, TTHH*}
    (d) $\frac{6}{16} = \frac{3}{8}$
3. (a) {$P_1, P_2, P_3, P_4, P_5, P_6, P_7$}; $\Pr(P_1) = .25$; $\Pr(P_2) = \Pr(P_3) = \Pr(P_4) = \Pr(P_5) = \Pr(P_6) = \Pr(P_7) = .125$
    (b) 1 to 3; 1 to 7
5. (a) 17,576,000   (b) 15,818,400   (c) 11,232,000
7. (a) 40,320   (b) 40,319
9. (a) {red, blue, yellow, purple, orange}
    (b) {red, blue, yellow}
    (c) {purple, orange}
    (d) $\Pr(\text{red}) = \Pr(\text{blue}) = \Pr(\text{yellow}) = 0.1$; $\Pr(\text{purple}) = \Pr(\text{orange}) = .35$
11. { }, {*A*}, {*B*}, {*C*}, {*A, B*}, {*A, C*}, {*B, C*}, {*A, B, C*}
13. (a) {*A, B, C, D, E*}
    (b) {*AB, AC, AD, AE, BA, BC, BD, BE, CA, CB, CD, CE, DA, DB, DC, DE, EA, EB, EC, ED*}
15. (a) $\frac{1}{12}$ (b) 1 to 11   (c) 11 to 1   (d) $\frac{11}{12}$
17. (a) $S = \{$ ⚁⚁, ⚁⚂, ⚁⚃, ⚁⚄, ⚂⚁, ⚂⚂, ⚂⚃, ⚂⚄, ⚃⚁, ⚃⚂ $\}$

    (b) $\frac{5}{18}$

    (c) 5 to 13

19. (a) $\frac{1}{36}$

    (b) $\Pr(ss) = \frac{1}{36}$; $\Pr(sf) = \frac{5}{36}$; $\Pr(fs) = \frac{5}{36}$; $\Pr(ff) = \frac{25}{36}$

■ **Jogging**

21. $1 - (\frac{35}{36})^{25} \approx .51$

23. (a) $(.98)^{12} \approx .78$

    (b) $(.98)^{12} + 12(.02)(.98)^{11} \approx .98$

25. (a) 0, 1, 2, 3, 4, 5   (b) $\frac{5}{18}$ (c) $\frac{1}{6}$

27. $2^6 = 64$

29. (a) 24,360   (b) 4060

31. (a) Let m be the number of heads tossed. Then $8 - m$ is the number of tails tossed and $W = m - (8 - m) = 2m - 8$. So, W is even and for $m = 0, 1, 2, 3, ..., 8$, we have $W = -8, -6, -4, -2, 0, 2, 4, 6, 8$.

    (b) $W = 1.50m - 1.75(8 - m) = 3.25m - 14$. So, for $m = 0, 1, 2, 3, ..., 8$, we have $W = -14, -10.75, -7.50, -4.25, -1, 2.25, 5.50, 8.75, 12$.

CHAPTER 16

■ **Walking**

1. (a) 52   (b) 50%   (c) 68%   (d) 16%

3. (a) 44.6   (b) 59.4   (c) 14.8

5. (a) 1900   (b) 50

7. (a) approximately the 3rd percentile

    (b) the 16th percentile

    (c) around the 22nd or the 23rd percentile

    (d) the 84th percentile

    (e) the 99.85th percentile

9. (a) 95%   (b) 47.5%   (c) 97.5%

11. (a) 13   (b) 80   (c) 250   (d) 420   (e) 488   (f) 499

13. $\mu = 75$, $\sigma = 3$

15. $\mu = 80$, $\sigma = 10$

17. (a) 16th percentile   (b) 97.5th percentile   (c) 18.6 lbs

19. (a) 97.5th percentile   (b) 99.85th percentile   (c) 8 lbs

■ **Jogging**

21. Extra large: 125; large: 2375; discount: 2500.

23. (a) .5   (b) .475   (c) .025   (d) 0

25. For an honest coin $p = 1/2$. Using the dishonest coin principle with $p = 1/2$, one gets $\mu = n\cdot(1/2) = n/2$ and $\sigma = \sqrt{n\cdot(1/2)\cdot(1/2)} = \sqrt{n/4} = \sqrt{n}/2$.

27. (a) $\mu + 0.25\sigma = 55 + 0.25 \times 12 = 58$ points

    (b) 45 points corresponds to the 20th percentile. $(55 - 0.84 \times 12 \approx 45)$

    (c) 83 points corresponds to the 99th percentile. $(55 + 2.33 \times 12 \approx 83)$

    (d) Approximately 50 students (1%)

    (e) Approximately 2000 students. (52 points corresponds to the 40th percentile.)

Index